The W. K. Kellogg Foundation and the Nursing Profession: Shared Values, Shared Legacy

Books Published by the Honor Society of Nursing, Sigma Theta Tau International

Pivotal Moments in Nursing: Leaders Who Changed the Path of a Profession, Houser and Player, Volume I, 2004, and Volume II, 2007.

Daily Miracles: Stories and Practices of Humanity and Excellence in Health Care, by Briskin and Boller, 2006.

When Parents Say No: Religious and Cultural Influences on Pediatric Healthcare Treatment, by Linnard-Palmer, 2006.

Healthy Places, Healthy People: A Handbook for Culturally Competent Community Nursing Practice, Dreher, Shapiro, and Asselin, 2006.

The HeART of Nursing: Expressions of Creative Art in Nursing, Second Edition, Wendler, 2005.

Reflecting on 30 Years of Nursing Leadership: 1975-2005, Donley, 2005

nurseAdvance Collection. (Topic-specific collections of honor society published journal articles.) Topics offered are: Cultural Diversity in Nursing; Disaster, Trauma, and Emergency Nursing; Gerontological Nursing; Health Promotion in Nursing; Implementing Evidence-Based Nursing; Resources for Implementing Evidence-Based Nursing; Leadership and Mentoring in Nursing; Maternal Health Nursing; Oncology Nursing; Pediatric Nursing; Psychiatric-Mental Health Nursing; Public, Environmental, and Community Health Nursing; and Women's Health Nursing; 2006.

Technological Competency as Caring in Nursing, Locsin, 2005.

Making a Difference: Stories from the Point of Care, Volume 1, Hudacek, 2005.

A Daybook for Nurses: Making a Difference Each Day, Hudacek, 2004.

Making a Difference: Stories from the Point of Care, Volume 2, Hudacek, 2004.

Building and Managing a Career in Nursing: Strategies for Advancing Your Career, Miller, 2003.

Collaboration for the Promotion of Nursing, Briggs, Merk, and Mitchell, 2003.

Ordinary People, Extraordinary Lives: The Stories of Nurses, Smeltzer and Vlasses, 2003.

Stories of Family Caregiving: Reconsideration of Theory, Literature, and Life, Poirier and Ayres, 2002.

As We See Ourselves: Jewish Women in Nursing, Benson, 2001.

Cadet Nurse Stories: The Call for and Response of Women During World War II, Perry and Robinson, 2001.

Creating Responsive Solutions to Healthcare Change, McCullough, 2001.

Nurses' Moral Practice: Investing and Discounting Self, Kelly, 2000.

Nursing and Philanthropy: An Energizing Metaphor for the 21st Century, McBride,2000.

Gerontological Nursing Issues for the 21st Century, Gueldner and Poon, 1999.

The Roy Adaptation Model-Based Research: 25 Years of Contributions to Nursing Science, Boston Based Adaptation Research in Nursing Society, 1999.

The Adventurous Years: Leaders in Action 1973-1999, Henderson, 1998.

Immigrant Women and Their Health: An Olive Paper, Ibrahim Meleis, Lipson, Muecke and Smith, 1998.

The Neuman Systems Model and Nursing Education: Teaching Strategies and Outcomes, Lowry, 1998.

The Image Editors: Mind, Spirit, and Voice, Hamilton, 1997.

The Language of Nursing Theory and Metatheory, King and Fawcett, 1997.

Virginia Avenel Henderson: Signature for Nursing, Hermann, 1997.

For more information and to order these books from the Honor Society of Nursing, Sigma Theta Tau International, visit the society's Web site at www.nursingsociety.org/publications, or go to www.nursingknowledge.org/stti/books, the Web site of Nursing Knowledge International, the honor society's sales and distribution division, or call 1.888.NKI.4.YOU (U.S. and Canada) or +1.317.634.8171 (Outside U.S. and Canada).

The W. K. Kellogg Foundation and the Nursing Profession: Shared Values, Shared Legacy

Joan E. Lynaugh • Helen K. Grace • Gloria R. Smith • Roseni R. Sena • María Mercedes Durán de Villalobos • Mary Malehloka Hlalele

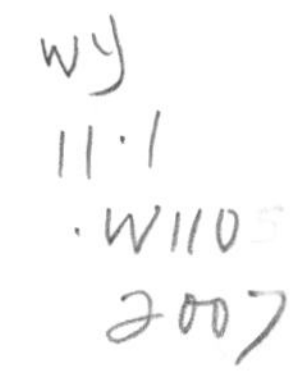

Sigma Theta Tau International

Editor-in-Chief: Jeff Burnham
Project Manager: Carla Hall
Development Editor: Eric Cox
Proofreader: Billy Fields

Cover Design by: Rebecca Harmon
Interior Design and Page Composition by: Rebecca Harmon

Printed in the United States of America
Printing and Binding by V. G. Reed & Sons

Sigma Theta Tau International
550 West North Street
Indianapolis, IN 46202

Visit our Web site at **www.nursingknowledge.org/stti/books** for more information on our books.

ISBN-10: 1-930538-52-9
ISBN-13: 978-1-930538-52-8

Library of Congress Cataloging-in-Publication Data

The W. K. Kellogg Foundation and the Nursing Profession : Shared Values, Shared Legacy / contributing authors, Joan E. Lynaugh ... [et al.].
p. ; cm.
Includes bibliographical references and index.
ISBN-13: 978-1-930538-52-8 (pbk.)
ISBN-10: 1-930538-52-9 (pbk.)
1. W. K. Kellogg Foundation--History. 2. Nursing services--United States--History. 3. Nursing--Developing countries--International cooperation. I. Lynaugh, Joan E. II. Sigma Theta Tau International.
[DNLM: 1. W. K. Kellogg Foundation. 2. Foundations--history--Africa, Southern. 3. Foundations--history--Latin America. 4. Foundations--history--United States. 5. Nursing Services--history--Africa, Southern. 6. Nursing Services--history--Latin America. 7. Nursing Services--history--United States. 8. Education, Nursing--history--Africa, Southern. 9. Education, Nursing--history--Latin America. 10. Education, Nursing--history--United States. 11. History, 20th Century--Africa, Southern. 12. History, 20th Century--Latin America. 13. History, 20th Century--United States. WY 11.1 W1105 2007]

RT4.W55 2007
362.17'3--dc22

2007003022

Dedicated to the board members and staff of the W. K. Kellogg Foundation and the many nurses across the globe who shared a vision of helping people to improve their health and productivity through community-driven solutions.

Table of Contents

Introduction

Helen K. Grace

For more than 75 years the W. K. Kellogg Foundation has worked with nurses and nursing to improve health care. No other foundation has provided as much support over such a prolonged time. Shortly after joining the program staff at the Foundation, I became intrigued with the Foundation's connection to the nursing profession. As a nurse and as a sociologist, I was fascinated with the concept of social change. There is no better place to study this phenomenon than in a philanthropic organization with a mission to foster social change. In my role as a program director and later as a vice president for programs, I was not a student, but a participant in decision-making underlying the grant-making process.

In my early years at the Kellogg Foundation, I began to read some of the old files, and marveled at the tales of the "Flying Squadron," public health nurses who worked as part of the Michigan Community Health Project (MCHP) in the thirties. That led me to wonder what had happened to nursing through the years, and in what ways the Kellogg Foundation had contributed to the paths taken. Day-to-day work kept me from indulging this fascination, so the Kellogg Foundation contracted with a professional nursing historian, Joan Lynaugh, to study the connection between the nursing profession and the Foundation. She chose to confine her study to the period of 1930 (the beginning of the Kellogg Foundation) up to 1982. Her historical study comprises the first section of this book.

As I neared retirement and Gloria Smith assumed leadership for health programming, I took on the task of organizing a historical database for the Foundation covering 1930 to 1994. As part of this project, I reviewed all board of trustees minutes over this time period, as well as all program staff reports to the board, and identified grants that were made in keeping with the board's decision-making. This allowed me to see nursing within the context of Foundation programming through the time period covered by Dr. Lynaugh's analysis up to 1994. The second section of this book includes my reflections—gathered from this historical database—on Foundation decision-making from 1930 to 1982 that influenced the funding for nursing.

The history of nursing and the Kellogg Foundation from 1982 to 1994 was one that I lived. During this time there were no particular initiatives targeted specifically toward the nursing profession. The health program goals were multidisciplinary, and this included support for nursing. Nurses stepped up and provided leadership by building partnerships with communities and with other health professions in education, among other community-based health projects across the nation. Nurses in community-based health services for families, children, adolescents and the elderly are highlighted in Chapter 3.

In 1997, Gloria Smith became vice president for programs and provided leadership for health for the next five years. She moved health programming ever closer to identifying obstacles to obtaining health services and giving voice to underserved populations. In a major initiative to rebuild education that focused on community-based, public health professions, the Foundation returned to its roots in the 1990s as a locally based Community Health Project in which public health departments were the cornerstone. The "Community Voices" initiative focused on the delivery of health services in a wide array of communities across the country, particularly for underserved populations. Dr. Smith reports on this part of the history in Chapter 4.

The story of the Kellogg Foundation and nursing would be incomplete if it focused only on the United States, and if it were only our story. The Foundation's work with nursing in Latin America and the Caribbean began in 1940 and continues to this day. The first grant made in southern Africa was in support of a nursing project in Botswana in 1986. In both geographical areas, nursing garnered considerable support from the Kellogg Foundation, though the approaches used in these two regions were similar in some aspects and different in others. There is much to learn from these examples that extends far beyond these settings. In telling these stories, we have included participants from these regions who reflect on their own experiences. Drs. Roseni R. Sena and María Mercedes Durán de Villalobos have provided leadership to nursing in Latin America and the Caribbean for nearly 20 years, and they share their analysis of the Foundation's support for nursing in that region and lessons for the philanthropic community. Dr. Mary Malehloka Hlalele, a former program director from southern Africa for the Kellogg Foundation, worked with nurses to assess the impact of funding for that region.

From the perspective of the Foundation, Dr. Smith and I report the rationale behind international programming and our assessment of the experience. The international story is told in Chapters 5 and 6.

The final chapter in Part I analyzes the relationship between the nursing profession and the Kellogg Foundation over its more than 75-year history. This chapter offers a chance to look back over this long road of collaboration and summarizes observations that might be useful in moving toward the future.

In the early years, the nursing profession and the Foundation worked closely in the initial Michigan Community Health Project. The Foundation's goals to improve the health of children in rural Michigan and foster the development of public health departments were highly compatible with those of the nursing profession, and the MCHP became an excellent vehicle to provide leadership in community health nursing. Prominent nursing leaders were on the staff of the Foundation, and many nurses came to Michigan as part of their educational programs to learn from these leaders.

World War II disrupted this relationship. The attention turned from community health to a focus upon nursing's contribution to winning the war. The National Committee for Nursing Services brought together the leadership of all of the nursing organizations at that time. The Foundation supported the Committee by preparing adequate numbers of nurses for the war effort. This was the first of a number of efforts on the part of the Kellogg Foundation to bring unity to the nursing profession's organizational structure. With the end of the war, the Foundation continued to support the Committee, hoping the professional organizations could remain unified. That was not to be. When the Foundation ended funding, the Committee disbanded and the organizations returned to their separate entities.

When the war ended, attention once more turned to the home front. During the war, the development of community hospitals had gained momentum. With this development came a need for additional nurses to staff the hospitals. Meanwhile, the nursing profession focused on upgrading the nursing educational system. A large number of nurses were still trained in hospital-based schools of nursing, and the profession pushed to move nursing into educational institutions instead. With guidance from the nursing advisory committee, the Kellogg Foundation provided support to develop associate degree nursing education in the community colleges. These efforts led to an increase in numbers of nurses in the field to meet the growing demand at hospitals, and successfully moved nursing education into the educational mainstream. At the same time, the nursing profession pushed for baccalaureate education as the entry point to the profession, and the large influx of associate degree nurses subverted this movement.

The burgeoning hospital field shifted the focus of both the Foundation and nursing from the community to the hospital. Their goals—the preparation of nursing leaders through nursing administration programs and the support of associate degree nursing to supply hospitals—were complementary.

Over the years, a lack of consensus on how nurses are to be educated has been a major concern of the Kellogg Foundation. At least five attempts have been made to bring the professional nursing organizations together with one unified voice, starting with the National Committee for Nursing Services, the Esther Lucille Brown study of Nursing and Nursing Education, the Lysaught Commission on Nursing Education and Nursing Practice, the Joint Practice Commission, and the National Commission on Nursing Implementation (NCNIP). Unfortunately, that vision is yet to be realized.

More recently, the nursing profession increasingly has allied itself with the hospital industry and has distanced itself from the community. Seeing prestige in the big-business aspects of health care, nursing has centered primarily on hospital-based care settings. While the Kellogg Foundation also followed the path away from community-based health care to the hospital, in the past 20 years, it has rediscovered the community as the center for health care.

Some perceive that the Foundation's support for the profession of nursing per se has diminished. But in reality, support for the work that nurses do at the community level has never been stronger. Within the nursing profession, however, the work of these nurses is neither valued nor visible. The most successful efforts have been those that supported the work of nurses in a wide array of roles in practice and educational settings. Areas that received critical Foundation support include development of graduate education in nursing, nursing service administration education, associate degree nursing education, non-traditional education that facilitates educational mobility, fostering of regional organizations such as Western Inter-institutional Compact on Higher Education and Nursing (WICHEN) and Southern Regional Educational Board (SREB), and nursing education-service collaborations such as the model at University of Rochester.

Ironically, the Kellogg Foundation's contemporary nursing vision is more compatible with that of nurses in Latin America, the Caribbean, and southern Africa than with the profession in the United States. In these regions, nursing is the backbone of primary health care, and nursing leadership has wisely drawn upon Foundation resources to help provide quality care.

Despite this longstanding history, the relationship between philanthropy, and in particular the Kellogg Foundation, and the nursing profession has been fraught with tension.

Private funding is a precious resource that must be used to foster change. It should not be used to maintain the status quo, to fill gaps in funding from other sources, or to improve the competitive edge of one organization over another. The most effective strategies use private support to leverage change on a broad scale.

I hope this review of the historical connection between the nursing profession and the Kellogg Foundation informs both the nursing and the philanthropic communities. One of the biggest challenges for the philanthropic community is how to use valued private resources to promote positive social change. And the challenge to professions such as nursing is to understand how private funding can accelerate desired changes. This book provides a detailed picture of the interplay between a philanthropy focused on investing its resources in people and a professional field that has as its mission caring for people—and how they served both interests as they saw fit.

About the Authors

Joan E. Lynaugh, PhD, RN, FAAN

Joan E. Lynaugh is professor emerita and director emerita of the Barbara Bates Center for the Study of the History of Nursing at the University of Pennsylvania School of Nursing. Long interested in improving the delivery of health care to all Americans, she worked to develop the nurse practitioner role, redesign nursing curricula and teach nurses to meet that challenge. For more than 25 years she has encouraged the development of historical research on nursing and health care. Believing that historical research offers context and perspective on contemporary problems, she was one of the founders of Penn's Center for the Study of the History of Nursing in 1985. Her own research and writing focuses on hospital development, the role of women in institution building, labor supply and demand phenomena, nurse-physician relationships, international nursing, and the evolution of higher education in nursing. She is the author or co-author of 10 books and 59 chapters and articles. In 1992, she became the first editor of *Nursing History Review,* the official journal of the American Association for the History of Nursing. She consults often on matters of historical scholarship in the United Kingdom, Ireland, and Canada as well as throughout the United States.

Helen K. Grace, PhD, MPH, FAAN

Helen K. Grace has served as a leader in nursing and primary health care nationally and internationally. Her academic roles over the years included faculty positions primarily in psychiatric nursing, associate dean for graduate study, and dean of the College of Nursing, all at the University of Illinois. After completing a diploma at West Suburban Hospital and her BSN at Loyola, Dr. Grace received her master's degree in mental health nursing at the University of Illinois, and her PhD in sociology from Northwestern. In 1982, Dr. Grace took a position with the W. K. Kellogg Foundation serving in multiple roles, including program director and vice president of programs at the Foundation. Her work with the Foundation positioned her well to use her big picture vision and emphasis on systems to impact nursing and primary health care on a very broad level. Dr. Grace is known nationally and internationally for her leadership in community-based primary health care. She is a pioneer and change agent whose underlying focus became larger systems and public policy. She has been described as a devoted advocate for nursing who continues to position nursing as part of the solution to health care. Through her strong leadership, she has been able to facilitate the funding and advancement of community-based care around the world including nurse managed health centers in the United States.

Gloria R. Smith, PhD, MPH

Gloria R. Smith started her career as a public health nurse in Detroit in 1955 after completing her bachelor's degree at Wayne State University. She received her MPH from the University of Michigan, and an MA and PhD in anthropology. She was dean at the University of Oklahoma and from there became the first nurse to head the State of Michigan's Department of Public Health. In 1988, Dr. Smith became the dean of Wayne State University College of Nursing. In 1991, she began her work at the W. K. Kellogg Foundation in Battle Creek, Michigan, as coordinator for programs in health, and became vice president for programs in 1995. Dr. Smith was instrumental in a number of major health initiatives funded by the W. K. Kellogg Foundation to redirect resources to strengthen community-driven and community-based health services delivery. These included Community Partnerships in Health Professions Education, Community-Based Public Health, and Comprehensive Community Health Models of Michigan. The ultimate objective of these projects was to identify and demonstrate strategies for improving access to high quality, cost-effective, and culturally competent health care to communities, particularly medically underserved communities. The projects underscored the importance of multidisciplinary collaboration and further highlighted the critical roles and contributions of nurses in expanding access to vulnerable and underserved populations. Dr. Smith has worked internationally in southern Africa and South America to expand access to community-based health care through increasing the numbers of adequately prepared nurses in severely underserved regions.

Roseni R. Sena, PhD

Roseni R. Sena was born in Belo Horizonte, State of Minas Gerais, Brazil. She received her degree in nursing in 1976, at the Federal University of Minas Gerais (UFMG). She did specialization studies in public health and earned a master's degree in epidemiology. She concluded her PhD at the University of São Paulo in 1996. During her doctoral studies, she received a scholarship at the California State University, at a distance education center, in 1995. She has been on the faculty of the School of Nursing of the UFMG since 1977, where she was dean from 1998 to 2002. She was national coordinator of professional education of the Ministry of Health in 2003 and 2004. Dr. Sena is a member of the Brazilian Nursing Association, where she is engaged in formulating innovative strategies for the teaching of nursing. She participated in the social movements for democracy in the country during the military regime between 1964 and 1984, and also in the movement for sanitary reform in Brazil and in the struggles to assure the right to health. She has been a consultant to the W. K. Kellogg Foundation in Brazil and Latin America for 20 years. During this period she has participated in various initiatives, including projects for the integration of nursing teachers and service nurses, community, rural and urban development and the Project for the Innovation in Training Health Professionals (UNI). She coordinates the Nucleus for Study and Research on Nursing Education and Practice (NUPEPE), as a researcher of CNPq (a federal research funding agency) since 1998. She has written books, chapters of books, articles and lectured regionally, nationally and internationally. She has been the advisor for many masters' dissertations and doctoral theses.

María Mercedes Durán de Villalobos, MNS

María Mercedes Durán de Villalobos is currently titular and emeritus professor at the Universidad Nacional de Colombia, and chair of the research committee at Universidad de La Sabana, Bogotá. She has taught undergraduate and graduate students in the areas of chronic and cardiovascular care, nursing theory and research for more than 25 years. After graduating from Univeridad Nacional de Colombia, she earned a master's degree in nursing sciences and a certificate of advanced graduate studies at Boston University, as well as from the Universidade Federal do Rio de Janeiro, Brazil. Professor Durán de Villalobos has been dean of the School of Nursing and vice president of students at Universidad Nacional de Colombia; visiting professor at Universidad Nacional de Rosario, Argentina; Benemérita Universidad Autónoma de Puebla, and Universidad Nacional Autónoma de México, México. She worked as a consultant for the W. K. Kellogg Foundation for Latin America and Santo Domingo, on community-based and educational projects.

Mary Malehloka Hlalele, MD

Mary Malehloka Hlalele is currently with Management Science for Health (MSH), a private, nonprofit educational and scientific organization working to close the gap between what is known about public health problems and what is done to solve them. Before joining MSH, she was Africa region health and nutrition advisor for World Vision International. She also has worked as faculty coordinator through the office of the dean of the faculty of health sciences, University of Witwatersrand. She was a founding member on the board of trustees of the Tropical Institute of Community Health and Development in Africa. In 1998/99, she was a member of the executive planning committee for the W. K. Kellogg Foundation's Leadership Forum. Prior to that, she was the program director for health in the Africa region for the W. K. Kellogg Foundation. Besides having worked as a physician in Morija and coordinator of community health services in Bophuthatswana, Dr. Hlalele pioneered the establishment of non-racial nursery and primary schools in the Thaba-Nohu district. Dr. Hlalele did her pre-medical training at Bath Technical College, City of Bath, England, and medical training at the College of Medicine, University of Lagos, Nigeria. Her post-graduate work includes a diploma in primary health care.

Part I

The Kellogg Foundation and American Nursing: 1930–1980

Joan E. Lynaugh

The Kellogg Foundation and American Nursing: 1930–1980

Joan E. Lynaugh

Education, then, beyond all other devices of human origin, is a great equalizer of the conditions of men—the balance wheel of the social machinery . . . it does better than to disarm the poor of their hostility toward the rich, it prevents being poor.

—Horace Mann, Report to the Massachusetts State Board of Education, 1848.

The inspiration for Part I of this book was the suggestion by Helen K. Grace that someone should explore the intriguing historical relationship between nursing and the W. K. Kellogg Foundation. I took her suggestion seriously, and with generous support from the Foundation and vital information from nurses and former Foundation officers, pursued their parallel but linked histories. Little did I realize in 1988, when Helen Grace made her suggestion, that this project would absorb my attention for the next eight years. The case of nursing and the Kellogg Foundation is a window on a tumultuous period; investigating its myriad events sent me down several research paths. I am grateful for excellent source materials and advice from the librarians and staff at the Foundation in Battle Creek. My nursing and medical history colleagues have taught me much and tried to correct my errors. I am grateful to them. I am keenly aware of the hazards of writing contemporary history, but I continue to believe this case study can help shape understanding of our present and, perhaps, give added meaning to the future.

1 The Context for the Case

This section chronicles 50 years in the relationship between two 20th century American institutions—nursing, a new field of work, which ultimately became indispensable to modern life, and the W. K. Kellogg Foundation, which grew from a local, beneficent project to become one of the nation's largest philanthropic foundations. The W. K. Kellogg Foundation, part of the second wave of great American foundations, was founded in 1930.

These modern American foundations were founded by affluent, conservative, mostly elite or self-made citizens who believed that the private sector should take charge of social change. Their brand of specialized, focused, and corporate philanthropy supplanted the more global, idiosyncratic, and often reactive benevolence of the 19th century. Their attitude was certainly less coercive than 19th century adherents of the ideals of "scientific charity," who sharply discriminated in their treatment of the deserving and undeserving poor. Instead, the foundation-oriented philanthropists of the 20th century sought to control and give direction to what they saw as inevitable, sometimes threatening, but necessary impulses toward social change. No doubt they hoped to formally and permanently enhance American culture. Carnegie's libraries, Rockefeller's medical schools, and Kellogg's child health goals reflected their self-conscious and confident vision of 20th century philanthropy.

Nursing owes its very existence to philanthropy and voluntarism, two quintessentially American approaches to social problem-solving.

Organized nursing appeared on the American scene about 60 years earlier than the Kellogg Foundation. Then, during the Great Depression, nursing and the Kellogg Foundation came together. Linking their stories offers a special historical vantage point on fundamental, sweeping changes in American society between 1930 and 1980. There is another historical logic in this juxtaposition: Nursing owes its very existence to philanthropy and voluntarism, two quintessentially American approaches to social problem-solving (Stevens, 1989; Rosenkrantz & Buck, 1990). Now, of course, nursing is part of the gigantic health care business so important and problematic in today's world. How this largest of the health occupations adapted and grew in the dynamic social environment during the middle decades of the 20th century, and how it was influenced by Foundation leaders, offers another view of the complicated, multifaceted history of health care in America.

Nursing's Origins

A brief overview of American nursing's origins puts the 20th century partnership between nursing and Kellogg in perspective. Nineteenth century nursing schools and hospitals, where the first trained nurses learned the skills of mending and nourishing the ill and wounded, were often underwritten by philanthropic benefactors. For instance, the 1861 nursing school at Women's Hospital in Philadelphia was aided by funds from local philanthropist Pauline Henry, who felt that Philadelphia should invest in the civic improvement that the Nightingale-trained nursing reform promised (O'Brien, 1987, p. 2-17). More than 40 years later, in 1906, Helen Hartley Jenkins sponsored the first postgraduate nursing program at New York's Columbia University Teacher's College, which produced nursing leaders for the next 50 years (Roberts, 1954, p. 67).[1]

In the years between these two nursing benchmarks, educational programs and trained nurses derived much of their financial support from the charitable impulses of wealthy community leaders. Visiting Nurse Societies, in particular, were popular charities of late 19th century women who,

[1] *Teachers College faculty would become quite involved with the Kellogg Foundation in the 1940s and later.*

with their husbands, raised or donated money to subsidize home care of the urban poor. Women like Mrs. William Jenks of Philadelphia and men like Jacob Schiff of New York understood that nurses were essential to implementing reformers' philanthropic and reform impulses. The visiting nurses, who went from door to door giving care to immigrants and impoverished people, were, in effect, the agents of these social reformers who pledged their private wealth to projects such as the visiting nurses, hospitals, or settlement houses.[2] All across the country, benefactors who had, in the burgeoning American economy, accumulated wealth beyond their own needs, found in health care and nursing a suitable outlet for morally or religiously inspired generosity (Buhler-Wilkerson, 1989; Ginsberg, 1985).

But the dependence of the new field called nursing on individual benevolence and voluntarism proved problematic for a fledgling occupation. Paying for hospital and home care of the poor plus the education and wages of nurses soon exceeded the capacity of private beneficence alone. Those nurses who relied on these sources found themselves working long hours for low pay, in effect subsidizing their benefactors' charitable impulses. Schools for nurses, exclusively based in hospitals which were also heavily dependent on charity, were financially vulnerable. Hospitals quickly began to exploit their pupils as a means of curtailing rising labor costs. The normal pattern for nurses in training was to trade substantial amounts of their own labor in the hospital for a modest level of instruction. In many hospitals volunteer Boards of Lady Visitors tried to ameliorate the working conditions of the pupils by paying for better residences for nurses and supporting part of the educational expense.

Nurses and their educational system became caught up in the remarkable late 19th century surge of hospital building funded by group or individual charity and inspired by ethnic, religious, or community pride and progressive ideology. Between the years 1875 and 1900, more than 4,000 hospitals opened in cities and towns from coast to coast. Although some of these were tax-supported public hospitals and a moderate number were

[2] *For an overview and analysis of the visiting nurse movement in the United States, see Buhler-Wilkerson, 1989. A helpful series of essays illuminating the prevalence of this movement can be found in McCarthy, Kathleen D. (Ed.). (1990). Lady Bountiful Revisited: Women, Philanthropy, and Power. New Brunswick and London: Rutgers University Press.*

proprietary hospitals, the majority (70%) were nonprofit, voluntary institutions dependent on a combination of philanthropy and income from paying patients. By 1900 more than 500 of these new institutions operated nursing schools (Oderkirk, 1985, p. 30–37). This system of hospital training remained the dominant model for nursing until after World War II.[3]

Transferring a portion of the responsibility for care of the sick from home and family to professional nurses and hospitals was a popular American health care idea that was rapidly implemented in the four to five decades after the Civil War.

Transferring a portion of the responsibility for care of the sick from home and family to professional nurses and hospitals was a popular American health care idea that was rapidly implemented in the four to five decades after the Civil War. But putting the sick together in a designated place invited chaos and the spread of disease. Nurses were expected to not just dress wounds, give medications, and instruct in nutrition and hygiene, but also to maintain order. With infections sweeping through hospitals, nurses were critical in keeping them stable and clean. Neither hospital building nor the development of nursing was subject to any overall planning at the state or national levels. Development of the hospital and nurse education enterprise was firmly situated in local hands. Although leaders in nursing, led by Adelaide Nutting, the first head of nursing at Teacher's College, Columbia University in New York, struggled to set up standards for education and practice, the common reality of nursing was on-the-job training and the lowest possible expenditures for nurses' services.

Meanwhile, the nation's general health came under criticism and scrutiny after the revelation that significant numbers of recruits to the military services in World War I were unfit for duty. And vexing problems in finding sufficient and qualified nurses to serve during the war focused attention on nursing issues.

Even before the war was over, demand for public-health nursing improvements stimulated a conference on the matter under the auspices of the Rockefeller Foundation. The research done by Josephine Goldmark and Haven Emerson, who collaborated on an in-depth study of Cleveland's hospitals and health system for the Cleveland Hospital Council in

[3] *For the best discussion of nursing's 19th century origins and 20th century conflicts see Reverby, Susan. (1987). Ordered to Care: The Dilemma of American Nursing, 1850–1945. Cambridge and London: Cambridge University Press.*

1920, impressed decision makers at the Rockefeller Foundation. Goldmark was selected to focus on the nursing problem and direct a national study. This meeting led to the 1923 report, "Nursing and Nursing Education in the United States." A few years later, Haven Emerson, a leading proponent of improvements in public health, began his long involvement at the new W. K. Kellogg Foundation as advisor and board member.

The Goldmark report, as the 1923 Rockefeller study came to be called, produced a list of recommendations addressing the nursing situation as it was seen in 1922-23 (Goldmark, 1923). It called for

- Establishing an endowment for schools of nursing to be controlled by boards or committees independent of hospitals;
- Shortening hospital school programs from 36 to 28 months to eliminate noneducational and repetitive work;
- Preparing and licensing subsidiary workers for specified populations of patients;
- Offering special training built on basic nursing education to prepare administrators, teachers, and public health nurses; and
- Developing and strengthening university schools of nursing.

In spite of the Rockefeller Foundation's efforts, their widely distributed report prompted little change. Private-duty nursing remained the dominant form of nursing practice. Nurses completed training only to find themselves competing with an ever-growing number of other nurses for a diminishing number of jobs. Some public health observers would later look back on the twenties as the era in which public health became "nationalized."

Control of nursing education, however, was still dispersed among thousands of local hospitals and relatively weak nursing organizations. One interesting note on "Nursing and Nursing Education in the United States" is how very similar its 1923 recommendations would be to the post-World War II nursing plans of the not-yet-created W. K. Kellogg Foundation (see Chapter 3).

As the culture of the twenties, the "decade of prosperity," tended to hide the deep problems of economic instability and threats of social dislocation, advocates for pubic health education were pushing a disease-prevention agenda with special focus on improving the health of women and children. In contrast to the Progressive rhetoric of the era before World War I, this campaign relied on conservative, even nationalistic arguments. At a 1926 meeting of the American Child Health Association, for example, Herbert Hoover argued for placing greater emphasis on the health of the American child. Weaving together themes of nationalism, eugenics, and public health, he exhorted his listeners with: "Our work is racial defense . . . if we want this civilization to march toward higher economic standards, to moral and spiritual ideals, it will march only on the feet of healthy children" (Rogers, 1990).[4]

Voluntary organizations such as the National Tuberculosis Association and the American Red Cross public-health nursing program, as well as the federally sponsored Children's Bureau, all supported the renewed public health campaign. At the time there seemed to be a growing faith in the power of health education to combat the perceived weakening effects of unhealthy lifestyles and immigration on American vitality. The link between nursing and the 1920s public health campaign was, of course, the need for professional messengers to carry the message and work of public health to the public (Stuart, 1989, p. 111-131).

At the Rockefeller Foundation the argument that better public health was linked to better education for these nurse-messengers seemed sound and worth supporting. Mary Beard, a capable and influential public-health nursing leader, was recruited by the Rockefeller Foundation to oversee their nursing initiatives. During the twenties the Foundation underwrote two new university schools of nursing: first at Yale University under the direction of Annie Goodrich, and later, at Vanderbilt University, where Shirley Titus became the first dean. In each case, the Foundation supported five years of demonstration, apparently to see if women and nurses could survive on all-male campuses, then followed up by

[4] *For an analysis of child health, see Merke, Richard A. (1990). Save the Babies: American Public Health Reform and the Prevention of Infant Mortality, 1850–1929. Baltimore and London: The Johns Hopkins Press.*

endowing each school with a million dollars. In the late 1920s, however, the Rockefeller Foundation lost interest in the nursing aspect of public health reform, and ceased its funding (Abrams, 1993, p. 119-137).

In Cleveland, Western Reserve University created a new nursing school by absorbing Maternity Hospital School and Lakeside Hospital School. After five years of testing, the new college program was endowed, this time for $1.5 million by philanthropist Frances Payne Bolton. In 1930, a different scenario played out as the Chicago-based Illinois Training School, an unusual free-standing, philanthropically sponsored hospital school, transferred its assets of $500,000 to the University of Chicago to found a university nursing program. Never developed as an undergraduate program, the Chicago effort, led by Nellie X. Hawkinson, offered degree-completion programs for graduates of hospital schools, and later, master's degrees. It closed in 1959. The University of Chicago story, with its underdeveloped and short-lived nursing programs, symbolizes the difficulties inherent in establishing university-based education for nurses (Baer, 1992, p. 43-48).

University nursing programs, including Nutting's program at Columbia, and early schools at the Universities of Minnesota and Cincinnati, were training nurses in public health theory and skills. The small cadre of nurses who obtained advanced education during this decade and the next were to play an important role in the upcoming collaboration between the Kellogg Foundation and nursing.[5]

Other university schools opened in the mid-twenties: the Universities of Virginia and California at the instigation of state nursing societies, and the University of Rochester (New York) as part of a new medical-hospital complex partially underwritten by the Rockefeller Foundation. The development of nursing in American universities must, however, be kept in perspective. In 1926 there were still only 25 schools of nursing offering college degrees; altogether, they enrolled 328 students. Clearly this could have only a minimal impact in the context of the 2,286 schools of nursing in the country—an all-time high reached in 1927.

[5] *Some nursing leaders who were educated in these schools later worked closely with the Kellogg Foundation, e.g., Katherine Faville, Lucille Petry (Leone), Mary Kelly Mullane, and Rozella Schlotfeldt. And, of course, Mildred Tuttle came from this tradition. Her school, Peabody College in Nashville, played a significant role in public health education during this era.*

Reform-minded nurse educators, undaunted by the lack of public response to the Goldmark report and encouraged by philanthropic and other educators' support, continued to document the problems in nursing. Just a few years after the Goldmark study, Frances Payne Bolton, the Rockefeller Foundation, the Commonwealth Fund, and nurses themselves funded still another study. The first reports of the Committee on the Grading of Nursing Schools quantified the situation. May Ayers Burgess, who led the study (1928 to 1934) noted that in 1926, 37% of "schools" in nursing had no teacher at all, 43% had one teacher, and only 2% had four or more teachers. In 1928 her report indicated that nearly 50% of all schools/hospitals employed six or fewer graduate nurses for all purposes in the school/hospital combination (Burgess, 1928, p. 385).

Burgess coined a phrase that summed up the problem: "preparing too many [nurses], but too few" (Roberts, 1954, p. 179). In spite of several decades of reform rhetoric, the hospital-based nursing education system continued to turn out a large number of graduates into the health marketplace, where the private-duty jobs for which they were prepared were scarce. Meanwhile, hospitals complained that shortages of student recruits curtailed their ability to care for patients. Public health reformers, on the other hand, complained that they could not find competent nurses to staff their health care demonstration projects, clinics, schools, and industrial nursing jobs.

The stark limits on hospital-based nurse education are evident in the 1917 standard curriculum, which called for less than 600 hours of instruction in three years, or about a day a month or two hours a week. Ten years later, in 1927, the standard curriculum called for an increase to three hours a week, a little more than one day per month. Nurses were expected to learn as they worked, although, as Burgess noted, there were only a few experienced, fully trained nurses in the hospitals working with the pupils.

Historian Mary Roberts said that before 1925 she could find no evidence of a clearly defined and vigorous nursing movement to replace students with trained nurses in hospitals. Nurses seeking opportunities for postgraduate education were limited to courses in public-health nursing and administration. And finally, nursing's professional organization, the American Nurses Association, which obtained its members via the alumnae associations of the hospital schools, was necessarily ambivalent about any educational reform which threatened the hospital schools from which virtually all its members were drawn.

Trying Times

The years leading up to the creation of the W. K. Kellogg Foundation were a time of increasing difficulty and frustration for health care and community leaders trying to manage the development of nursing and provide safe nursing care to the American public. They faced a dual dilemma: The existing system called on them to try to improve the education of nurses inside impoverished, non-educationally oriented American hospitals, and to deliver effective, safe nursing care using student nurses in hospitals and ill-prepared graduates in public health. Social scientists studying the problem, social welfare reformers, and philanthropic staff as well as individual philanthropists recognized the inherent futility in the traditional system of nursing education and practice.

The years leading up to the creation of the W. K. Kellogg Foundation were a time of increasing difficulty and frustration for health care and community leaders trying to manage the development of nursing and provide safe nursing care to the American public.

Nursing changed significantly in the 20th century. Although first viewed as a promising reform that addressed both the care needs of a modernizing society and the occupational needs of women, nursing in the twenties and thirties increasingly experienced, in the words of Isabel Stewart of Teacher's College, "stress and turmoil . . . experimentation and painful self-examination" (Stewart, 1931, p. 601-611). The number of nurses per capita constantly increased, but the organizations in which they practiced, such as private-duty registries, hospitals, and public health agencies, functioned chaotically. Nursing enjoyed a positive social image in the thirties; at the same time its ambiguous social role as an all-female occupation and its peculiar educational system hindered its maturation.

Inseparable from this story are rapid developments affecting American institutions and beliefs about health. Nurses would rarely dominate the debate about what or how much health care would be delivered and who would get it. In fact they had to struggle to participate at all. Nevertheless, their essential role in health care delivery was assumed by virtually everyone. Far less clear, however, were answers to the questions about who would enter nursing, how they would be educated, how much of the work of health care they would do, and how nursing costs would be paid.

The W. K. Kellogg Foundation would enter this scene of "stress and turmoil" and, in a remarkably short time, become a significant voice in these fundamental debates about the American health care system. Almost two decades would pass, however, before initiatives on education and problems in nursing could be launched. In the meantime the campaign for public health reforms, especially those focused on child health, attracted the interest of Will Keith Kellogg of Battle Creek, Michigan.

Notes and References

Abrams, Sarah E. (January 1993). Brilliance and Bureaucracy: Nursing and Changes in the Rockefeller Foundation, 1915-1930. *Nursing History Review.*

Baer, Ellen D. (January/February 1992). Aspirations Unattained: The Story of the Illinois Training School's Search for University Status. *Nursing Research.*

Buhler-Wilkerson, Karen. (1989). *False Dawn: The Rise and Decline of Public Health Nursing, 1900–1930.* New York: Garland Publishing, Inc.

Burgess, May Ayers. (1928). *Nurses, Patients, and Pocketbooks: Report of a Study of the Economics of Nursing Conducted by the Committee on the Grading of Nursing Schools.* New York City: Committee on the Grading of Nursing Schools.

Ginsberg, Lori D. (1985). *Women and the Work of Benevolence: Morality and Politics in the Northeastern States, 1820–1885* (Doctoral dissertation, Yale University, 1985).

Goldmark, Josephine A. (1923). *Nursing and Nursing Education in the United States: Report of the Committee for the Study of Nursing Education and a Report of the Survey*. New York: The Macmillan Company.

O'Brien (D'Antonio), Patricia. (January-February, 1987). All a Woman's Life Can Bring: The Domestic Roots of Nursing in Philadelphia, 1830–1885. *Nursing Research, 36.*

Oderkirk, Wendell. (November 1985). Setting the Record Straight: A Recount of Late Nineteenth Century Training Schools. *Journal of Nursing History.*

Roberts, Mary. (1954) *American Nursing History and Interpretation.* New York: The Macmillan Company.

Rogers, Naomi. (May 1990). *Vegetables on Parade: Child Health Education in the 1920s.* Presented at American Association for the History of Medicine Annual Meeting, Baltimore, Maryland. Professor Rogers's paper reviews the changing rhetoric of public health reform during the twenties.

Rosenkrantz, Barbara, & Buck, Peter. (1990). *Philanthropic Foundations and Resources for Health: An Anthology of Sources.* New York and London: Garland Publishing, Inc.

Smillie, Wilson G. (1952). *Preventive Medicine and Public Health.* 2nd Ed. New York: The Macmillan Company, p. 10.

Stevens, Rosemary. (1989). *In Sickness and in Wealth: American Hospitals in the Twentieth Century.* New York: Basic Books.

Stewart, Isabel. (May 1931). Trends in Nursing Education. *The American Journal of Nursing, 31.* Stewart succeeded Adelaide Nutting as head of Teacher's College nursing programs.

Stuart, Meryn. (1989). Ideology and Experience: Public Health Nursing and the Ontario Rural Child Welfare Project, 1920-25. *Canadian Bulletin of Medical History, 6.* This reports the efforts of the Province of Ontario to educate mothers in an effort to lower unacceptably high infant mortality. Stuart calls attention to the futility of educational efforts in the presence of severe economic disadvantages.

2 Creating the Kellogg Foundation

By 1930, when Herbert Hoover invited him to be a delegate to the White House Conference on Child Health and Protection, Will Keith Kellogg had already begun to turn his interest away from his cereal empire to search for better ways to amplify and organize philanthropic distribution of some of his accumulated wealth. Sometimes called "The King of Cornflakes," Kellogg, over a period of 25 years, built an enormously successful business based on dried and flaked wheat and corn meal, a health food he popularized. Now almost 70, he worried that the Fellowship Corporation, the entity he created in 1925 to distribute $1,000,000, was not going to be equal to the task of effectively reinvesting "his money in people" (Powell, 1956, p. 303).

Biographers agree that the conservative, even frugal Kellogg was deeply moved by the central message of his friend Herbert Hoover's 1930 White House conference, which was the well-publicized "Children's Charter" announcing an expansive new concept of the American child's right to health (Powell, 1956, p. 312; Nielsen, 1985, p. 18; W. K. Kellogg Foundation, 1987, p. 18). The economic downturn that led to the Great Depression accelerated the obvious and growing need Kellogg already saw.

The Charter pledged full access to health care and education for American children, as well as establishment of organized child welfare services in every locality. It dramatically expanded previous public positions on Americans' public health rights.

Included in its 19 points stipulating, among others, prenatal care, periodic exams, adequate schooling, protection against excessive labor, and access to voluntary youth organizations, was a demand for local health, education, and welfare organizations staffed by public health officials, including nurses. Reprinted in the first bound and published report of the W. K. Kellogg Foundation in 1942, the Children's Charter came to be the guide for the Michigan Community Health Project, one of the new Foundation's first health experiments (W. K. Kellogg Foundation, 1942).[1]

The conservative, even frugal Kellogg was deeply moved by the central message of his friend Herbert Hoover's 1930 White House conference, which was the well-publicized "Children's Charter" announcing an expansive new concept of the American child's right to health.

The W. K. Kellogg Foundation, which was, for its first two months, actually called the W. K. Kellogg Child Welfare Organization, intended to "receive and administer funds for the promotion of the welfare, comfort, health, care, education, feeding, clothing, sheltering, and safeguarding of children and youth, directly or indirectly, without regard to sex, race, creed, or nationality, in whatever manner the Board of Trustees may decide" (Powell, 1956, p. 306). But, Kellogg added, the greatest good for the greatest number can come only through the education of the child, the parent, the teacher, the family physician, and the community in general. "Education," he was fond of saying, "offers the greatest opportunity for really improving one generation over another" (W. K. Kellogg Foundation, 1987, p. 18).

Foundation documents indicate that Kellogg tried to give his staff a free hand and not restrict them in any way. He wanted their use of his money to be flexible so long as it promoted the health, happiness, and well-being of children. He funded the new Foundation with income from a $45 million trust, mostly in Kellogg company stock.

---

[1] *For details on the conference and the charter, see White House Conference, 1931.*

Will Keith Kellogg's Early Years

Will Keith Kellogg was born in 1860; his parents, John Preston Kellogg and Ann Janette Stanley Kellogg, were active in the Seventh Day Adventist Church in Battle Creek. His formal education stopped in the sixth grade; for the next 30 years he worked, first for his father selling brooms, and then later for his brother as manager of the Battle Creek Sanitarium and its associated businesses.

Founded in 1866 as the Western Health Reform Institute, the famous Sanitarium successfully based its therapies on hydropathic and health reform ideas of the late 19th century. It became the pulpit from which Will Keith's flamboyant and influential physician brother, John Harvey Kellogg, preached his own brand of good health with vegetarianism, exercise, and clean living at its nucleus. John Harvey Kellogg and his Sanitarium are historically important and fascinating in their own right, but for the purposes of this story, one aspect of his health program is most fundamental. The grain-based diet offered at the Sanitarium, modified and made more palatable by John Harvey's brother Will, ultimately became Kellogg's Corn Flakes, the key product of the W. K. Kellogg Company and the basis of Will Kellogg's fortune.

John Harvey Kellogg and Will Keith Kellogg had a productive but very tumultuous relationship. Will, the younger brother, felt unfairly treated by John Harvey, who insisted on taking three quarters of their income, the lion's share, for himself. Will, married, raising three children (two others died), and working constantly to manage the myriad enterprises of the Sanitarium, despaired of his future. He finally left his brother in 1901 and, in the midst of an acrimonious fraternal conflict, created his own new company in 1906. The split between the brothers was never healed.

W. K. Kellogg spent his young adulthood in his brother's shadow. After age 40, the next phase of his working life was devoted to building a national and international food company using with great advantage the advertising and transportation strategies of 20th-century commerce. His foundation, he felt, needed to be created quickly and well because he did not expect to live long enough to oversee the distribution of his wealth himself. In fact, of course, he lived to be 90, and so, in what could be called his third life, he experienced some sense of what his largesse could accomplish.

Twentieth-Century American Foundations

Kellogg's decision to create a foundation to distribute his personal wealth was based on distinguished precedents in the United States. The nation's first foundation was incorporated by Congress in 1847 to accommodate the gifts of James Smithson, thereby creating the Smithsonian Institution. Of course, by the time Kellogg was pondering his choices, the Rockefeller Foundation, Carnegie, and others were already demonstrating how to responsibly manage huge individual benefactions.

Private foundations were, by the late 1920s, a distinct component of the so-called "independent" sector of the American system. They were readily distinguished from business and industry on the one hand and government on the other.

Private foundations were, by the late 1920s, a distinct component of the so-called "independent" sector of the American system. They were readily distinguished from business and industry on the one hand and government on the other. Their freedom of action, although challenged from time to time, seemed to be sustained by the public because of their usefulness in addressing social problems or needs ignored by government and uninteresting to private enterprise. The foundations' ability to act quickly and to deal with societal issues in a flexible, responsive manner also avoided the ponderousness inherent in democratic public decision-making.

More than 40 years after Kellogg began to assemble his foundation, in 1970 the federal Commission on Foundations and Private Philanthropy defended American foundations with this argument: "If foundations and individuals do not support heterodoxy, who will? If private-sector philanthropy will not support worthwhile activities that are neither self-supporting nor have sufficient political support for public funding, who will?" (Commission on Foundations and Private Philanthropy, 1970, p. 126).

Historians and others commenting on philanthropy often speculate on the ways in which foundations reflect both the attributes or weaknesses of their founders and the capitalist system that makes enormous private wealth possible (Wheatley, 1988; Rosenkrantz & Buck, 1990; Smith & Chiechi, 1974). Waldemar Nielsen, a contemporary critic and interpreter of foundations, particularly approved of the "earthy excellence" of Kellogg's legacy. He ranks Kellogg with John Rockefeller, Sr., Andrew Carnegie, and Julius Rosenwald as among the great philanthropists in American life: "One of those rare individuals who had both a vision for the social use of his great wealth and the capacity to organize an instrument to realize that vision" (Nielsen, 1985, p. 265).

The New Foundation

From its founding in 1930 to the beginning of World War II, Kellogg's new foundation confined itself, with few exceptions, to local health and education issues in Michigan. In fact, it began as an operating, rather than grant-making, foundation. As the inaugurator of Kellogg's concept, the Michigan Community Health Project (MCHP), so named in 1935, illustrated both the possibilities and limitations of public health reform and private-public collaboration. It enacted the Kellogg trustees' decision to expend resources on the application of knowledge rather than investing in research or simply giving funds for relief (W. K. Kellogg Foundation, 1942).

The trustees eschewed "expert" demonstrations, especially those imposed by those not involved in the situation being reformed. They wanted to focus on locally identified problems solvable through education and local ingenuity. The basic question the trustees asked was whether local leadership, stimulated by education and temporary extra money, could and would develop effective ways to meet community needs and improve child health, education, and welfare.

The trustees eschewed "expert" demonstrations, especially those imposed by those not involved in the situation being reformed. They wanted to focus on locally identified problems solvable through education and local ingenuity.

The newly created staff at the Kellogg Foundation opened their offices at 250 Champion Street, Battle Creek, Michigan. They worked hard to make their vision of reform in public health and health education come to life. The Foundation's first full-time president, Stuart Pritchard, M.D., assumed the post in 1933. Long an advocate of better education for physicians and himself an experienced public health physician, he joined the Foundation to start a program of continuing education to improve his colleagues' diagnostic and therapeutic capabilities. His approach and ability helped him move quickly into the Foundation's leadership role.

Pritchard's arrival at Kellogg was as successor to two part-time presidents, Wendell Smith and Dr. A. C. Selmon. Selmon, an Adventist missionary and the founding president of Kellogg's philanthropy, impressed Kellogg through his occupational health work in the Battle Creek cereal plant. But he had first gained Kellogg's confidence by treating him for pneumonia in China. Kellogg's success in recruiting Pritchard, however, assured a leader capable of building a system to carry out Kellogg's version of the modern public health agenda. Other staff for the new Foundation were drawn from the surrounding area or recruited for their special skills from other parts of the country.

The Michigan Community Health Project

At first, Pritchard and the Foundation focused on local health projects. The most famous of these, the MCHP, assembled local political and professional leadership and encouraged interdisciplinary collaboration to set up and run public health services at the town and county level. Starting in one county in 1931, the Foundation subsidized local health projects until the citizens could be persuaded to take them over. The MCHP became a national model, widely influential on the development of public health services. In its design and rhetoric, the Michigan Community Health Project reflected Kellogg's emphasis and faith in education (Nielsen, 1985, p. 269).

The MCHP became a national model, widely influential on the development of public health services. In its design and rhetoric, the Michigan Community Health Project reflected Kellogg's emphasis and faith in education.

Lulu St. Clair (Blaine), the Foundation's first nurse and the MCHP director of health education from 1932 to 1936, organized a corps of public health nurses called the Flying Squadron. These nurses traveled the counties organized under the MCHP using their preparation in public health and education to teach not only children but their teachers the ways of hygiene, good nutrition, detection of health problems, and preventive health services.

In 1932 local dentist Emory Morris joined the Kellogg health project as a volunteer. Nurse Mildred Tuttle, a graduate of Peabody College in Nashville, Tennessee, accepted a position as public health nurse in the Hastings, Michigan, part of the project in that same year. Both would rise to leadership roles, but in the meantime, they were instrumental in the success of the MCHP experiment. Morris and Tuttle remained with the Foundation for the rest of their careers and saw astonishing growth from the 1940s to the late 1960s.

Marguerite Wales (Norton) joined the Foundation in 1935 as its first director of Nursing. An evaluator, author, and veteran of Lillian Wald's famous public-health nursing program at Henry Street in New York City, she was the architect of many of the innovative clinical programs in the MCHP. Because of her extensive connections in the East, she linked the Kellogg model to the national and international public-health nursing scene.

Kellogg's Michigan Community Health Project was local, to be sure, but it was sweeping in its intent. Foundation plans called for creating public health care in selected Michigan counties where no such authority had previously existed. Foundation staff reviewed the procedures of other foundations, considered their own priorities, investigated resources in Michigan, and from their findings, developed a set of project principles.

First, the new health care system would offer services to all the population, not just the poor or unserved. And, the MCHP would stay in the philosophical middle ground, between controversial "state medicine" and medical conservatives, who advocated leaving all responsibility for health care to private medical practitioners. Finally, planners would rely on education and persuasion to achieve their goals rather than seeking legislation or mandated change (W. K. Kellogg Foundation, 1942, p. 15).

With these principles in mind, Foundation grants eventually went to seven southern Michigan county boards of supervisors willing to establish health departments. Foundation funds supplemented the modest amounts of public money from the United States Public Health Service and the State of Michigan available for the same purpose.

Barry County was first, receiving its grant in 1931. The county supervisors heard proposals from Kellogg staffers Drs. Selmon and Pritchard explaining the project. The supervisors' vote to go along was close, 11 to 9 (Ramos, 1986, p. 14). This is not surprising given the fundamental political changes the supervisors confronted. Implicit in the Kellogg plan was more authority for the county and less control for each town. Combined with consolidation of rural schools, which was going on at the same time, the move to centralize health care must have made the supervisors worry that their town constituents would react negatively to so much centralization.

One out of four Americans would be unemployed by 1933. A county, much less a town with an insufficient tax base, could not institute or sustain even basic health services.

But those were Depression years, and families were already experiencing economic strain in an era where one out of four Americans would be unemployed by 1933. A county, much less a town with an insufficient tax base, could not institute or sustain even basic health services. Thus, the money that Kellogg stood ready to invest would have been hard to turn down. And, Kellogg staff smoothed the process using great care in introducing their ideas and respecting local feelings. Staff participated in intensive community meetings and sought consultation to try to bring everyone who was a stakeholder into the planning. They were especially careful of local physicians and dentists, funneling all direct patient care through their private offices so as to avoid the taint of "state medicine." In particular, they eschewed any public health nurse usurpation of patient confidence previously reserved for private physicians (W. K. Kellogg Foundation, 1942, p. 10).[2]

[2] *Commentary on public health nurses and private physicians stresses collaboration between public systems of health education and autonomous private medical practice. The extreme sensitivity of medicine to any perceived infringement on its prerogatives is detailed in Rosen, 1983, p. 41, 43, 97.*

Physicians were reassured that public health nurses would not substitute for them or steal their patients by providing medical services. Local physicians received detailed accounts of the preventive, health education strategies that Kellogg's staff wanted to start through the new county health departments. Ruth Tappan, an MCHP public health nurse writing for the professional nursing press, enthusiastically described the child-focused care system that the MCHP tried to organize around the newly centralized public schools in the counties (Tappan, 1932, p. 215-217).

Each county health department included a medical director, public health engineer, one public health nurse for every 5,000 people, and clerical staff. Budgets were calculated at $1 per capita of each county's population. Michigan paid 10 cents of that dollar, the county paid 25 cents, and Kellogg paid 65 cents. The overall goals of the health departments were to establish health data on birth, death, and disease; to get control of infectious diseases; to improve maternal and child care; to improve school hygiene; to set up food and milk sanitation systems; and to improve general sanitation in each county.

Special educational programs were organized for physicians, dentists, community leaders, and nurses. For physicians and dentists the emphasis was on preventive medicine and children's dentistry. The idea was to try to get individualistic practitioners to visualize a context in which health problems developed, and to think of their responsibility, beyond their own private practice, to the whole population (W. K. Kellogg Foundation, 1942, p. 9). The goal of these training sessions was to shore up and expand the abilities of known and accepted leaders in the community. Using pragmatic strategies, they worked on practical, everyday health problems such as getting children immunized.

The idea was to try to get individualistic practitioners to visualize a context in which health problems developed, and to think of their responsibility, beyond their own private practice, to the whole population.

For instance, school superintendents and principals wanted courses in "problems affecting the health of the child," so Kellogg staff set them up. Schoolteachers and school board members took special classes to learn how to disburse health knowledge through the schools. To establish the maternity nursing service, Kellogg funded postgraduate courses for MCHP nurses in obstetrical nursing and care of the newborn at maternity centers around the country. By 1942, it had spent $152,000 on this special nurse training alone.

By 1935, the seven projects were all functioning in Barry, Allegan, Eaton, Van Buren, Hillsdale, Branch, and Calhoun counties. About 280,000 people in those counties became involved. Notwithstanding the trustees' stated belief that giving actual relief was a public, not a private responsibility, the Foundation subsidized medical and dental services for children who could not pay private fees (W. K. Kellogg Foundation, 1942, p. 18). Physicians and dentists simply certified to the child's county health department that the care was given and Kellogg picked up the bill and paid them.

Kellogg's financial subsidy markedly expanded the practice income of local physicians and dentists; and these practitioners and their patients benefited from Kellogg investments in laboratory and diagnostic equipment for local hospitals.

This private welfare did contradict the trustees' original reluctance to give relief. But it reflected the fundamental reality of the Great Depression voiced in the complaint of the 1932 Committee on the Costs of Medical Care: "Between the physician, dentist, nurse or hospital—able and willing to provide service—and the persons who need [them] stands a barrier compounded of ignorance and of inability to pay." The committee's complaint, based on a survey of 9,000 families from all income groups and regions of the country, helped sustain the idea that Americans had a right to some minimum of health care (Davis, 1937, p. 75-99).[3] Kellogg's financial subsidy markedly expanded the practice income of local physicians and dentists; and these practitioners and their patients benefited from Kellogg investments in laboratory and diagnostic equipment for local hospitals.

The staff at Kellogg were on the road all the time during those MCHP years. In one year, 1937, they logged 626,206 miles in seven counties, making 37,070 home visits, 10,294 school visits and 53,569 counselor visits for tuberculosis, maternal and infant care, and preschool health checks. The "counselors" were public health nurses who did general health teaching and looked for persons needing medical or dental follow-up (Seven Michigan Counties Vote to Preserve Health, 1938, p. 518-519).[4] Public health nurses, some with special maternity training, staffed school and home-health programs. Engineers consulted on water, sewage, and school construction projects to improve overall sanitation.

[3] *See also Roberts, Mary. (1954). American Nursing: History and Interpretation. New York: The Macmillan Company, p. 236. Roberts, who edited the American Journal of Nursing and participated in nursing affairs from 1930 to 1955, devotes considerable space to the Kellogg Foundation in her history.*

[4] *There was one counselor for every 5,000 persons.*

The maternity nursing service, which provided nurse visits for pregnant women and home delivery support, was popular with both citizens and physicians. As it became better known among county physicians, they began to refuse cases unless the women were willing to be cared for by the maternity nursing delivery service either in their own home or at the hospital (Burnett, 1941, p. 1365-1372).

In the mid-thirties about 40% of babies born in the MCHP area were delivered in hospitals. Nurses and physicians used the maternity program to encourage women to seek professional care during pregnancy, then they encouraged them to use the hospital for subsequent deliveries. The Foundation staff planned that home maternity care would be transitory. Nurses and physicians applauded the gradual acceptance of hospital delivery by Michigan's rural women. Although certainly not entirely attributable to the MCHP, by 1955, 90% of area babies were born in hospitals.

In the mid-thirties about 40% of babies born in the MCHP area were delivered in hospitals ... although certainly not entirely attributable to the MCHP, by 1955, 90% of area babies were born in hospitals.

As the MCHP matured, it proved to be an invaluable training ground for other health professionals interested in public health organization and care delivery. The Foundation underwrote fellowships to encourage graduate and undergraduate students to spend time observing MCHP projects. Thirty universities sent students for experiences ranging from a week to a year of concentrated study. Managing field visits and producing seminars for observers became part of the routine at the MCHP.

Between 1930 and 1940, the W. K. Kellogg Foundation spent $2,971,670 on MCHP direct services. This figure does not include funds for fellowships, the youth camps operated by the Foundation, and capital funds for school and hospital renovation. When those are included, overall expenditures during the entire MCHP rose to nearly $8 million. In some years, the project consumed more than two-thirds of the Foundation's endowment income.

Creation of the W. K. Kellogg Trust in 1935 conveyed to the Trustees $82 million, to which William Keith Kellogg later added more funds. This yielded a large reservoir of interest income to expend each year: that responsibility would ultimately add impetus to the need to reorganize the Foundation. The Trust and the grant-making opportunities it created for the Foundation, in other words, helped nationalize what began as a local, hands-on philanthropy.

Changes at the Foundation

By 1940 the Foundation was ready to turn to other projects and, in fact, was considering altering its entire approach to philanthropy. Evaluation of the MCHP indicated that its intended goals were being met. As had been hoped, the counties, with help from the state of Michigan, began to assume financial responsibility for some essential health programs previously offered through the MCHP (Smilie, 1937, p. 544-556; W. K. Kellogg Foundation, 1942).

The pragmatic pattern of demonstration on a small scale accompanied by education and followed by expanded funding of successful elements of the original project would appear again and again. Preference for locally driven, non-elite programming also persisted as part of the Foundation culture.

By this time, Kellogg trustees had adopted the common American philanthropic idea that innovations supported by the Foundation, should, if they were worthwhile, find long-term support from their own communities, institutions, or other agencies. The trustees adopted a general policy of limited-term funding.[5] However, they did continue to subsidize educational leaves, fellowships, and special programs to support MCHP concepts.

Looking back toward their beginnings, the staff at Kellogg had good reason to feel great satisfaction.[6] In just 10 years their new foundation proved its worth by creating an acclaimed and broadly influential community health project; what is more, they did it in the midst of the profound social disruptions of the Great Depression. Finally, the MCHP would prove to be something of a prototype for subsequent Kellogg funding approaches. The pragmatic pattern of demonstration on a small scale accompanied by education and followed by expanded funding of successful elements of the original project would appear again and again. Preference for locally driven, non-elite programming also persisted as part of the Foundation culture.

[5] *It is important to note that the Kellogg Foundation, although using this policy of limited term funding after it became a grant-maker, did fund certain categories of projects over long periods of time. The persistent attention to nursing and to hospital administration is a good example of this tendency.*

[6] *The Foundation published an account of its beginnings in The First Eleven Years, a low-key but satisfying account of the Michigan Community Health Project and its accomplishments.*

World events, culminating in the entry of the United States into war in 1941, however, overtook any self-congratulations on the Michigan Community Health Project, or, for that matter, any hoped-for extension of the Foundation's prized Michigan initiatives. Before December 7, some staff hoped that the Michigan public health program could be expanded and further tested on a national scale (W. K. Kellogg Foundation, 1944). But both the war and internal changes diverted attention from the MCHP. Partial MCHP funding continued for a few more years but, once the war began, the Project lost its place as the centerpiece of the Foundation. And, in fact, the chaos of World War II would alter the lives of all the staff at Kellogg.

Notes and References

Burnett, Elizabeth. (December 1941). A Rural Home Maternity Service. *American Journal of Nursing, 41.*

Commission on Foundations and Private Philanthropy. (1970). *Foundations, Private Giving, and Public Policy—Report and Recommendations of the Commission on Foundations and Private Philanthropy.* Chicago: University of Chicago Press.

Davis, Michael M. (1937). Eight Years Work in Medical Economics, 1929–1936. In B. Rosenkrantz & P. Buck (Eds.), (1990), *Philanthropic Foundations and Resources for Health—An Anthology of Sources.* New York and London: Garland Publishing, Inc.

Nielsen, Waldemar. (1985). *The Golden Donors: A New Anatomy of the Great Foundations.* New York: Dutton Books.

Powell, Horace B. (1956). *The Original Has This Signature: W. K. Kellogg.* Englewood Cliffs, N.J.: Prentice Hall, Inc.

Ramos, Mary Carol. (1986). *The W. K. Kellogg Foundation and the Michigan Community Health Project.* University of Virginia: unpublished manuscript, 14; presented at the American Association for the History of Nursing, Annual Meeting, Charlottesville, Virginia, 1984.

A Report upon the Organization and Activities of the W. K. Kellogg Foundation. (May 14, 1937). In B. Rosenkrantz & P. Buck (Eds.), (1990), *Philanthropic Foundations and Resources for Health—An Anthology of Sources.* New York and London: Garland Publishing, Inc.

Roberts, Mary. (1954). *American Nursing: History and Interpretation.* New York: The Macmillan Company.

Rosen, George. (1983). *The Structure of American Medical Practice, 1875–1941.* Philadelphia: University of Pennsylvania Press.

Rosenkrantz, Barbara, & Buck, Peter, (Eds). (1990). *Philanthropic Foundations and Resources for Health—An Anthology of Sources.* New York and London: Garland Publishing, Inc.

Seven Michigan Counties Vote to Preserve Health. (December 1938). *Trained Nurse and Hospital Review, 101,* 518-519.

Smilie, W. G. (May 10, 1937). A Confidential Report to the Director of the Kellogg Foundation.

Smith, William, & Chiechi, Carolyn. (1974). *Private Foundations—Before and After the Tax Reform Act of 1969.* Washington, D.C.: American Enterprise Institute for Public Policy Research.

Tappan, Ruth. (April 1932). Activities of the W. K. Kellogg Foundation. *Public Health Nursing, 24.*

W. K. Kellogg Foundation. (November 1944). *Annual Report.* Battle Creek, Michigan: WKKF archives.

W. K. Kellogg Foundation. (1987, 3rd ed.). *I'll Invest My Money in People: A Biographical Sketch of the Founder of the Kellogg Company and the W. K. Kellogg Foundation.* Battle Creek, Michigan: W. K. Kellogg Foundation.

W. K. Kellogg Foundation. (1942). *The First Eleven Years, 1930-1941.* Battle Creek, Michigan: Trustees of the W.K. Kellogg Foundation.

Wheatley, Steven. (1988). *The Politics of Philanthropy.* Madison, Wisconsin: University of Wisconsin Press.

White House Conference. (1931). *White House Conference on Child Health and Protection.* New York: The Century Company.

3 Grant-Making and Professionalizing

The unexpected death of W. K. Kellogg Foundation president Stuart Pritchard in August 1940, followed by the total confusion of war, prompted a series of significant staff changes at the Foundation. Pritchard's interests had complemented the public health and educational commitments of Will Keith Kellogg. As a trusted associate of Kellogg, first at the Battle Creek Sanitarium and later throughout his tenure as Foundation president, Pritchard tested and developed their shared ideas within the framework of the Michigan Community Health Project (MCHP). In some senses, Pritchard's death marked the end of an era at Kellogg.

For a while afterward, the Foundation was directed in tandem by President George B. Darling and General Director Emory Morris. But, after World War II began, Darling left to join the National Research Council in Washington.[1] Morris took over as both Foundation president and general director in 1943.

Emory Morris officially joined the Foundation in 1933 when he was a 28-year-old dentist just starting out in practice. From a small Michigan town himself, he became acquainted with the Foundation as a volunteer, helping in the dental education programs of the MCHP. After accepting a full-time position as the Foundation's director of Dental Education, he rose rapidly in the Foundation leadership.

[1] *Darling had served as assistant treasurer, assistant secretary, and comptroller before assuming the presidency of the Foundation.*

Morris presided over the reorganization of the Foundation from an operating, hands-on, local enterprise into a grant-making, more nationally focused Foundation. He brought in strong leaders in nursing, hospital management, education, and agriculture. Witnesses from both inside and outside the Foundation all concur that the Kellogg Foundation between 1942 and 1967 mirrored the character and style of Emory Morris.

The War Years: The Foundation Reorients Itself

His tenure, however, began in an atmosphere of war-related turmoil. About 35 Foundation staff left Battle Creek to join the uniformed services. The services of Director of Nursing Marguerite Wales were loaned to the American Red Cross. The Foundation donated her salary for two years while she organized the Red Cross wartime programs for the East Coast: she never returned to the Midwest. Mildred Tuttle, who by then was senior counselor at the Hillsdale unit of the MCHP, took a leave of absence from her post to direct the public-health nursing program at Wayne State University in Detroit. Nurses left the Foundation both for active service in the military and for critical domestic projects. By 1942, two thirds of the previous year's MCHP nursing staff was gone.

Foundation programs immediately began to change focus to war-related problems. One of these, which later became a major theme of Kellogg philanthropy, arose from the travel limitations imposed on Latin Americans by the European war.

Foundation programs immediately began to change focus to war-related problems. One of these, which later became a major theme of Kellogg philanthropy, arose from the travel limitations imposed on Latin Americans by the European war. Initiated by a request to Emory Morris from the U.S. Office of Inter-American Affairs (directed by Nelson Rockefeller), the Foundation brought physicians, dentists, and nurses from Latin American countries to the United States for graduate studies in their respective disciplines (W. K. Kellogg Foundation, 1941, p. 13). Cut off from their customary European study opportunities, these scholars pioneered a program of new fellowships in American medical, public health, and nursing schools. Twenty-three fellowships in public-health nursing were given in the first year. Extensive Kellogg support for health and other projects in Latin America continues today (W. K. Kellogg Foundation, 1980, p. 20; Arnett, 1986, p. 5).

In an effort to cope with the instant shortage of health personnel created by military demands for doctors and nurses, the Foundation began granting educational support to health personnel across the country. It funneled nursing scholarship money to collegiate schools and to hospital nursing programs associated with accredited medical schools.[2] Between 1943 and 1944 the Foundation expended $247,643 on nursing scholarships. The scholarship money made it possible for nursing students to stay in school, complete their education, then fill needed nursing jobs, rather than delay graduation as they worked to earn tuition or support themselves. Part-time study combined with work was a common pattern for nurses studying for baccalaureate degrees, since their low salaries precluded savings and they could get almost no educational aid.

Medical students too, got scholarship money to live on so they could continue to study through the summer rather than working; in turn, student availability for summer study made wartime three-year medical school programs possible. Encouraging nursing and medical schools to rapidly increase their output of graduates went to the top of the private and public agenda as war became a reality.

The Foundation funded other programs to compensate for the loss of nurses to overseas service, including home nursing courses for family members, high school nurses' aide training programs, and refresher courses intended to bring retired nurses back into practice. Most of these programs were focused on Michigan's immediate problems. In another sense, however, these small projects served as "practice programs" for later initiatives aimed at increasing access to health care by creating more providers. Local projects probably were all the constricted Foundation staff could manage during the early war years. Similar home nursing courses, aid training, and refresher courses for nurses spread across the country during the war as localities strained to find ways to meet civilian health needs.

---

[2] *The Kellogg Foundation did not usually support hospital-based schools of nursing. These early grants are a rare exception. The issue of what schools to support with nurse education funding, i.e. which, if any, hospital schools to include, was a significant political problem for public funding which followed soon after.*

Beginning with the 1942 Appropriations Act, precedent-setting public investment in nursing education through the United States Public Health Service (USPHS) helped attract recruits into nursing by reducing the financial barriers to past high school education for applicants who would otherwise have to go to work.

Beginning with the 1942 Appropriations Act, precedent-setting public investment in nursing education through the United States Public Health Service (USPHS) helped attract recruits into nursing by reducing the financial barriers to past high school education for applicants who would otherwise have to go to work.[3] Federal dollars also helped prepare faculty to staff rapidly expanding nursing schools.

Kellogg also responded quickly to the war-caused nursing shortage, but its resources were miniscule compared to the $4.5 million eventually appropriated by Congress for the USPHS nurse-recruitment program. The nation's demand for more nurses was relentless on both civilian and military fronts. Although small in proportion to public spending, Kellogg's investments in nursing were targeted and persistent and its influence would grow throughout the war.

The National Nursing Council

The war posed both a crisis and an opportunity for the national nursing leadership. As it began to seem imminent in late 1939 and early 1940, two nursing veterans of the World War I era tried to rouse nursing organizations to face up to the war emergency. Julia Stimson, president of the American Nurses Association (ANA), sought to avoid the factionalism she believed seriously weakened nursing's response to World War I.[4] Mary Beard, director of the Red Cross Nursing Service, knew that the traditional source, the Red Cross First Reserve, was no solution to the insufficient number of nurses available for military service.

In September 1939, as Hitler invaded Poland, Beard urged the National League for Nursing Education to help recruit graduate nurses for the First Reserve and try to increase enrollment in nursing schools. Beard

[3] *This federal investment was the first support for entry-level training for nurses. The 1935 Social Security Act carried a provision for subsidizing the education of nurses for public health practice; this money was available to graduate nurses seeking further education, but not to persons seeking basic nursing education.*

[4] *Julia C. Stimson was a veteran of World War I who served in France. Her memoir of the war, Finding Themselves (New York: The Macmillan Company, 1927), attracted attention to her abilities. She was Superintendent of the Army Nurse Corps, then assumed a series of leadership posts in nursing organizations culminating in the presidency of the American Nurses Association (ANA).*

knew that most Red Cross First Reservists were administrators and teachers needed on the home front, and, in any case, their numbers were insufficient to the task. Even at that late date, with everyone thinking of war, the authorized strength of the Army Nurse Corps was just 675 nurses. War planning estimates would soon stipulate six nurses for every thousand soldiers. Most problematic of all, there was no official body of the federal government charged with the task of recruiting nurses from civilian life into the military. But, as Lucille Petry Leone recalls it, the invasion of the Low Countries in 1939 finally "got attention in spite of isolationism" (Personal communication, November 11, 1989).

In her memoirs, Stella Goostray, who was Superintendent of Nurses at Boston Children's Hospital when the war began, told how nurses coped with the crisis. On July 10, 1940, Isabel Stewart wrote to Goostray: "I believe we should have a Commission . . . representative of the nursing profession as a whole and that it should be at work now, . . . [on national defense needs]" (Goostray, 1969, p. 120). Goostray agreed; she quickly called on trusted colleagues. Two weeks later, on July 29, Julia Stimson chaired a meeting of representatives of all five major nursing organizations plus the Red Cross and the six nursing units of the federal government.

That day they created the Nursing Council on National Defense, a new, voluntary organization charged with recruiting nurses for the war and students for nursing schools (Goostray, 1969, p. 114-116). Almost two more years would pass before the federal government could establish a military nurse recruitment system. In the interim the war nursing crisis would remain in the hands of this small, self-selected band of women.

Given their general lack of preparedness and the monumental task they faced, the first thing the National Nursing Council needed was full-time, energetic leadership. Elmira Bears Wickenden, formerly of the National Organization of Public Health Nursing (NOPHN) and the Red Cross, took over as executive director in October 1941. Wickenden, according to Lucille Petry Leone, was a masterful person who had great judgment. "She would be the model for the well-dressed executive woman . . . a powerful person" (Personal communication, November

11, 1989). Stella Goostray succeeded Julia Stimson as president of the renamed National Nursing Council for War Service (NNCWS) early in 1942.

American Nursing and the Kellogg Foundation

The Foundation's interest in nursing changed from providing local public health funding to influencing national health policy.

It is here that the story of the longstanding relationship between the Kellogg Foundation and American nursing really begins. The Foundation's interest in nursing changed from providing local public health funding to influencing national health policy. The Foundation was drawn into the nurses' problems because the National Nursing Council had little or no financial capability to fulfill its mission to recruit nurses for the war effort. The Council—representing the American Nurses Association (ANA), National Organization of Public Health Nurses (NOPHN), National League of Nursing Education (NLNE), Association of Collegiate Schools of Nursing (ACSN), National Association of Colored Graduate Nurses (NACGN), and a cross-section of national nursing service agencies including the American Red Cross—accepted the task of finding 3,000 nurses per month for the Armed Forces plus recruiting more students into nursing schools. During its first 18 months the member organizations paid the Council's bills and loaned it space, but after the war started its volunteer staff was swamped by the recruitment workload. The "National Nursing Council was a miracle to bring unity in such diversity," according to Lucille Petry Leone, but the organizations were poor and could not keep up its funding" (Personal communication, November 11, 1989).

It was Mary Beard, an experienced foundation hand from her days with the Rockefeller Foundation, who went to the little-known foundation in Battle Creek early in 1942 hoping to get money to actually operate the Council (W. K. Kellogg Foundation Annual Report, 1944, p. 16). The Council, according to historian and contemporary observer Mary Roberts, had lofty ambitions but suffered under its uncertain budget. The relatively unstinting support of the Foundation alleviated the budget problem and became, after 1942, the Council's principal source of income (Roberts, 1954, p. 355). An ambitious agenda for reforming nursing grew out of this philanthropic-professional collaboration. Wickenden's warm and frequent correspondence with President Morris reveals their community of purpose and mutual trust (Microfilm collection, WKKF archives).

Between 1942 and 1946 the Foundation contributed $331,500 to keep the many activities of the Council going.[5] Foundation dollars gave NNCWS badly needed autonomy to execute its mission. The Council enjoyed the freedom and flexibility of a private, voluntary organization funded mainly by private philanthropy rather than by the memberships of its constituent groups.

Elmira Bears Wickenden seized on the war crisis as an opportunity to mobilize the various nursing groups. She saw a chance to work out some of nursing's perennial problems: its dysfunctional educational system, the poorly organized systems in which nurses worked, and their low pay. In a very real sense, she argued, the need for the National Nursing Council was symptomatic of nursing's difficulty in dealing with any national issue. Its founding nurses had realized that some new entity was necessary as a means to circumvent the cumbersome and divided decision-making of the numerous but weak professional and federal nursing organizations.

In effect, they used the war as impetus and the Council as a bridge between the old system and the as yet unformulated new national nursing movement. Wickenden and NNCWS found support for their comprehensive reform ideas at the Foundation in the persons of President Emory Morris and Mildred Tuttle, who, in 1942, returned to Battle Creek to direct the Nursing Division at Kellogg.

So the reach of the Council began to extend far beyond its original mission of war nurse recruitment. Collaborating with the United States Public Health Service (USPHS) and the Office of Civilian Defense, the Council took the broadest possible view of their special task of balancing military and civilian needs for nurses. The broad mandate implied

[5] *The United States Public Health Service took over most of the national nurse recruitment campaign in 1943; it spent $569,933 in three years.*

They launched an ambitious array of projects: recruit more women into nursing, reform the educational system, desegregate nursing practice and education, and strengthen the national nursing organizations.

in their charge offered an opportunity to explore new approaches to long-term problems afflicting nursing practice and nursing education. They launched an ambitious array of projects: recruit more women into nursing, reform the educational system, desegregate nursing practice and education, and strengthen the national nursing organizations.

Because there was no accurate count of American nurses on which to base recruitment plans, the NNCWS, with help from the USPHS and state nursing organizations, quickly surveyed graduate nurses. They found about 290,000; of these, 173,000 were actively practicing. According to the survey, 100,000 nurses would be eligible, i.e. unmarried and under 40, for military service if they could pass the physical exam (McIver, 1942, p. 769). Military quotas required NNCWS to identify 3,000 nurses a month for potential service. Young, single nurses in private duty practice or doing nonessential hospital or public health work were urged to enlist, but NNCWS guidelines discouraged nurse administrators, teachers, and supervisors from going into the military. These senior nurses were needed on the civilian front.

The Council aimed to raise enrollment in nursing schools by 65,000. They found recruiting enough students to fill these quotas to be difficult since young high school graduates tended to prefer the adventure and high pay of war work. To solve this dilemma, the U.S. Cadet Nurse Corps Program (the Bolton Act) was passed in July 1943 to pay for students' education, uniforms, maintenance, and incidental expenses. In exchange for these government benefits, Cadet Corp students agreed (for the duration of the war) to join the military or accept essential civilian work after graduation. Lucille Petry Leone, director of the new USPHS Division of Nurse Education, ran the $160 million cadet program, which ultimately graduated 125,000 nurses.[6] With the recruitment problem now in the hands of the federal government, the Council could move on to its other projects.

As they worked together, the NNCWS evolved a method which diluted control of individual nursing organizations over the projects and

[6] *This discussion of the National Nursing Council for War Service is drawn, in part, from Lynaugh, Joan. (March-April 1991). Stepping In. Nursing Research, p. 39, 126-127. The Bolton Act was sponsored by Frances Payne Bolton, the philanthropist from Cleveland, who was at the time a member of the U.S. Congress from Ohio.*

decisions of the Council. Called "committees of interest," small groups of nurses selected from various nursing organizations were created by the Council to deal with specific problems such as student recruitment or analyzing the economic problems of nurses. Usually led by deans or leaders in public-health nursing, these committees reported to the Council or its planning committee, not back to the nursing organizations of their members. Katherine Faville, dean at Wayne State in Detroit, Michigan, chaired the Committee on Recruitment of Student Nurses and also consulted with the federal subcommittee on nursing of the Health and Medical Committee of the Federal Security Agency. Katherine Tucker, head of the nursing program at the University of Pennsylvania, chaired the Supply and Distribution Committee. And, Estelle Massey Riddle (Osbourne) headed the Coordinating Committee on Negro Nursing.

As the full-time consultant to this latter committee, Riddle launched a vigorous campaign to recruit Negro students and provide better nurse education opportunities for them. Later this committee focused on getting Negro nurses access to full participation in the military nursing services. An intense, well-orchestrated public information effort finally broke down the restriction on black military nurses being appointed to officer status similar to their white counterparts.

The National Nursing Council became something of a magnet for the conflicts and tensions aroused by all these diverse interests. Historian Mary Roberts, who, as editor of the *American Journal of Nursing,* followed these developments closely, referred frequently to this dissention, albeit with oblique, tactful comments. References to "misunderstandings," "the great difficulties of keeping [everybody] informed of changes in a fast moving program," "terrific pressures and considerable criticism," and "many stresses and strains" are sprinkled through her detailed account of the workings of the NNCWS (Roberts, 1954, p. 354-367). Fundamentally, the Council agenda intended to insert nurse participants into national planning groups where they had not been before.

NNCWS president Stella Goostray recalled "pretty spirited discussions," especially in Council dealings with representatives from the American Hospital Association (AHA). The AHA had insisted that it be a member of the Council; decisions involving nurses were, of course, vitally important to the hospital association's constituents. Goostray reported AHA's objection to the plan to shorten hospital-based nursing programs receiving funds under the Cadet Nurse Corps program (reducing training to 30 months instead of 36). It meant "the loss of senior students during the last six months of training, when they would be of the most service" to the hospital. James Hamilton, then president of the AHA (1943), was well-liked and respected by the nurses but, Goostray reported, "We did not always agree with him or with the efforts of the AHA to direct our ways." Hamilton irritated Goostray one day with complaints that nurses were trying to dictate all the changes. She replied, "We [had] been in bondage long enough, and like the children of Israel we were on the march to the Promised Land" (Goostray, 1969, p. 122).[7]

The National Council members gradually worked out an internal consensus on the future of nursing education. For instance, in 1945, they agreed that any federal aid for education which might follow the Cadet Nurse Corps Program should be restricted to accredited nursing schools and that no aid should go to vocational schools. After some debate they concurred that the federal auspices under which nursing and aid for nursing should be administered would be the United States Public Health Service. Some argued that the Office of Education would be more appropriate but the political base built in the USPHS before and during the war made it the majority option (Goostray, 1969, p. 122, 123).

Kellogg staff were urging the nursing leadership to get organized, to consolidate, to be able to guarantee a reliable supply of nurses to the country.

Even as they struggled to finish war-related projects, staff at Kellogg and nursing leaders worked on postwar plans. All was not always well between the National Nursing Council and Kellogg, either. As the correspondence makes plain, Kellogg staff were urging the nursing leadership to get organized, to consolidate, and to be able to guarantee a reliable supply of nurses to the country (W. K. Kellogg Foundation, Tentative Outline of a Composite Program for Nationwide Action in the Field of Nursing). During one exchange, in 1946, Emory Morris refused to grant more funds for Council projects unless they could verify progress toward

[7] *Hospitals always resisted shortening nurses' training time or adding more classes. This debate arose in the deliberations leading to the 1923 Goldmark report, the "Grading Committee" reports of the 1930s, and in the series of NLNE curriculum guides published between 1917 and 1937.*

restructuring and merger of the national nursing organizations. Responding to an inquiry from Marion Sheahan, chair of NNCWS's powerful Planning Committee, Morris said, "We are delaying any assistance . . . until after the final structure of the professional nursing organizations of this country has been established."[8] Six more years would go by before restructuring the American Nurses Association, the National League of Nursing Education, the National Organization of Public Health Nurses, the National Association of Colored Graduate Nurses, and the rest would be accomplished, although not quite in the way that Morris and Tuttle hoped. The Foundation, in the meantime, funded many other nursing projects notwithstanding Morris' impatience (W. K. Kellogg Foundation, 1946-47, Michigan Council on Community Nursing, p. 103-105).[9]

Summing up their relationship after the war, Wickenden selected three landmark nursing initiatives she attributed to Kellogg's support of the NNCWS from 1943 to 1948: (1) work leading to the *Study of Nursing Schools,* i.e. the Brown Report; (2) the *Socio-Economic Status of Nurses Study* (a project with the Bureau of Labor Statistics and the *American Journal of Nursing),* which compared nurses' income and career prospects to other fields; and, (3) the Council's far-reaching accreditation investigations. This latter project helped form the National Nursing Accreditation Service, set up in 1949. As two personally gratifying Council achievements, Wickenden cited Council clarification of responsibilities of professional nurses versus practical nurses and aids, and the ultimately successful plan to integrate black and white professional nursing organizations (Nelson, 1946, p. 816-819; National Nursing Council, 1948, p. 756-758; Roberts, 1954, p. 666-668.).

---

[8] *Emory Morris to Marion Sheahan. (August 23, 1946). Battle Creek, Michigan: W. K. Kellogg Foundation archives. Sheahan was, at the time, chair of the National Nursing Council Planning Committee, which was developing postwar plans. Later she became head of the National Committee for the Improvement of Nursing Service (NCINS), the postwar successor to the National Nursing Council. Sheahan enjoyed the confidence of Emory Morris and Mildred Tuttle.*

[9] *A prototype for what Tuttle and Morris wanted was set up and funded by Kellogg in Michigan. The Michigan Nursing Center Association, Inc. brought together the Michigan Nurses Association, the Michigan League of Nursing Education, the Public Health Nurses, the Industrial Nurses Association, and the Michigan Association for Practical Nurse Education into one organization. The plan implemented the suggestions of the Raymond Rich Associates' so-called "Structure Study of National Nursing Organizations," which, however, did not become the national pattern.*

And, returning the compliment in her 1945 "Report to the Board" on nursing's wartime programs, Mildred Tuttle quoted Elmira Wickenden, "Perhaps in the end the NNCWS's function is to stimulate other people to do their best for nursing in order that nursing may do its best for the nation" (W. K. Kellogg Foundation Annual Report, 1944-1945, p. 11-12).

Kellogg's decision to invest in the National Council and its agenda during these stressful years worked to help direct the fundamental changes in the education, work, and organization of nursing.

Kellogg's decision to invest in the National Council and its agenda during these stressful years worked to help direct the fundamental changes in the education, work, and organization of nursing. Nursing was transformed from a segmented, local, loosely held occupational group, guided by a small band of leaders, to a more united, national, ambitious, and self-conscious entity (Correspondence between Mildred Tuttle and Elmira Wickenden; Roberts, 1954, chap. 34; Wickenden, 1945, p. 508-512). The Foundation, during this time, selected a focus for its benevolence, developed staff expertise, and began an enduring commitment to health care and quality nursing services for the American people.

Mildred Tuttle and the Foundation

The crucial first link and negotiator between the Foundation and nursing was Mildred Tuttle, director of the Nursing Division. She joined the Foundation as a public health nurse in 1932 and stayed 36 years until her retirement in 1968.

Born in 1903 in Madison, Ohio, Mildred Tuttle experimented briefly with piano—studying at a conservatory in Berea, Ohio. Then she turned to nursing, earning a bachelor's degree from Western Reserve University in Cleveland. She practiced as a public health nurse in her hometown for four years—then studied at Peabody College and Vanderbilt University in Tennessee, completing her Master of Arts in 1932. Thus, when she went to Kellogg she was far better educated than most women, and part of a tiny band of nurses holding advanced university degrees.[10]

[10] *As late as 1948, when Tuttle was building programs to prepare nurse faculty and upgrade schools, only two of all nurses held the master's degree, a factor which frustrated hopes for rapid expansion of nursing education in universities. The Ph.D. also was a rarity among nurses.*

Her first Kellogg appointment, as a public health nurse in the community health education project at Hastings, Michigan, led to a series of nursing positions in the MCHP. She took a leave in 1938 to study at the East Harlem Nursing and Health Service Project in New York City (Reverby, 1985).[11] Later she earned her master's in Public Health Nursing at her alma mater, Frances Payne Bolton School of Nursing, Western Reserve University. Tuttle's professional orientation was public-health nursing and her preparation and experience qualified her for leadership in the field. Her own philosophy emphasized clinical practice and highly responsible roles for nurses. When she accepted her leadership post at the Foundation she sought the counsel of public-health nursing colleagues such as Katherine Faville, Ruth Freeman, and Lucille Petry Leone.[12] Although her personal style seems to have been low-key and consultative, she was persistent, perhaps even dogged in pursuit of her goals. Tuttle and Morris shared a reputation for congenial, responsive relationships among their associates in the national health care community.

Shifting Focus to Hospital Nursing

Notwithstanding Tuttle's background in public health, the Foundation's major nursing initiatives during her tenure as director of the Nursing Division mainly addressed the formidable problems in nursing education and hospital nursing. By 1944, the Foundation staff was concentrating on postwar planning for the quantity and types of health personnel, surveying the quality and role of community hospitals, considering specialization in nursing, and preparing administrators for health services.

By 1944, the Foundation staff was concentrating on postwar planning for the quantity and types of health personnel, surveying the quality and role of community hospitals, considering specialization in nursing, and preparing administrators for health services.

Graham Davis, for instance, was committed to the centrality of hospitals in the postwar health care system. Unlike Morris and Tuttle, Davis did not work his way up in the Foundation. He was recruited, probably

[11] *From 1922 to 1941 the East Harlem Project offered a highly respected program of field training in public health. It promulgated consultation between specialists and generalists, an advanced level of clinical knowledge for nurses, and the concept of broad public responsibility for health.*

[12] *In 1944, the first Nursing Advisory Committee at the Foundation, selected by Mildred Tuttle and approved by Emory Morris, included Leah Blaisdell Bryan of the National Organization of Public Health Nursing (NOPHN); Ruth Freeman, a public health educator from the University of Minnesota; and Lucille Petry (Leone) and Minnie Pohl of the United States Public Health Service (USPHS).*

by Emory Morris, from the Duke Endowment to head the new Hospital Division in 1940. Originally from North Carolina, Davis early established himself as an expert on hospital reform. His direct style comes through in his contributions to the WKKF annual reports and his published papers. Described by a contemporary as a "very earthy person," Davis wanted to upgrade community hospitals; he was a strong proponent of the idea of regionalization and hospital accessibility. After serving as president of the American Hospital Association in 1947-48, he left Kellogg in 1951. He was chosen to direct the Commission on Financing of Hospital Care under the aegis of the American Hospital Association's Hospital Research and Education Trust; after only a year in this role he resigned because of illness.

Graham's associate and successor, Andrew Patullo, began his long career with Kellogg partially because of Graham Davis. Patullo sought a fellowship in hospital administration at Kellogg because Davis was there, arriving in Battle Creek in 1943. After working with rural hospitals in Michigan and on studies directed by the School of Public Health at the University of Michigan, Patullo officially joined the Foundation in Davis's division. When Davis left in 1952, Patullo took over as Hospital Division head. He would spend nearly 40 years at the Foundation, ultimately serving as vice president and senior vice president and a member of Kellogg's board. Patullo was the voice of modern hospital administration at Kellogg—well-regarded and trusted in the professional communities linked with the postwar expansion of hospitals.

But for Mildred Tuttle, in particular, concern about the number of nurses and the quality of postwar nurses' education became paramount. She recognized that the postwar decision to expand hospitals would create unprecedented demands on nursing.

Tuttle's Agenda for the Trustees

The postwar situation, as Tuttle saw it, is outlined in a 1945 "Report to the Trustees of the Kellogg Foundation" (W .K. Kellogg Foundation, 1944-1945). Here the influence of the wartime NNCWS is clear; parts of Tuttle's report are drawn almost word for word from a wartime publication written by Isabel M. Stewart, chair of the NNCWS Committee on Educational Policies.

Tuttle called for combining professional nurse training with "general cultural" courses in baccalaureate programs. She dismissed the common, traditional route to nursing, the hospital-based diploma school, as too limited in scope and quality for the modern era. Tuttle outlined baccalaureate nursing education as a layer of general education in the first two years topped by professional theory and practice courses.

Most baccalaureate nursing programs at the time were five years long; students usually enrolled in standard freshman and sophomore college courses and then moved into the three-year clinical training programs. Their three-year clinical training was often identical to diploma students' program.[13] Tuttle urged integrated, stronger, university-controlled clinical training courses.

Although strenuous wartime efforts to increase nursing student numbers succeeded, the number of students could have grown even faster, Tuttle argued, had the schools and faculties been better prepared.[14] She told the trustees about successful, war-related efforts to centralize resources for nursing education, recruit black women into nursing, and the growing trend toward collegiate instead of hospital-based schools. Colleges were glad to recruit nursing students, she somewhat optimistically noted, as they tried to build up enrollments depleted by the war.

Although strenuous wartime efforts to increase nursing student numbers succeeded, the number of students could have grown even faster, Tuttle argued, had the schools and faculties been better prepared.

On the other hand, some problems in nursing were intensified by the war. Students were often left in charge of patient care, especially on nights and weekends. There were too few qualified faculty, and disturbingly poor working conditions for both staff nurses and faculty. Tuttle's state-by-state review of problems was a monotonous dirge of

[13] *Tuttle's discussion of baccalaureate nursing education and the other routes to preparation for nursing was drawn from a National Nursing Council for War Service publication of 1942, A Guide for the Organization of Collegiate Schools of Nursing, 26. This guide was prepared by Isabel Stewart of Teacher's College, Columbia University, with collaboration from Roy Bixler, a well-known educational administrator whose wife, Genevieve Bixler, often consulted at Kellogg on nursing projects in the 1940s. It was not uncommon for baccalaureate and diploma programs to co-exist on the same university campus.*

[14] *In January 1944 there were 112,249 students registered in 1,307 schools, an increase of 24,661 over January 1941.*

shortage: too few faculty, students, head nurses, and assistive personnel. Hospital personnel vacancies in 1944 reportedly ranged from 10% to 25% in these four categories. Tuttle emphasized this shortage data to refute the then-prevalent fear that nurses returning from the military would flood the field. To bolster her argument she added projections for numbers of future nurses who would be needed in public health and to staff the Veteran's Administration hospitals.

Tuttle was also critical of the quality of wartime emergency nurse education programs, pointing out that they offered nursing students only minimum preparation. As a partial remedy for this she advocated flexible college-based educational programs to accommodate returning military nurses and civilian nurses who needed upgrading to qualify for "adequate community service."

In this particular proposal to the trustees, Tuttle did not elaborate on master's education for nurses other than to refer to the unique Yale program. Established in 1923 with money from the Rockefeller Foundation, the Yale School of Nursing required the baccalaureate for admission and granted the Master of Nursing as the first professional degree. Drawing on the consensus of her colleagues at the time, she implied that the Yale approach to educating nurses was too radical for widespread implementation.

Tuttle summed up this important first nursing proposal by urging the trustees to fund 10 carefully selected university baccalaureate nursing programs. She proposed implementing a step-by-step agreement with the universities to ensure the long-range stability of the funded schools and the commitment of their universities to nursing. To fund the three-year program she sought $270,000, planning for grants of $12,000 per year per school, totaling a modest $60,000 total for each school.

Tuttle called the Kellogg trustees' attention to the Association of Collegiate Schools of Nursing (ACSN), founded in the mid-1930s. Characterizing the ACSN as "the first voluntary association [promoting] higher standards . . . enforced by the schools themselves," she urged the trustees to accept her plan to make ACSN membership a prerequisite for Kellogg funding for higher education (W .K. Kellogg Foundation, 1944-1945, Proposed Post War Plan).

The trustees' approval launched a key piece of Kellogg's postwar nursing initiative. Working from a list of schools approved by the Association of Collegiate Schools of Nursing, Emory Morris wrote to the university presidents to inform them of the program. When the presidents responded, Mildred Tuttle invited their deans of nursing to Battle Creek to discuss their ideas about the project; then the deans submitted proposals. Ten American university schools of nursing ultimately received funding: Boston, Chicago, Colorado, Minnesota, Oregon, Pennsylvania, Pittsburgh, Teachers College at Columbia University (New York City), Wayne State in Detroit, and Western Reserve (Cleveland, Ohio), along with McGill University in Toronto. The project had several names but eventually came to be known as the Graduate-Post Graduate Nursing Education Program.

Influence of the National Nursing Council

There is no question that staff at the National Nursing Council played an important role in Tuttle's selection of participants. The creation of Boston University School of Nursing, for example, can be traced from Council staff recommendations to Mildred Tuttle followed by Kellogg funding for Boston University in 1947. According to Stella Goostray, the ". . . school was voted [by the Boston University trustees] when the first W. K. Kellogg grant was made" (W. K. Kellogg Foundation Plan, 1946-1949).[15] The postwar plans of American nursing leaders were built on the premise that money to university-based nursing should go to the strongest, best led schools. They were eager to inform Mildred Tuttle which schools were best as Kellogg's postwar funding decisions were made. They had learned that they could increase their influence through collective action in their

[15] *It should be noted that the entire budget of the Boston University Division of Nursing (immediate predecessor of the School) was $65,021 in fiscal year 1946. A $12,000 grant would be nearly a fifth of their budget.*

experience with the National Nursing Council. At the United States Public Health Service, similar views supporting the expansion of the university nursing education strategy prevailed. According to Lucille Petry Leone, the public and private immediate postwar agendas for nursing were indistinguishable from each other (Personal communication, November 11, 1989).

Most nursing school deans used their Kellogg grants to hire young, clinically trained faculty. At Wayne State in Detroit, for example, Dean Katherine Faville recruited Irene Beland for medical-surgical nursing, Esther Reid for pediatrics, Evelyn Johnson for obstetrics, and Anne Connelly for public health (Personal communication, November 14, 1989).[16] Mary Kelly Mullane, who knew and admired Tuttle, believed she handpicked strong deans for funding, enabling them to hire faculty they would not otherwise have had (Personal communication, November 4, 1989). In New York, Dean R. Louise McManus recruited Bernice Anderson, Eleanor Lambertsen, and several other new faculty to improve the Teacher's College (Columbia University) program, perhaps the largest in the nation. Martha Ruth Smith at Boston University developed an advanced clinical nursing education project and a new program for administration in nursing, bringing in four new faculty to staff the programs.

Although Tuttle originally proposed the postwar education program for only three years, it stretched on to eight years in some cases. Both this pioneering program and its participating schools played an important role in influencing events in the 1950s.

The schools all tried to improve their nursing course offerings by upgrading science content. Some used the case method of teaching to encourage critical thinking and problem-solving behavior in their students. Courses for practicing nurses and for nursing instructors were underwritten with Kellogg funds (W. K. Kellogg Foundation Annual Report, 1946-1947). Generally, the deans seemed to accept the liberal arts courses on their campuses as adequate and concentrated on bringing professional courses up to their universities' standards (Hanson, 1991, p. 341-350; Russell, 1959).

Although Tuttle originally proposed the postwar education program for only three years, it stretched on to eight years in some cases. Both this

[16] *Dr. Schlotfeldt, later dean at Case Western Reserve University School of Nursing, was at Wayne State University School of Nursing herself at the time.*

pioneering program and its participating schools played an important role in influencing events in the 1950s. The Graduate-Post Graduate Nursing Education Program represented Mildred Tuttle and the Foundation's interest in better-quality faculty in nursing and thus better education. Another initiative launched almost at the same time addressed their goal of satisfying persistent demands for more caretakers.

Practical Nurses—Toward Differentiating Nursing Practice

In 1946 staff at Kellogg began to experiment with vocational or practical nurse training programs. The 1946-47 annual report described the practical nurse as "a person trained to care for the sub-acute, convalescent, and chronic patients in their own homes or in an institution, one who works under the direction of a licensed physician or a registered professional nurse and is prepared to provide household assistance when necessary" (W. K. Kellogg Foundation Annual Report, 1946-1947, p. 102).

Based on their wartime aide training experience and relying on recommendations in a Tuttle-commissioned report called "Nursing Resources and Needs in Michigan," the Foundation, as its first step, created a Michigan-based project. By linking hospitals, the public school system, the Michigan State Nurses Association, and the National Association for Practical Nurse Education, the Foundation sought to clarify and find consensus on just what the practical nurse could do and how to train people for the role. The projects were intended to demonstrate how assistive nursing personnel could extend the capabilities of the professional nurse to relieve the chronic shortage of more highly trained nurses. In the Michigan practical nurse project, classroom instruction became the responsibility of the state vocational education department; it thus was infinitely expandable through the public school system. Students spent four months in the classroom and eight months in clinical training. Teachers for practical nurses were prepared in a Kellogg-supported training program at Wayne State University School of Nursing in Detroit.

By linking hospitals, the public school system, the Michigan State Nurses Association, and the National Association for Practical Nurse Education, the Foundation sought to clarify and find consensus on just what the practical nurse could do and how to train people for the role.

After three years of experience with the Michigan project, Kellogg expanded its practical nurse initiative into five southern states. Arguing that the problem of adequate nursing personnel was more acute in the South than in other regions, the Foundation awarded grants to Alabama, Arkansas, Florida, Illinois, Louisiana, and Mississippi.

With resources from Kellogg, Genevieve Bixler, author of "Nursing Resources and Needs in Michigan," was recruited by the Michigan Nursing Council for War Services to do the study.[17] Her report was based on a statewide Michigan survey paralleling the 1944 state survey of hospitals conducted by the Commission on Hospital Care. Bixler's survey, which became the template for dozens of subsequent state nursing surveys, investigated whether the number of nurses was adequate, evaluated capabilities of nursing schools, and asked whether nurses were actually trained for the jobs they were doing (Roberts, 1954, p. 525).

Data on hospitals and nurses were urgently needed and quickly assembled to establish a basis for creating the postwar health care system. Tuttle and Bixler realized how important their data-rich nursing survey would be in putting through the comprehensive nursing program they and their colleagues wanted. However, for reasons not clear in Tuttle's reports or elsewhere, Bixler's survey did not investigate "the financial aspects" of the nursing schools or judge the "quality of the educational programs." Tuttle mentioned these omissions each time she referred to the Michigan study in internal Foundation documents, leaving the distinct impression that she felt constrained from doing an in-depth study in these areas.

Conflicts between hospitals' need for student labor and nursing demands for higher quality education remained a constant theme all during this period.

The continuing dependence of the hospitals on their schools of nursing invariably created barriers to criticism of hospital school conditions. Lucille Petry Leone recalls discovering that, in fact, "hospitals were against nurses" in the course of her work setting up the Cadet Nurse Corps during the war (Personal communication, November 11, 1989). Conflicts between hospitals' need for student labor and nursing demands

[17] *Genevieve Bixler was a 1913 graduate of the University of Iowa and earned her MA at Chicago. She was on the Teacher's College faculty until 1941, when she went with the new, expanded staff at USPHS. Mildred Tuttle had great confidence in Bixler. She and the Foundation frequently retained her as a consultant. Bixler and her husband Roy were later killed in a plane crash.*

for higher quality education remained a constant theme all during this period.

The 21 recommendations in Bixler's report seemed to cover every other aspect of nursing in Michigan. Among them, they called for establishing objectives for a "program of nursing for the state"; standard qualifications for nurse supervisors and staff nurses; centralization of small schools of nursing; elimination of unqualified personnel; concentration on baccalaureate-level nursing programs, including support for the new programs at Wayne State University in Detroit, University of Michigan at Ann Arbor, and other selected schools; better training in nurse specialties; and better preparation for administrators, supervisors, and teachers. The Michigan survey, as with all similar successor surveys, verified the acute shortage of nurses and nurse faculty.

Bixler's "Recommendation 16" called for a better definition of the functions of the nonprofessional nurse. Kellogg responded with the licensed practical nurse project outlined earlier and then carried the practical nurse training idea beyond the borders of Michigan.

This technique of funding surveys, commissions, and study groups and then building grant-making programs around the findings came to be a classic Kellogg Foundation strategy. Bixler's Michigan study, another byproduct of deliberations at the National Nursing Council, became the basis of several years of funding initiatives.

This technique of funding surveys, commissions, and study groups and then building grant-making programs around the findings came to be a classic Kellogg Foundation strategy.

Another technique seen here is testing of new ideas in Michigan before developing national programs. Although the Foundation did not take its MCHP concept outside of the state, as it might have if not for the war, postwar staff continued to use Michigan as a bellwether as they experimented in early grant-making.

The Foundation and Mildred Tuttle moved ahead with their funding for university professional nursing programs and high school-based practical nursing programs. In doing so, they began to define a two-tier approach to nursing education. This concept, which called for differentiation of nursing work based on education, as well as ideas of efficiency and economy in nursing training, gave a new twist to the most intractable nursing controversy. How and where should nurses be educated and how much did they need to know? Many nurses thought that removing nursing education from domination by hospital-run schools was *the* central objective. But their goal was frustrated by the fact that baccalaureate programs produced too few nurses to meet the needs of the nation's hospitals, mental institutions, and home nursing.

In 1945, even the most avid advocate of baccalaureate preparation for nursing did not envision college-educated women taking over first-level, bedside care in hospitals. The immediate problem was finding enough nurses to do the work while building a cadre of better-prepared administrators, teachers, and head nurses.[18] For some, the licensed practical nurse seemed to fill the description, but broader experimentation was in the offing.

A cascade of postwar commissions and reports poured out of the late 1940s and continued through the 1950s.

Worries about an impending national nursing crisis extended far beyond the nursing leadership and the Kellogg Foundation, of course. A cascade of postwar commissions and reports poured out of the late 1940s and continued through the 1950s. Two early studies spawned by the National Nursing Council on War Service led off: the first Brown Report, *Nursing for the Future* (1948), and the committee to implement the Brown Report, called the National Committee for the Improvement of Nursing Services (NCINS). The Brown Report helped justify and sustain the 1948 National Nurse Accreditation Services, which differentiated adequate from inadequate nursing schools. Brown's study was underwritten by the Carnegie Foundation, and the Russell Sage Foundation purchased 500 copies and circulated them nationwide.

Even as Brown's report was being published, a national committee was formed to try to push through her recommendations. At the Kellogg Foundation, Mildred Tuttle hoped to find support for Marian Sheahan's plans for the Committee. In the 1948-49 annual report, Tuttle mentioned the Rockefeller Foundation startup support of the "Committee to

[18] In 1948 only one nursing student out of 10 was enrolled in a collegiate program; furthermore, half of these students were enrolled in programs leading to a nursing diploma only. At the time it was common for diploma and degree programs to coexist on college campuses. Twenty-nine out of 77 schools offered both; eight colleges offered a diploma only. Extrapolating from this situation, probably about 5,000 out of 100,000 students enrolled in schools were in baccalaureate programs.

Implement the Brown Report," and suggested that the trustees might be interested in giving [the plan] further consideration. Soon renamed the National Committee for the Improvement of Nursing Services (NCINS), the group was charged with figuring out the nursing needs of hospitals, from defining the functions of nurses and other care personnel to estimating how many nurses would be needed, to encouraging hospitals and educational institutions to improve nursing and nursing education.[19]

As a first step, NCINS sponsored a program called the Interim Classification of Schools of Nursing Offering Basic Programs (1949). A year later, it published *Nursing Schools at Mid-Century,* a detailed description of nursing education.

Somewhat earlier, the Hospital Survey and Reconstruction Act of 1946 (Hill-Burton) had provided funds for state surveys of hospitals; similar state surveys of nursing needs modeled after Bixler's Michigan study followed. Not surprisingly, these parallel surveys proved that more and bigger hospitals were needed and, for nursing, that more and better nurses were needed, along with the schools and teachers to prepare them. By 1949, 35 states and Hawaii had surveyed nursing needs using Bixler's model, encouraged by Esther Lucille Brown's recommendations in *Nursing for the Future,* and helped by USPHS money.

In New York, at Teacher's College, Columbia University, R. Louise McManus convened another panel, the Committee on the Function of Nursing. This one, headed by economist Eli Ginzberg, produced a report called "A Program for the Nursing Profession," calling for establishing two nursing levels: the professional, college-trained nurse and the practical nurse. Ginzberg and his group thought that by 1960, a ratio of two practical nurses to every professional nurse in a total population of 625,000 nurses would be about right.

In spite of all the attention nursing received, however, the actual hospital nursing situation in 1950 was well summed-up almost 40 years later by

---

[19] *The National Committee for the Improvement of Nursing Services (NCINS) was first mentioned in the 1948-1949 Annual Report, and thereafter appeared in each report until it was disbanded in 1953. The Rockefeller Foundation invested $10,500 to get it started; thereafter Kellogg provided substantial support each year, ultimately contributing about $200,000.*

Eleanor Lambertsen: "In 1938 we were still studying whether we needed graduate nurses in hospitals; after the war there was no question—but also no nurses" (Personal communication, December 13, 1989).

Notes and References

Arnett, Jan Corey. (1986). *For the People of Latin America: The W. K. Kellogg Foundation and Its Partnership with the People of Latin America for the Improvement of Human Well-Being.* Battle Creek, Michigan: W. K. Kellogg Foundation.

Goostray, Stella. (1969). *Memoirs: Half a Century in Nursing.* Boston: Nursing Archives.

Hanson, Kathleen S. (November-December 1991). An Analysis of the Historical Context of Liberal Education in Nursing Education from 1924 to 1939. *Journal of Professional Nursing, 7.* Hanson provides helpful background for the development of thinking about appropriate liberal arts preparation for professional nurses.

Lynaugh, Joan. (March-April 1991). Stepping In. *Nursing Research, 39.*

McIver, Pearl. (February 1942). Registered Nurses in the U.S.A. *American Journal of Nursing, 42.*

Emory Morris to Marion Sheahan. (August 23, 1946). Battle Creek, Michigan: W. K. Kellogg Foundation archives.

National Nursing Council. (December 1948). *American Journal of Nursing, 48.*

Nelson, Sophie. (December 1946). National Nursing Council Reports. *American Journal of Nursing,* 46

Reverby, Susan (Ed). (1985). The East Harlem Health Center Demonstration: An Anthology of Pamphlets. *The History of American Nursing.* New York and London: Garland Publishing, Inc.

Roberts, Mary. *American Nursing.* (1954). New York: The Macmillan Company.

Russell Charles H. (1959). *Liberal Education and Nursing.* New York: Teachers College, Columbia University for The Institute of Higher Education.

Mildred Tuttle to Elmira Wickenden. W. K. Kellogg Foundation archives.

W. K. Kellogg Foundation. (1941). *Annual Report* (unpublished). Battle Creek, Michigan: WKKF Archives.

W. K. Kellogg Foundation. (1944). *Annual Report.* Battle Creek, Michigan: WKKF Archives. Unpublished.

W. K. Kellogg Foundation. (1944-1945). *Annual Report.* Battle Creek, Michigan: WKKF archives, 11-12. Unpublished.

W .K. Kellogg Foundation. (1944-1945). Proposed Post War Plan in Nursing Education. *Reports of the Officers to the Trustees, 1944-1945.* Battle Creek, Michigan: WKKF Archives.

W. K. Kellogg Foundation. (1946-1947). *Annual Report.* Battle Creek, Michigan: WKKF archives.

W. K. Kellogg Foundation. (1946-47). Michigan Council on Community Nursing. *Annual Report.* Battle Creek, Michigan: WKKF archives.

W. K. Kellogg Foundation Plan, 1946-49, Mugar Special Collections, Boston University.

W. K. Kellogg Foundation. Tentative Outline of a Composite Program for Nationwide Action in the Field of Nursing. Battle Creek, Michigan: W. K. Kellogg Foundation microfilm collection, reel p-9.

W. K. Kellogg Foundation. (1980). *The First Half-Century, 1930-1980: Private Approaches to Public Needs.* Battle Creek, Michigan: W. K. Kellogg Foundation, p. 20.

Wickenden, Elmira Bears. (July, 1945). The National Nursing Council Reports. *American Journal of Nursing, 45.*

4 Hospitals, Nurses, and Education—Eternal Triangle

Postwar hospital development and expansion accentuated the concern and debate about the supply and quality of nurses. The Kellogg staff's simultaneous goals of moving nurse preparation into the educational mainstream and expanding hospitals placed them squarely in the middle of this complex and long-running argument about nursing's role in the health care system.

Educators actively debated the appropriateness of professional education alongside liberal arts studies in universities. For academics, the idea that nursing, which many doubted was even a profession, could be a suitable subject for university study seemed strange. Nursing's educational history, long embedded in the operations of hospitals, significantly impeded its acceptance as a suitable academic subject (Russell, 1959).

The argument centered on how much general education nurses needed, what should be the content of professional education, and where nurses' training should be based. The fact that most nurses were women and that many universities still resisted the idea of women as students and faculty was crucial to the context of this debate. Nursing, along with social work and education, upset the traditional patterns of university life, not only because they were professional schools, but also because they promised to bring many women into campuses.

For academics, the idea that nursing, which many doubted was even a profession, could be a suitable subject for university study seemed strange.

And, importantly though not always explicitly, the impact of nursing's educational decisions on hospital services and costs always played a part in the discussion. Implicit in this linking of nurse education to hospital expansion was the understanding that hospitals primarily supply nursing or personal-care services; so anything that affected nursing immediately affected hospitals.

Another complication was changing opinion about the role of the community hospital in an increasingly science-based, medically driven health care system. Postwar medical education was centered primarily in university-based medical complexes. Much of nursing education, based in "home" hospitals in local communities, was distanced from medical science influences. Tensions rose among planners as they tried to balance the goal of improving nurse education and practice against the local prerogatives and nursing traditions of community hospitals.

Between 1944 and 1952 the Foundation spent nearly $1 million on its postwar Graduate-Post Graduate Nursing Education Program, which gave 13 American universities grants to improve and expand baccalaureate education for nurses. In roughly the same time period the Foundation invested $895,402 in programs testing the licensed vocational or practical nurse concept. Thus, the 1950 Kellogg Foundation view of nursing education combined the licensed practical nurse and baccalaureate nurse to replace the hospital-trained diploma nurse who, for the 60 previous years, staffed hospitals and supplied the nation's private duty nurses. But hospital expansion and change throughout the social system in the country would accelerate and ultimately shake these early, carefully laid Foundation plans.

American Hospitals after World War II

The Foundation was much involved in private and public strategies to improve and expand hospital services during the war, and stepped up its commitment afterward. Both Graham Davis and Andrew Patullo, who successfully led the Hospitals Division at the Foundation, played major roles in forging the new national consensus on upgraded and expanded hospital services. As historian Rosemary Stevens notes in her definitive study of 20th century hospitals, the Commission on Hospital Care (1944-1947) found the formula for its state-by-state quantitative hospital analysis in a Kellogg-assisted pilot study carried out in Michigan hospitals (Stevens, 1989, p. 214).[1] The Commission ultimately concluded that all health services at the local level should be coordinated around the community hospital, and that voluntary insurance plans be relied on to finance care. Its dramatic recommendations for hospital expansion called for an additional 195,000 general hospital beds plus 45,000 beds for tuberculosis patients. The Commission also called for further study of the shortage of facilities for the mentally ill (Commonwealth Fund, 1947).

The Hill-Burton legislation of 1946 funneled federal money into the states to support hospital and public health surveys and planning, constructing and/or renovating hospitals. New construction estimates for hospital beds were based on the idea that there should be 4.5 beds per 1,000 population, an estimate double that in Kellogg's 1946 annual report. Over the next two years all the states in the nation (except Nevada) applied for Hill-Burton funds by submitting plans for hospital construction. A series of amendments appropriating more funds expanded the concept of Hill-Burton throughout the 1950s and 1960s. By 1971, the federal government had contributed $3.7 billion of a total of $13 billion spent on hospital development.

The Commission on Financing Health Care, a successor to the Commission on Hospital Care, was set up in 1951 by the American Hospital Association (AHA), and charged to "study the costs of providing adequate hospital services and determine the best systems of payment for such services" (Weeks & Berman, 1985, Appendix G). Initially chaired by Kellogg's Graham Davis until he resigned because of illness in late 1952, the new Commission began where the Commission on Hospital

[1] *The Michigan project is reported with pride in Kellogg's 1945-46 Annual Report, where the "heterogeneous" and uncoordinated hospital system of the time is sharply criticized. The text in the report is unsigned but the style seems to be that of Graham Davis.*

Care left off. It studied voluntary prepaid insurance plans with an eye toward strengthening them, ways to finance care for those unable to purchase insurance, and the problem of rising hospital costs.

In 1952 the American College of Surgeons banded with the American Hospital Association, the American College of Physicians, and others to form the Joint Committee on Accreditation of Hospitals (JCAH). The JCAH eventually assumed central and sole responsibility for hospital accreditation. In 1953 the American Hospital Association convened still another conference to define the difference between hospitals, nursing and convalescent homes, and domiciliary institutions in an effort to clarify which institutions qualified for construction funds (Weeks & Berman, *Shapers,* 1985, Appendix F, p. 286-287).[2] Throughout this period voluntary insurance expanded from covering 30% of the population in 1946 to 75% of the population by 1960.

In general, all these public and private health initiatives shored up local, as opposed to federal, decision making in health care and carefully avoided any commitment to national or compulsory health insurance.

In general, all these public and private health initiatives shored up local, as opposed to federal, decision making in health care and carefully avoided any commitment to national or compulsory health insurance. At the same time, two distinct trends in hospital development emerged: the university-based medical complex, which valued research and medical education, and the new and up-to-date community hospital, which primarily emphasized patient care and depended on local approval.

Both university medical centers and community hospitals called for expansion and improvement, issuing promises that catered to the public's desire for better quality of hospital care and access for all. However, although postwar hospitals focused on care for acutely ill and surgical patients, a growing proportion of their patients actually suffered from chronic illnesses, some of which were associated with an aging population. Many of these patients experienced repeated admissions, surviving ever more severe episodes in the trajectory of their illnesses until death. Most nursing care was given by student nurses or private-duty nurses paid by the

[2] *An explanation for the American Hospital Association decision to define these various institutions is in George Bugbee's testimony before the Senate Committee on Labor and Welfare in March 1954. Bugbee laid out the complex layers of facilities in the American system and the problems of unequal access based on lack of insurance or limited insurance. He also represented the hospitals' concern that funds for construction might be dispersed beyond the control of hospitals.*

patients. This arrangement encouraged hospital use and helped assure hospitals' economic survival. It was inherently unstable, however, because it could not sustain the hospitals' promise to assure safe and better care.

The Meaning of More Hospitals for Nursing

In the 1950s, a combination of postwar social, cultural, political, and economic factors influenced nursing practice. Americans, fresh from war and experiencing an economic strength previously unknown, desired the best health care money could buy. Nursing was part of that health care package. Scientific and technological improvements in diagnosis, pharmaceuticals, and surgical techniques fueled expectations for rights to more and better health care. As interventionist treatment began to dominate in hospitals, nurses became even more necessary to the institution as skilled technicians and caretakers. In many ways, the community hospital was a microcosm of postwar America. More and more it seemed to be a place where citizens could be restored to health as the country restored itself to peace and prosperity.

Americans, fresh from war and experiencing an economic strength previously unknown, desired the best health care money could buy. Nursing was part of that health care package. Scientific and technological improvements in diagnosis, pharmaceuticals, and surgical techniques fueled expectations for rights to more and better health care.

Higher expectations placed on hospital nurses, however, were not yet matched by change in their education or status. Just the same, nurses needed to follow their patients into the hospitals, and they accepted staff nursing positions instead of pursuing careers in private duty. As care became ever more centralized in hospitals, nursing employment there increased, continuing the trend begun just before the war.

A nagging truism troubled hospital and nursing leaders. Nurses are supposed to make hospitals safe for sick people. In the years after 1946 and through the fifties, the expression "below the margin of safety" appeared very frequently in studies of nursing practice. In 1949, for instance, Mildred Tuttle sponsored a study of nursing practices in eight small Michigan hospitals by nurse Thelma Ryan of Teacher's College, Columbia University. Ryan's analysis of nursing revealed the problems experienced by these hospitals, whose patients were forced to rely on nurses with limited basic education and scant opportunity for clinical updating. Tuttle urged the creation of a statewide system of in-service education for hospital personnel (W. K. Kellogg Foundation, 1948-1949, p. 156-157).

The poor education of nurses, as is evident from the plethora of studies and commissions on the subject, created deep-seated and widespread problems. Lucille Petry Leone recalled the wartime dilemma of trying to decide which schools to support with funds administered by the Cadet Nurse Corps program. "We all knew what was going on there in the small schools," and, she went on to explain, "we would have preferred to give money only to the best schools" (Personal communication). Small, hospital-based schools often lacked teachers or libraries, and many had low admission and graduation standards. It was estimated that student nurses made up 80% of the workforce in the nation's hospitals; most worked 44 to 48 hours per week in addition to attending classes. As director of the Cadet Nurse Corps experiment, Petry argued, however, "When you are trying to get 65,000 students in nursing schools you have to use the plant, such as it is" (Personal communication).

Once the war was over, however, many nurses thought it was time to close the old nursing education plant and build a new one. The commentary on schools of nursing in Esther Lucile Brown's *Nursing for the Future* helped make the reformers' point. Brown characterized the postwar nursing education system as incompetent to produce the requisite number of *qualified* (italics hers) nurses needed because of expanding health services. Her bluntness attracted widespread attention, both positive and negative. One negative reaction came from Graham Davis, then retiring president of the American Hospital Association and Kellogg Foundation staffer.

Davis was especially critical of Brown's assertion that many small hospital schools of nursing were "socially undesirable." He defended these schools, pointing out that they made it possible for hospitals to deliver nursing service throughout the war and during postwar shortages. He charged that Brown's book "ignores the facts of life" (Graham Davis Attacks Brown Report on "Nursing for the Future," 1948, p. 71, 138).

The nursing leadership thought the facts of life were already spelled out when it came to creating schools of nursing. Most accepted the approach that Teacher's College head Isabel Stewart laid out in 1942. (This was the same document copied by Mildred Tuttle in proposing her postwar Graduate-Post Graduate Nursing Education Program [Stewart, 1942]). Stewart believed the essentials for a good school were:

- A viable relationship with a high-quality hospital;
- A liberal, supportive university administration;
- A reasonable budget that could sustain expensive clinical training;
- A faculty capable of self-governance.

Stewart saw two immediate tasks for nursing education; first, it must attract and educate generic students in nursing, and second, hospital nursing-school graduates needed to be upgraded to "specialty" practice. By "upgrading," she meant training for education, nursing care management, and public-health nursing. "It is not practical at the present time to establish academic requirements for nursing school personnel equivalent to those generally regarded as standard in colleges and universities," she said, pointing out that the baccalaureate degree was generally accepted as the minimum for faculty (Stewart, 1942, p. 22). Stewart was generally cautious in her claims, and she made no argument that baccalaureate education improved the safety of patients or productivity of nurses in practice. Instead, she and many commentators who followed her focused on "building the educational plant" to train nurses and produce nurse managers who, they assumed, would direct less well-prepared nurses.

After the war, economist Eli Ginzberg was the least willing of the reformers to compromise and delay change for nursing. He and his 1948 Committee on the Function of Nursing recommended that the hospital school-trained nurse be phased out, "since it is contemplated that the entire nursing mission will eventually be no place for the registered (hospital graduate) nurse" (Committee on the Function of Nursing, 1950, p. 101). Starkly phrased, but clear enough.

Efforts at Reform

In Kellogg's 1951-52 annual report, Mildred Tuttle summarized the effects of the Foundation's investment in postgraduate and graduate education in nursing. The money supported new faculty, improved the schools' financial status in their universities (Boston University's School of Nursing owed its existence to Kellogg funding), helped set up advanced clinical training programs, provided faculty training, stimulated research productivity, and helped nursing faculty become involved in the broader university.

The limitations of small budgets and low enrollment coupled with inexperienced, ill-prepared faculty and university presidents and provosts who were skeptical of nursing programs plagued all but a handful of the initiatives.

On the other hand, she reported, the schools were judged to be very expensive by university officials, suffered from lack of qualified clinical and research faculty as well as weak administrative and educational leadership, and had problems establishing essential learning opportunities with hospitals and other clinical affiliates. The limitations of small budgets and low enrollment coupled with inexperienced, ill-prepared faculty and university presidents and provosts who were skeptical of nursing programs plagued all but a handful of the initiatives. Foundation investments in schools at the University of Pennsylvania and the University of Chicago, for example, disappointed Tuttle at the time. The objectives of funding were not met at either school because of decisions by university leaders that prevented the schools of nursing from growing. Since all these schools had been handpicked by Morris and Tuttle as most likely to succeed, their problems probably reflected the national scene.[3]

[3] Schools selected and funded by the Foundation to develop graduate nurse education were Boston University, University of Chicago, University of Colorado, University of Minnesota, University of Oregon, University of Pennsylvania, University of Pittsburgh, Teacher's College at Columbia in New York City, Wayne State University in Detroit, and McGill University in Toronto, Canada. Of these schools, seven would also be funded to develop nurse administrator programs in a later Foundation effort. These were Teacher's College, Columbia University; Boston University; the Universities of Chicago, Colorado, Minnesota, Pittsburgh, and Wayne State University in Detroit.

Given the state of baccalaureate education, the 1951 recommendation by the National Committee for the Improvement of Nursing Service (NCINS) that 30,000-35,000 graduates of baccalaureate programs be prepared to fill team-leader and head-nurse positions in the nations' hospitals seemed utopian. Head nurses usually took overall administrative responsibility for a ward or unit, anywhere from 30 to 60 patients, and reported to a supervisor or, in small hospitals, to the director of nurses. "Team leaders," a term coined in 1948, were registered nurses who worked under the direction of the head nurse and assumed responsibility for a group of patients assisted by students or non-professional personnel.

NCINS calculated that to achieve their goal, there would have to be one baccalaureate graduate nurse for every 18 hospital patients. Furthermore, they called for continuing education for 75% of the nurses who were managing care. NCINS acknowledged this would require tripling admissions to the nation's colleges, but argued that colleges were only graduating about 2,000 nurses each year. They believed and recommended that the college and university system could raise admissions from 4,000 to 12,000 annually.[4]

At the same time, NCINS and the nursing leadership also pushed for another approach to improved nursing education and practice. NCINS was the first home for the National Nursing Accreditation Service, a five-year program funded by the Commonwealth Foundation, the National Foundation for Infantile Paralysis, the Rockefeller Foundation, and later by income from accrediting services. The impetus behind accreditation, of course, was to improve the existing schools of nursing and try to bring them into conformity with a national standard.

NCINS reported in a 1949 survey that fewer than 10% of the nation's schools could meet its "good school" standard. Among other things, NCINS's "good school" controlled its own faculty, who held at least baccalaureate degrees; its students worked and attended class no more than 44 hours per week with four weeks of annual vacation; and the curriculum

[4] *This recommendation was part of NCINS report to the Foundation in 1951. The number graduating from baccalaureate nursing programs in the United States reached 35,550 in 1985-86. This figure includes graduates of two-year and three-year programs who returned to college for their baccalaureate degrees. See Nursing Data Source, vol. 1 (1991). New York: National League for Nursing.*

included at least 250 hours of biological and physical sciences, 165 hours of social sciences, and 1,000 hours of medical sciences, including clinical practice. Very wide variation was found among postwar nursing schools on all these measures (West & Hawkins, 1950, p. 4-5). NCINS planned to shore up the existing nursing education system and build baccalaureate programs in the nations' colleges and universities.

Efforts to write a national standard for nursing education effectively began in 1917, when the first curriculum guide was published; it was revised in 1927 and 1937, but there was no means for enforcing the standards other than through the various state boards regulating nursing.[5] Throughout the 1930s and during the war, a number of committees tried to create lists of approved nursing schools. Objections from the American Hospital Association and internal conflicts impeded and delayed progress.[6] Now the goal was for the profession to develop voluntary accreditation for all schools of nursing.

Efforts to write a national standard for nursing education effectively began in 1917, when the first curriculum guide was published; it was revised in 1927 and 1937, but there was no means for enforcing the standards other than through the various state boards regulating nursing. Throughout the 1930s and during the war, a number of committees tried to create lists of approved nursing schools.

The decision to try to improve all schools was contradictory to the idea of moving nursing education into colleges and universities because it meant committing resources to hospital-based schools. As Esther Lucile Brown later thought of it, the "raising all boats" strategy may have inhibited what she termed "the better institutions" from real change (Bear, 1986). Her concern, however, was expressed 35 years after the fact. In *Nursing for the Future* (1948), Brown herself recommended a gradualist strategy to tide the nation over while, at the same time, trying to create a new nursing education system based on the baccalaureate. Mildred

[5] *The Curriculum for Schools of Nursing, written in 1917 by Adelaide Nutting, was supplanted by a 1927 and 1937 version prepared by Isabel Stewart. Regulation of nursing practice by state boards was voluntary until the late 1930s; that is, there was no penalty for practicing or naming oneself as a nurse regardless of training or licensure.*

[6] *For a detailed account of the debate over accreditation between 1930 and the NCINS report, see Goostray, Stella. Memoirs: Half a Century in Nursing (1969). Boston: Nursing archives, Boston University.*

Tuttle's postwar Graduate-Post Graduate Education Program's awards to university nursing schools helped build the new system. In the next decades she would try another strategy for building a nursing education system across the country.

Regional Planning for Nursing

One of the Foundation's premises for reform rested on the idea of regional strategizing for change. The practice of demonstrating and testing ideas in Michigan and then exporting the ones that worked formed the background for Kellogg's development of regional funding techniques. In the 1950s, '60s, and '70s, the nursing version of this was WICHE, SREB, and MAIN.

In 1956 the Western Interstate Commission on Higher Education (WICHE) received funds to begin a program with fairly general goals. The Commission intended to encourage planning and growth for nursing in the West, identify health care problems and determine solutions, and improve the nursing care of patients through education and better faculty. This program of cross-state cooperation among institutions was mediated by the Western Council on Higher Education in Nursing (Tuttle, Annual Program Report to the Board of Trustees, 1967-1968, p. 24-25). In 1962 the Foundation again supported Council programs that increasingly emphasized shared planning and research development across schools. Some 83 colleges and universities (93 different programs) belonged to the Council, which met at least twice a year. The Council also attracted funds from the United States Public Health Service and from the respective states of its members.

Similarly, in 1962 the Southern Regional Education Board (SREB) was funded to establish a Council on Collegiate Education for Nursing. The nursing council represented 142 nursing programs in 86 colleges and universities in the Southwest. Over time, SREB served as a conduit of funding to start a series of Foundation programs ranging from statewide surveys of nursing needs to nursing care needs of minority groups to programs enabling registered nurses to earn baccalaureate degrees. The Midwest Alliance in Nursing (MAIN), founded in 1979, also planned to facilitate cooperation between nursing education and nursing service in the Midwest (State, Regional, National Education—Nurses, p. 37).

Regional nursing organizations proved quite durable, especially since the Foundation was willing to provide some operating support and use them as clearinghouses to administer a wide variety of funding projects. In this way the organizations became influential. WICHE and SREB, in particular, were able to stimulate and direct change in nursing education in their respective spheres of influence.

Regional nursing organizations proved quite durable, especially since the Foundation was willing to provide some operating support and use them as clearinghouses to administer a wide variety of funding projects. In this way the organizations became influential.

In the 1950s and 1960s, still another approach, which moved nursing education out of the hospital but not into baccalaureate, began to gain momentum.

The Associate Degree in Nursing

The idea of the Associate Degree in Nursing probably took root with President Harry Truman's 1950 Commission on Higher Education, followed by Margaret Bridgman's study *Collegiate Education for Nursing* in 1953.[7] The President's commission stressed the idea of expanding education norms to two years post-high school or at least 14 years of public education, setting a new educational goal for the nation. National enthusiasm grew for the community college (formerly called junior college) as a place to prepare for technical and specialized jobs. Plans for nursing education reform were quickly linked with this popular new educational strategy. The two-year idea for preparing the nation's nurses attracted much attention in discussions at the Foundation.

During most of her tenure at the Foundation, Mildred Tuttle employed a technique for consulting with nurses around the country called

[7] *The 1947 Truman Report was officially titled Higher Education for American Democracy: A Report of the President's Commission on Higher Education. Washington, D.C.: U.S. Printing Office. Bridgman was dean at Skidmore College when she was recruited by the Russell Sage Foundation to conduct this study of baccalaureate nursing education. She proposed two-year programs in community colleges as well as reform and expansion of baccalaureate nursing programs. Bridgman, Margaret. (1953). Collegiate Education for Nursing. New York: The Russell Sage Foundation.*

the Nursing Advisory Committee (NAC). At least once each year Emory Morris and Tuttle invited a small group of four or five well-known nurse leaders to Battle Creek to "appraise current activities of the [Nursing] Division and review opportunities for program development" (W. K. Kellogg Foundation Annual Report, 1944-45). Foundation president Emory Morris took great interest in these meetings, selecting or at least approving invitees, socializing with them and giving advice.

In 1952, for instance, he asked the NAC to consider the merits of what he called a "ladder plan" in education and the implications of the concept for basic nursing education.[8] The responses of NAC members indicated they were discouraged with the practical nurse-baccalaureate nurse concept, as they found it difficult to integrate into an educational ladder. Tuttle's minutes report that the 1952 group expressed "positive emphasis on the two- and four-year programs" (Minutes of the Nursing Advisory Committee Meeting, January 10-11, 1952). Two years later the NAC generally agreed with the philosophy of the two-year program assuming the associate degree nurse would replace the practical nurse.

But the minutes reveal a sample of problems to come. The "function of the professional nurse [the baccalaureate graduate] is [not clear] in view of changing medical care and complexities arising from an enlarged nursing service team" (Minutes of the Nursing Advisory Committee, January 28-30, 1954, p. 8). The question was always whether patients could be safely cared for if professional nurses oversaw and supervised a "team" of less-educated or trained workers instead of directly providing the more complex care themselves. Although the actual responsibility of the professional nurse remained vague with regard to the associate degree nurse (ADN), the Nursing Advisory Committee of 1955 was still ready to give a "unanimous opinion . . . that the hospital school of nursing will be replaced by the four year collegiate program. . . . In addition to the collegiate programs there will be the two-year programs" (Minutes of the Nursing Advisory Committee Meeting, March 1 and 2, 1955).

[8] *The Nursing Advisory Committee meeting of 1952 convened Lucile Petry, chief nurse officer, United States Public Health Service; Dean Ruth Kuehn, School of Nursing, University of Pittsburgh; Dean Katherine Faville, College of Nursing at Wayne State University in Detroit; and H. Lenore Bradley, Board of Nurse Examiners, New York State. Also attending were Mark Beach, executive director, Greater Detroit Hospital Fund; Margaret Bridgman, Russell Sage Foundation; and Vergil Slee, director of the local Barry County Health Department.*

In August 1957, Mildred Montag of Teacher's College, Columbia University in New York City, was invited to Battle Creek to report on her evaluation of two-year college programs in nursing. Montag had been working on the concept for some years; her first proposals appeared in 1951 in a book based on her doctoral dissertation called *The Education of Nursing Technicians* (Montag, 1951). Morris and Tuttle were deeply interested in Montag's work in spite of her explicit rejection of the "ladder" concept of career mobility from associate degree nurse to professional nurse, which had such appeal to Morris.

Her evaluation of two-year nursing education leading to the associate degree in nursing involved seven colleges and one hospital program; it extended over a five-year period ending in 1956. Her conclusions were that (1) nurses prepared in the new program could carry out the functions usually associated with the registered nurse and could pass licensing examinations; (2) the new programs attracted students; (3) the new programs did fit into the community college successfully; (4) community colleges could manage the costs of the programs; (5) hospitals were willing to provide essential clinical training without charging the students or the colleges; and (6) employers must realistically orient the new graduate and accept him/her as a beginning practitioner (Montag, 1959, p. 339-341).

Following Montag's presentation to Foundation staff, the Nursing Advisory Committee's suggestions for continuing to promote the concept included consultation services, work conferences, and developing instructional material, as well as sponsoring faculty preparation. Ultimately, the Foundation funded what came to be called the "Four States Project." The Four States Project was, to some degree, modeled after the Foundation program begun in 1947 to support practical nurse training.[9] New ADN educational projects were funded in New York, California, Florida, and Texas.

The Four States Project promoted statewide planning and was intended to improve faculty for ADN education, continuing education for nurses,

[9] *The practical nurse training programs were tested in Michigan and then extended to Alabama, Arkansas, Florida, Illinois, Louisiana, and Mississippi. The geography of both the practical nurse programs and the new ADN initiative was affected by recognition of smaller number of nurses being prepared in the Southeast and West, and postwar shifts in the population of the country.*

provide consultation about ADN education, set up demonstration centers, plan and evaluate new programs and graduates, and run in-service programs for employers of the new ADN nurses. Since the concept was quite different from traditional, hospital-based nursing education, state regulation and licensing for practice often needed to be changed. In Texas, objections from nurse leaders and educators effectively undermined the project. According to Bernice Anderson, college educators in Texas were not interested in terminal programs that offered courses at the freshman and sophomore level only. Although Texas seemed a suitable place for experimentation in 1958, the project there was abandoned by 1963 (Anderson, 1966, p. 231 and 248). During 1958 and 1959 the Foundation spent about $1.5 million on the Four States Project, bringing its total allocation to ADN education at that point to more than $3 million.

During 1958 and 1959 the Foundation spent about $1.5 million on the Four States Project, bringing its total allocation to ADN education at that point to more than $3 million.

At least two factors were paramount in the deliberations of the NAC and Foundation staffers during the 1950s. First, all were deeply worried by the need to increase the number of nurses to meet hospital staffing requirements. Some argued that the nation must have 350 nurses per 100,000 population; this would require a 40% growth in the number of nurses (Minutes of the Nursing Advisory Committee, November 21-22, 1957).[10] No one wanted to rely on the hospital schools to produce this increase. In spite of the earlier position taken by the NCINS, it did not seem to the Kellogg group that the baccalaureate programs could or should do it. Second, decision makers were convinced by Mildred Montag's evaluation of two-year programs. In short, the shortage of nurses was accepted as a critical problem and the solution of the ADN graduate seemed feasible.

Francis Payne Bolton, congresswoman from Ohio and the legislator responsible for the wartime Cadet Nurse Corps, seemed to agree. She published a clear statement on the subject in the Congressional Record

[10] *W. K. Kellogg Foundation archives. The National League for Nursing called for this level of increase by 1970. In 1950 it was estimated that the ratio stood at about 240 nurses per 100,000. Forty years earlier, in 1910, there were 10 nurses per 100,000. During the same 40-year period the number of physicians per capita fell from 160 per 100,000 to 140 per 100,000. Leaders in medicine reduced the numbers of schools of medicine and raised admission requirements in the first two decades of the 20th century. In fact, the projections of the National League were almost met; by 1970 there were 369 nurses per 100,000 according to the United States Public Health Service.*

in 1958 (Bolton, 1958). Arguing that what the country needed was "a good two-year trained bedside nurse, with dignity of status, and a living wage," she read into the record parts of an enthusiastic article from *Modern Hospital.* The graduate of the two-year program, she said, would not be a head nurse or supervisor, would not be a public health nurse, or a nurse anesthetist, or an educator—and she would not be eligible for a commission in the Armed Forces. Instead, she would have an in-between status of her own (like a warrant officer in the military), and work under the direction of the senior bedside nurse. Nursing below the professional level should be a career in itself, Bolton argued. If two-year nurses wanted to become professionals, they would have to return to the university and obtain the necessary education. So, the idea of the ADN nurse seemed to be taking hold even if Texas nurses and others resisted.[11]

By 1960, the Foundation and its advisors from nursing, the hospital sector, and education seemed to be committed to a new plan for national nursing. The ADN graduate could alleviate the nursing shortage and meet the hospitals' demand.

Kellogg's Nursing Advisory committee of 1959 did question some of the elements of the Four State Project and, to some extent, the whole idea that ADN education was unique. Faye Abdellah and Rozella Schlotfeldt joined the NAC at its June meeting.[12] They doubted whether faculty planning to teach in ADN programs should be prepared separately from other nurses pursuing graduate studies. The issue seemed to be the extent to which faculty prepared in graduate programs would become specialized and expert in ADN educational techniques as opposed to becoming expert in some specific clinical nursing specialty. In the end, the group sustained its earlier consensus that the ADN program would be a terminal program preparing a "semi-professional" worker in nursing, but they did not explain exactly what this would mean in practice

By 1960, the Foundation and its advisors from nursing, the hospital sector, and education seemed to be committed to a new plan for national nursing. The ADN graduate could alleviate the nursing shortage and

[11] *Nurses who were active during this period had strong objections to the ADN concept. Agnes Olsen, who was head of the Connecticut Board of Nurse Examiners at the time, resisted the development of two-year programs, as did the Texas State board. Mabel Wandelt, a longtime nurse educator and leader, in telephone commentary to author, June 15, 1991, confirmed the outright hostility of some nurse leaders to the idea. Verle Waters, a leader in the ADN movement, agreed that the concept had a polarizing effect. Personal communication, November 10, 1989.*

[12] *The NAC was Mildred Montag; Mildred Newton, head of the School of Nursing at Ohio State; Faye Abdellah, assistant chief, Bureau of Medical Services at the United States Public Health Service; and Rozella Schlotfeldt, associate dean at Wayne State University in Detroit.*

meet the hospitals' demand. The practical nurse no longer seemed so attractive as an assistant to the professional nurse and solution to the shortage issue. The practical nurse did not fit the Truman Commission's post-high school educational scheme and furthermore, seemed to be a "second-class" nurse. Unspoken in the de-emphasis on practical nursing was the sense that an ADN, even though a terminal degree, would draw more respect to nursing. Left unclear, however, was the identity of professional nurses—their scope of work, education, and relationship to the "semi-professional" ADN graduate. Moreover, the hospital training schools still graduated the majority of nurses and, although shrinking in number, remained a force to be reckoned with.

Supply and Demand by 1960

In the 1950s, planners and leaders in all sectors seemed oblivious to the negative effect of low wages and poor working conditions on the ultimate attainment of safe nursing in hospitals.

In the late 1950s, good nurses were hard to find and harder to retain in the nation's hospitals. The shortage crisis was exacerbated by high rates of turnover. Although by 1958, 291,500 of all 460,000 American nurses worked in hospitals (60%), hospitals claimed that 11% of their nursing positions were vacant. Interestingly, wages and benefits of nurses did not improve in response to this clamorous demand. In 1959 weekly wages of Buffalo and Boston nurses averaged $60, in Chicago and San Francisco nurses took home around $72, and in Philadelphia nurses earned the magnificent sum of $56.50 per week. These wages did not compete with factory work, much less teaching; health insurance, vacations, and overtime did not exist for most hospital nurses. Nurses moved from job to job or dropped out of practice for varying lengths of time since seniority made no difference to their income. The fact was that hospitals could easily and cheaply replace departing nurses with graduates of their own training schools or with the rapidly rising numbers of ADN graduates from local community colleges.[13] The great debate about educating nurses rarely included much reference to the reality of life for those who

[13] *For more details on the troubled economics and supply issues in nursing, see Fairman, Julie, & Lynaugh, Joan (in press). Intensifying Care: A History of the American Critical Care Movement, 1950-1980. Philadelphia: University of Pennsylvania Press; and Lynaugh, Joan, & Brush, Barbara (in press). American Nursing: From Hospitals to Health Systems. Cambridge, Massachusetts, and Oxford, United Kingdom: Blackwell Scientific.*

joined the ranks. In the 1950s, planners and leaders in all sectors seemed oblivious to the negative effect of low wages and poor working conditions on the ultimate attainment of safe nursing in hospitals.

The 1950s saw sweeping changes, such as the introduction of an ADN education for nurses, tougher standards applied to schools and enforced through accreditation, and importantly, the growing ratio of nurses to the population. But, the effects—compacted rather than expanded professional nurses' education, and bolstered rather than shuttered hospital schools—would continue to plague nursing in the decades to come.

And, remaining largely unanswered was the question of how to distinguish between professional nurses and technical nurses. As Rozella Schlotfeldt pointed out, before genuinely distinctive baccalaureate or higher degree programs could be developed, it was necessary to differentiate the scope of practice of professional nurses (Personal communication, November 14, 1988). Viewed in retrospect, the lack of a distinction in the world of nursing practice assured that diploma, ADN, and baccalaureate graduates would all be treated the same. Educators could not, by virtue of simply asserting a difference, convince employers that one existed.[14]

Moreover, the educational strategies of the 1950s misjudged the character of the future marketplace for nurses. The technical and knowledge imperatives of increasingly complex patient care in hospitals and the changes in nurse-physician work sharing demanded better educated, more flexible nurses. The system needed a capable and safe direct caregiver who could function independently in a variety of clinical situations. The strategies of the fifties resulted in too many under-educated nurses, the very same problem May Ayers Burgess characterized in the 1930s as "preparing too many, but too few" (May Ayers Burgess, quoted in Roberts, *American Nursing,* p. 179).

[14] *According to Verle Waters, a distinguished leader of the ADN movement, the distinctions among hospital staff nurses were all entirely hypothetical until specialties and areas of expert nursing practice emerged. There was "only so long you can keep making maps without territories—beautiful as the maps are," as she put it. Personal communication.*

Turning to the Practice of Nursing

By the early 1960s, however, new attention focused on the abilities of nurses in practice. The number of registered nurses rose from 246 per 100,000 in 1950 to 279 per 100,000 in 1960, and the number of practical nurses more than doubled. Yet, a sense of impending crisis in hospital nursing persisted. In 1963, a federally sponsored study, *Toward Quality in Nursing*, reported inadequate education for nurses, too few programs in colleges and universities, depressed social and economic status of nurses, poor utilization of nurses, and lack of research on nursing practice. The study's recommendations included balancing hospital nursing staffs at 50% professional nurses, 30% licensed practical nurses, and 20% nursing assistants. This balance was bound to be difficult to achieve since practical nurses and nursing assistants outnumbered professional nurses by more than 100,000. Recognizing the gap between their desires and reality, the study group members proposed a sweeping program of federal investment in nursing education, nurse utilization in hospitals, and research (Report of the Surgeon General's Consultant Group on Nursing, 1963).

At the Kellogg Foundation, Mildred Tuttle sought advice from a new Nursing Advisory Committee in 1964. In April she wrote to the four members asking them how the Foundation might make "a further contribution to the preparation of faculty for junior colleges," and, what type of program could improve the capabilities of the hospital head nurse in patient care supervision (Mildred Tuttle to Faye Abdellah, Elizabeth Giblin, Helen Graves, and Rozella Schlotfeldt, April 9, 1964). All four of the advisors' responses avoided or opposed Tuttle's suggestion of special programs to prepare educators for ADN programs, but all expressed interest and concern about the clinical knowledge and skills of head nurses.[15]

Faye Abdellah urged preparation of clinical nurse specialists while Elizabeth Giblin thought that head nurses badly needed access to

[15] *Head nurses usually were responsible for 30 to 60 patients and supervised a combination of professional nurses, students, practical nurses, and nursing assistants. Not infrequently, the head nurse might be one of only two or three professional nurses on the scene.*

better clinical teaching to improve their care of patients. Abedellah's position was consistent with her arguments in *Patient Centered Approaches to Nursing*, an influential work in which she argued for professional nurses' greater involvement in direct nursing of patients (Abdellah, Fay, et al., 1960, p. 182-190). Rozella Schlotfeldt wrote that head nurses were often ill-prepared and that graduates of ADN and hospital programs were routinely asked to lead patient care without being prepared to do so. Not specifying whether she was speaking of hospital school graduates, ADN graduates, or both, she argued, "we are trying to have graduates of technical education programs exert leadership in patient care when indeed they have only technical preparation in that and no preparation in leadership" (Rozella Schlotfeldt to Mildred Tuttle, April 21, 1964). Schlotfeldt pointed out that top hospital administrators lacked an appreciation for nursing and that nurse educators were handicapped by lack of authority in the clinical settings where they taught students. Complaining that "we are not masters in the house we use for student experience," she recommended that the Foundation invest in demonstrations of better clinical practice as a basis for better education (Schlotfeldt to Tuttle, April 21, 1964).

Notwithstanding the cool reception of her advisors, Tuttle informed them some months later that the Foundation had appropriated funds for summer workshops in four regions of the country to help prepare faculty for ADN teaching. She also urged them to think of and propose programs that might upgrade baccalaureate education. By this time all were aware of the Nurse Training Act of 1964, which ensured a major increase in federal support for nursing education. Tuttle sought advice on how the Foundation might fill in the gaps in the Nurse Training Act.

The NAC finally met in April 1965 to review a long list of new ideas of interest to the Foundation. On the list were demonstrations testing the concept of the new "nurse practitioner," looking at collaboration among health professionals, appropriate use of differently prepared health workers, cost-effective health care delivery, and computer management of patient care data.[16] It stressed educational technology, demonstration-teaching

[16] *Loretta Ford and Henry Silver, a nurse and physician, were demonstrating in Colorado how nurses with additional education could assume much of the general care of children in the community. Their new type of nurse was called a pediatric nurse associate, but the term nurse practitioner caught on.*

centers, and intensive teacher-training, especially for ADN programs. It also recommended intensive care units (coronary care) and experimental programs for the preparation of clinical specialists.

This 1965 NAC agenda signified a shift in the focus of the Foundation, not necessarily reducing its interest in educational issues, but expanding into projects that might alter the conditions of nursing practice. As University of Chicago scholar Odin Anderson put it, "the nursing profession, which was shaped by the apprentice system as an arm of administration rather than a clinical department during a time when relatively few women attended college, [was] now trying to overcome its past" (Anderson, 1968, p. 8).

The past to be overcome included not only an eccentric educational system but a persistent muddle about nursing's scope of practice and professional responsibility.

Contextual Change

One major force for change in nursing during the 1960s was the very active intervention of the federal government in higher education for nurses and then later in testing new nursing roles for them. The 1944 Public Health Services Act included provision for a Division of Nursing in the United Sates Public Health Service. Lucile Petry Leone, a frequent adviser at the Foundation, became its first head. Among many other things, the Division of Nursing oversaw a series of federal investments in nursing education. The 1956 Extramural Grants Program in Nursing Research helped underwrite doctoral education for nurses, as did the 1962 Nurse Scientist Graduate Training Grants. In 1964 the Nurse Training Act began to pour funds directly into schools and into the hands of students, ensuring the rapid growth of masters programs in nursing. At first, the money supported nurses preparing to be faculty members and nurse administrators, but soon schools of nursing began to shift the curricular emphasis to clinical programs. The Nurse Training

One major force for change in nursing during the 1960s was the very active intervention of the federal government in higher education for nurses and then later in testing new nursing roles for them.

Passage of the Medicaid and Medicare legislation in 1965 intensified demand for nursing services and created an even larger revenue stream for hospitals and other health care delivery institutions.

Act was re-authorized time after time; it ultimately became the largest ($4 billion) educational investment in higher education for women ever made.

Passage of the Medicaid and Medicare legislation in 1965 intensified demand for nursing services and created an even larger revenue stream for hospitals and other health care delivery institutions. These public programs, which finally, after long debate, guaranteed funding of care for those over 65 and for those too poor to purchase care on their own, opened up the health care system. Coupled with employment-based insurance, Medicare and Medicaid seemed to ensure that all Americans could have access to health care. One of the consequences of this promised access was growing recognition that nurses must be thought of and used differently if the expectations of the public were to be met.

Notes and References

Abdellah, Fay, et al. (1960). *Patient Centered Approaches to Nursing.* New York: The Macmillan Company.

Anderson, Bernice E. (1966). *Nursing Education in Community Junior Colleges: A Four-State, Five-Year Experience in the Development of Associate Degree Programs.* Philadelphia: J. B. Lippincott Company.

Anderson, Odin. (1968). *Toward an Unambiguous Profession?: A Review Of Nursing.* Chicago: Center for Health Administration Studies.

Bear, Elizabeth, Ph.D., RN. (1986). Oral History of Dr. Esther Lucile Brown, Tape #1 (Doctoral dissertation, University of Texas at Austin, 1986).

Bolton, Hon. Frances P. (1958). Solution to the Nursing Dilemma: Calls for Understanding and Action on the Part of Professionals and the Lay Public. Extension of remarks in the Congressional Record. Washington, D.C.: U.S. Government Printing Office.

Bridgman, Margaret. (1953). *Collegiate Education for Nursing.* New York: The Russell Sage Foundation.

Committee on the Function of Nursing. (1950). *A Program for the Nursing Profession.* New York: The Macmillan Company. Eli Ginzberg chaired the committee and wrote the report with the cooperation of R. Louise McManus of Teacher's College, Columbia University. McManus was Isabel Stewart's successor as head of the nursing programs at Teachers College.

Commonwealth Fund. (1947). *Hospital Care in the United States.* New York: Commonwealth Fund.

Fairman, Julie, & Lynaugh, Joan (in press). *Intensifying Care: A History of the American Critical Care Movement, 1950-1980.* Philadelphia: University of Pennsylvania Press.

Goostray, Stella. *Memoirs: Half a Century in Nursing* (1969). Boston: Nursing archives, Boston University.

Graham Davis Attacks Brown Report on "Nursing for the Future." (October 1948). *The Modern Hospital, 71.*

Lynaugh, Joan, & Brush, Barbara (in press). *American Nursing:From Hospitals to Health Systems.* Cambridge, Massachusetts, and Oxford, United Kingdom: Blackwell Scientific.

Mildred Tuttle to Faye Abdellah, Elizabeth Giblin, Helen Graves, and Rozella Schlotfeldt. (April 9, 1964). Nursing Advisory Committee files, W. K. Kellogg Foundation archives.

Minutes of the Nursing Advisory Committee Meeting. (January 10-11, 1952). W. K. Kellogg Foundation archives. The minutes for these meetings are not signed but the text suggests that Tuttle prepared them from her notes and shared them with President Morris.

Minutes of the Nursing Advisory Committee. (January 28-30, 1954), p. 8. W. K. Kellogg Foundation archives.

Minutes of the Nursing Advisory Committee Meeting. (March 1 and 2, 1955). W. K. Kellogg Foundation archives. The NAC of 1955 included Marjorie Bartholf, University of Texas at Galveston; Lulu Wolf Hassenplug, dean of Nursing, University of California at Los Angeles; Eleanor

Lambertsen of Teacher's College, Columbia University; and Marion Sheahan, National League of Nursing.

Minutes of the Nursing Advisory Committee. (November 21-22, 1957). W. K. Kellogg Foundation archives.

Montag, Mildred. (1951). *The Education of Nursing Technicians.* New York: G. P. Putnam's Sons.

Montag, Mildred. (1959). *Community College Education For Nursing.* New York: McGraw-Hill Book Company.

Nursing Data Source, vol 1. (1991). New York: National League for Nursing. Report of the Surgeon General's Consultant Group on Nursing. (1963). *Toward Quality in Nursing: Needs and Goals.* Washington, D.C.: U.S. Department of Health, Education, and Welfare. The Consultant Group was chaired by Alvin Eurich of the Ford Foundation. It became the blueprint for subsequent federal funding of nursing education.

Rozella Schlotfeldt to Mildred Tuttle. (April 21, 1964). The letter was four pages long.

Russell, Charles H. (1959). *Liberal Education and Nursing.* New York: Teacher's College Bureau of Publications. Reporting on his study of education for nursing for the Institute of Higher Education, funded by the Carnegie Foundation, Russell reviews the debate about the merits of professional and technical education versus traditional liberal studies and shows how nursing might be integrated with liberal studies.

State, Regional, National Education—Nurses. (July 1979). Indiana University Foundation Appropriation Request. Barbara Lee files. Thirteen states participated: Illinois, Indiana, Iowa, Kansas, Michigan, Minnesota, Missouri, Nebraska, North and South Dakota, Ohio, Oklahoma, and Wisconsin. Another regional organization called the New England Board for Higher Education held similar goals.

Stevens, Rosemary. (1989). *In Sickness and in Wealth: American Hospitals in the Twentieth Century.* New York: Basic Books.

Stewart, Isabel M. (1942). A Guide for the Organization of Collegiate Schools of Nursing. New York City: National Nursing Council for War Service and the Association of Collegiate Schools of Nursing.

Tuttle, Mildred. (Unpublished). Annual Program Report to the Board of Trustees, 1967-1968. Battle Creek Michigan: W. K. Kellogg Foundation archives.

W. K. Kellogg Foundation. (1944-45). Annual Report. WKKF archives.

W. K. Kellogg Foundation. (1948-49). Nursing Service Study. Annual Report. Battle Creek, Michigan: WKKF archives.

Weeks, Lewis, & Berman, Howard. (1985). *Shapers of American Health Care Policy.* Ann Arbor, Michigan: Health Administration Press. The Health Administration Press was established in 1972 with support from the W. K. Kellogg Foundation in cooperation with the University of Michigan. Assistance and promotion of professional hospital administrators was a continuous feature of Kellogg programming after World War II.

West, Margaret, & Hawkins, Christy. (1950). *Nursing Schools at Mid-Century.* New York: National Committee for the Improvement of Nursing Services.

5 Re-conceptualizing Nurses' Work

Caring for one sick person entails an assortment of tasks and requires skill and knowledge ranging from simple to complex. But most of all, it means that one person must commit time to help the other. Caring for constantly changing groups of sick people, as in hospital nursing and other organized nursing arrangements that promise safe care, calls for all of the above, plus a high degree of organization and coordination to achieve a civil, humane atmosphere, safety, and reliable, continuous delivery of therapeutic services.

With this in mind, Foundation staffers and their advisors tried to rethink the essence of care and the scope of practice of nurses and other health workers, so as to attain what Odin Anderson labeled their utopian ideas of health reform. The very concept of nursing underwent more or less constant scrutiny during this era. What would be the nature of nursing? The escalating impact of science-based medicine on standards of health care, new ways of framing care such as "comprehensive care" and "progressive patient care," debate about the responsibilities of hospitals to their communities, changes in the organization of medical care, and growing recognition of behavioral and psychological factors in patient care all impinged on nursing. Much of this discussion centered on upgrading and expanding the abilities and role of the nurse; the goal was to improve quality while also improving access to health care.

What would be the nature of nursing? The escalating impact of science-based medicine on standards of health care, new ways of framing care such as "comprehensive care" and "progressive patient care," debate about the responsibilities of hospitals to their communities, changes in the organization of medical care, and growing recognition of behavioral and psychological factors in patient care all impinged on nursing.

Of special significance to nurses was Virginia Henderson's international definition of nursing, first published in 1959. Henderson's simple language gained international credence and became a new basis for changing legislation controlling nursing practice:

> The unique function of the nurse is to assist the individual, sick or well, in the performance of those activities contributing to health or its recovery (or to peaceful death) that he would perform unaided if he had the necessary strength, will or knowledge. And to do this in such a way as to help him gain independence as rapidly as possible (the definition was popularized in Henderson, 1966).

The Henderson definition laid out the social responsibility of the nurse. Her use of the word *unique* specifically detached the functions of the nurse from physician oversight. When the International Council of Nurses adopted her definition, it was the first time the organization omitted the physician as overseer of all of nursing.

Nurses, physicians, politicians, and journalists all participated in the discussion permeating the professional and policy literature of the late 1950s and throughout the 1960s. Frances Reiter, who studied the content of practice and who coined the term "nurse clinician" to label nurses expert in special areas of practice, probably was the most articulate leader of the effort to re-conceptualize nursing practice. Hildegard Peplau, who formulated an innovative and influential clinical specialist program in psychiatric mental health nursing at Rutgers University, promulgated the importance of nurse-patient relationships based on psychodynamic principles. Edmund Pelligrino, a respected medical educator and dean, urged responsible collaboration between medicine and nursing.[1]

Sociologist Odin Anderson, in 1968, enlarged on the difficulty of finding balance between what he called "the essentially utopian" social goal of equal access to health care services and various elements of professional life, including career patterns, incentives, standards, and job descriptions. Equity in health services and for professionals, in turn, had to be balanced against their cost to society, changing demographics, and new knowledge and technology (Anderson, 1968, p. 31).[2] The Kellogg Foundation addressed the question by sponsoring an array of projects ranging from "progressive patient care," a way of reorganizing practice by grouping hospital patients according to need for services, to funding for demonstrations of new practice-oriented university schools of nursing.

Progressive Patient Care

Faye Abdellah, a highly accomplished nurse, worked for the federal government and often acted as a Kellogg advisor. She was deeply interested in the ideas behind progressive patient care (PPC).[3] Abdellah's earlier concept of patient classification, i.e. developing a typology of nursing problems, became linked with redesigning the way patients were assigned to hospital beds. During an interview done in 1981, she recalled the enthusiasm she and her colleagues felt for this idea in the late 1950s. Convinced that they could improve quality and efficiency by clustering patients together based on the amount of care each needed, Abdellah and

---

[1] *Dr. Edmund Pelligrino, MD, a respected medical scholar who served as a national and professional voice for medicine, wrote often on the subject. See The Nurse Must Know—the Nurse Must Speak, 1960; The Challenging Role of the Professional Nurse in the Hospital, 1961, p. 56-62; and The Ethical Implication in Changing Practice, 1964, p. 110-112. Reiter was the first dean of the Graduate School of Nursing, New York Medical College (1960-69), now the Lienhard School of Nursing at Pace University. See Frances Reiter, 1904-1977, 1988, p. 268. Also see Peplau, 1952.*

[2] *Writing before the rapid rises in health care costs of the 1970s, Anderson offered a view characteristic of the 1960s, writing, "The economy will continue to expand for the foreseeable future . . . money, as such, need be of diminishing concern (p. 32).*

[3] *Fay Abdellah was well known for her early work on patient classification which she continued in her position with the United States Public Health Service. Her "21 problems" appeared in Abdellah, 1955.*

her colleagues set up small experiments. She recalled that the now familiar categories of intensive care, intermediate care, and self-care "seemed to fall in place." They assumed that patients would "progress" from one stage of recovery to another; hence the term "progressive patient care" (Weeks, 1983, p. 24). As it turned out, patients did not usually "progress" through more than two levels of care, but the principle of assigning patients to hospital space by need for care endured (DeVries, 1970, p. 44).

Each patient's level of illness determined the needed amount of nursing care; theoretically, this made more economical use of available nursing staff. The slogan promoting the idea was "the right patient in the right bed at the right time, for the right cost."

First proposed by the Army in 1951, PPC was promoted by hospital and nursing planners as a way to reorganize facilities, service, and staff around the medical and nursing needs of patients (Fairman, 1992, p. 11). Patients were admitted to hospital units according to their degree of illness and their need for medical and nursing care rather than assigning them to rooms using the customary categories of medical diagnosis, sex, or ability to pay. There were five basic levels in the progression from most ill to most well: intensive care for the seriously or critically ill; intermediate care for the patient requiring "a moderate amount of care;" self-care for the "physically self-sufficient;" long-term care for those needing extensive chronic care; and home care for convalescents and less needy chronically ill persons (Fairman, 1992, p. 39). Each patient's level of illness determined the needed amount of nursing care; theoretically, this made more economical use of available nursing staff. The slogan promoting the idea was "the right patient in the right bed at the right time, for the right cost" (Halderman & Abdellah, 1959, p. 38-42; also see W. K. Kellogg Foundation, 1980, p. 61). Progressive patient care, noted Jack Haldeman and Abdellah, "shows promise of helping hospitals with their two most pressing problems, scarcity of trained personnel and continuing improvement of services without unduly increasing costs" (Halderman & Abdellah, 1959).

During the time PPC was being tested, the "team system" of assigning nurses to care for patients was in vogue, but it was beginning to attract criticism as unwieldy and distracting professional nurses from their clinical work. (See Chapter 3.)

The architectural design of each unit was also thought critical to optimal efficiency and care delivery. When intensive care units were built for the purpose, they were designed with centrally located nurses' desks to permit constant observation; one or two beds might be behind glass walls to control noise without obstructing the nurses' view.[4] Self-care units, on the other hand, were intended to afford patients privacy and quiet as they convalesced from surgery or awaited diagnostic testing. Nurses practicing in self-care would, it was thought, primarily teach and support people learning to adapt to new diagnoses, chronic illness, or functional impairments.

According to both Abdellah and Kellogg's Andrew Patullo, the concept of self-care units failed. It did not "fit in" with medical and hospital preferences for treating patients who had specific, acute problems requiring therapeutic interventions. Physicians had no interest in the need to deal with the whole person, as in teaching or adapting to illness, which was inherent in the self-care concept (Abdellah, p. 24; and personal communication with Andrew Patullo, November 1988). Also, hospital administrators found that they needed the space occupied by self-care beds for patients with more acute care needs.

However, the intensive care concept did spread rapidly, as did the idea of creating special units called coronary care units for patients suffering from myocardial infarctions (heart attack). The Kellogg Nurse Advisory Committee meeting of April 1965 recommended that the Foundation give major attention to these experimental units for the intensive care of coronary patients.

By this time, demonstration units in several parts of the country clarified the organizational and nursing care implications of clustering together acutely ill patients. Nurse Rose Pinneo and physician Lawrence Meltzer of Philadelphia's Presbyterian Hospital were showing how such a unit could function (Pinneo, 1965, p. 76-79). What they believed was fundamentally new about intensive and coronary care was the increased responsibility of the nursing staff. "Intensive coronary care is primarily, and above all, a system of specialized nursing care [and its] success is predicated almost wholly on the ability of nurses to assume a new and different role" (Meltzer, Pinneo, & Kitchell, 1965). One of the influential

[4] *At first, however, many of the intensive care units were roughly fashioned out of available space. It was a few years before better-planned renovations could be completed.*

aspects of the Foundation's support of the progressive patient care idea and the coronary care unit was demonstrating that the concept was feasible in small, rural hospitals (Holt, 1970; see also DeVries, 1970).

National Commission for the Study of Nursing and Nursing Education

In 1967, the Foundation, in tandem with the Avalon Foundation and a private donor, funded a three-year study and report on nursing.[5] Recommendations made by this group, called the National Commission for the Study of Nursing and Nursing Education, were destined to become an outline for Foundation funding in nursing during the 1970s (Personal communication with Barbara J. Lee, former program director, W. K. Kellogg Foundation, November 1, 1988). According to its director, Jerome Lysaught, the four nursing problems identified for study were: supply and demand for nurses, nursing roles and functions, nursing education, and nursing careers. These problems were considered in the context of trends in health care delivery, social and cultural changes, advances in medical science, and changed economics of health (Lysaught, 1970, p. 4-10). Stressing the independence of the Commission, Lysaught wrote that the objective of the study was

> [to] improve the delivery of health care to the American people, particularly through the analysis and improvement of nursing and nursing education. People, as individuals, are the subjects and objects of health care. Care is required wherever people are gathered together—in hospitals, clinics, nursing homes, factories. This is the *raison d'etre* for nurses and nursing. Our efforts to enhance nursing and to improve the education, roles and careers of nurses begin with the critical need to ensure that better health care is provided for the person who requires it (Lysaught, 1970, p. 11-12).

[5] *The Avalon Foundation later became the Mellon Foundation. Encouraged by a $50,000 commitment from the American Nurses Association, the Kellogg and Avalon Foundations granted $100,000 each, to which was added $300,000 from an anonymous donor.*

The National Commission was intended as a response to the health care problems identified in the 1963 Secretary's report *Toward Quality in Nursing*. The rapidly escalating rate of change and conflicts within nursing and between nursing, hospitals and physicians during the 1960s guaranteed controversy for the Commission whatever actions it recommended.

The rapidly escalating rate of change and conflicts within nursing and between nursing, hospitals and physicians during the 1960s guaranteed controversy for the Commission whatever actions it recommended.

By 1970, the Commission reported a series of familiar concerns about nursing practice in hospitals, introduced the long-neglected areas of community and long-term-care nursing, and then laid out its recommendations regarding nursing roles and functions, nursing education, and nursing careers. It proposed 58 recommendations in all. Among those that interested the Kellogg Foundation were: establishing a national joint-practice commission between medicine and nursing; strengthening the ability of hospital nursing chiefs to be on par with other administrators and physicians; directing regional bodies to work on nursing curriculum and educational issues; considering the idea of education as an "open-ended process" with "access to enlarged opportunities . . . a right of every individual"; establishing state "master-planning committees" for nursing education; and, of course, closing of hospital schools in favor of ADN and baccalaureate programs (Lysaught, 1970, p. 116).

Its least understood recommendation suggested two "related but differing career patterns—episodic and distributive" (Lysaught, 1970, p. 81-147). The Commission thought that basic nursing programs should offer distinct preparation for two different nursing careers; one would prepare nurses to care for people with acute or chronic illnesses, particularly in hospitals, while the second option would prepare nurses to work mainly in health promotion and disease prevention, particularly in community settings (Lysaught, 1980, p. 91-92). This idea met with little favor as it seemed to segregate education for nursing into illness and health categories.

The Commission drew heavily on the work of Odin Anderson and sociologist Fred Davis. Quoting Anderson, it argued, "If medical specialists become the model [for nursing practice] the behavioral aspects of patient care will languish and be preempted by lesser trained nurse types.

... If the patient-as-person orientation can be the model, then ... professional nursing has a viable role model in its educational program" (Lysaught, 1970, p. 40).

These observations from social scientists reflected the debate in nursing about the nature of specialization. Would nurse specialists simply become "junior doctors"? What would the character of nurse specialists become? One thing was clear by then. Nursing wanted master's programs that could prepare nurse clinicians "capable of improving nursing care through the advancement of nursing theory and science" (American Nurses' Association, 1969, p. 2). The earlier approach to graduate study that focused on administration and teaching was now thought to "devalue nursing care and practice" by assuming that no further education in clinical nursing was needed beyond the baccalaureate (American Nurses' Association, 1969, p. 2).

Would nurse specialists simply become "junior doctors"? What would the character of nurse specialists become? One thing was clear by then. Nursing wanted master's programs that could prepare nurse clinicians "capable of improving nursing care through the advancement of nursing theory and science."

A significant shift in thinking about the relationship of ADN nurses to the rest of nursing shows up in several Commission recommendations. The recommendations urging career mobility and "articulation" suggest that the ADN program should no longer be seen as a terminal program but instead, a feasible first step toward advanced education. The National Commission thus refuted, or at least chose to ignore, ADN originator Mildred Montag's basic premise that ADN education was not a route to professional nursing. The Commission's position helped accelerate a trend toward linking programs, the "ladder" concept suggested years earlier by Emory Morris. The Commission also refuted the supply-side, shortage-inspired solutions of earlier years, saying that nurses were disenchanted with hospital nursing and that, without improving the conditions of nursing practice, simply graduating more nurses solved nothing.

The Lysaught Report, as it came to be called (Director Jerome Lysaught, an educator, was from the University of Rochester in New York), attracted much attention via its publications and regional meetings in 1970; the Commission continued to issue pamphlets and report at conferences throughout the seventies. On the advice of staff, the Kellogg board made appropriations totaling more than $485,000 in 1970 and 1972 to help implement its recommendations (W. K. Kellogg Foundation, 1970 and 1972).

Much of the Commission's attention and the response to its report necessarily focused on the struggle to find consensus on the argument about what came to be called "entry into practice."[6] What was at stake was how well educated the nurse should be in order to be licensed by the state to accept final nursing responsibility for all nursing needed by patients. Hospitals accepted the idea that a professional nurse with a valid state license should be the one responsible. But the question still remained: who was the professional nurse? For 60 years, the baccalaureate degree had been proposed but not implemented as the entry standard for state licensure for professional nursing. Virtually all studies about nursing and hospitals were influenced by this enduring and unresolved impasse.

The Position Paper of 1965

In 1965, five years before the Lysaught Commission's report, the American Nurses' Association (ANA) published an unequivocal statement on educational preparation for nursing. The so-called Position Paper framed the debate from then on. The ANA stipulated three levels of nursing practice: professional, with minimum preparation at the baccalaureate; technical, with minimum preparation at the associate degree; and, assistive, prepared in short, pre-service courses in vocational schools. The ANA called for education, which would "provide an environment in which the nursing student can develop self-discipline, intellectual curiosity, the ability to think clearly, and acquire the knowledge necessary for practice" (American Nurses' Association, 1965). The Position Paper insisted that all "education for those who are licensed to practice nursing

The ANA stipulated three levels of nursing practice: professional, with minimum preparation at the baccalaureate; technical, with minimum preparation at the associate degree; and, assistive, prepared in short, pre-service courses in vocational schools.

[6] *For a brief history of the argument, see Lynaugh, 1980, p. 266-270.*

should take place in institutions of higher education" (American Nurses' Association, 1965). The Position Paper was silent on career mobility. As it became well-known, it accentuated the controversy about nursing education throughout all the layers of the health care system.

Notwithstanding the position of nursing's professional organization, movement toward the goal of relocating nursing education was extremely slow. Change was problematic on two levels: it required voluntary closure of local hospital training schools, and, in effect, disenfranchised the licensed practical nurse who usually had one year of hospital- or high school-based training. The Kellogg Foundation's agenda in the 1970s, which did include several elements of the Lysaught report, was influenced by the urgency and complexity of this protracted struggle.

Changes within the Foundation

In 1967 Emory Morris relinquished the president's role and became chairman of the board at the Foundation. In January 1968 Mildred Tuttle resigned; she had reached the mandatory retirement age of 65. Both Morris and Tuttle spent almost their entire professional careers with the Foundation and worked closely together when they were at the height of their capacities. Emory Morris became head of the Foundation in 1943 when he was only 38; Mildred Tuttle began directing the nursing work at the Foundation in 1944 when she was 41.

The agenda of the last Nursing Advisory Committee meeting called by Mildred Tuttle covered a wide range of projects from announcing funding for the National Commission on Nursing to searching for ways to enable teachers of nursing to attain or retain their clinical expertise. Joining veteran NAC member Dean Schlotfeldt at the March 1967 meeting were Helen Belcher of the Southern Regional Education Board; Jesse Scott of the U.S. Public Health Service; Lillian Harvey, dean of Tuskegee Institute's School of Nursing; and Dorothy Smith, dean of the School of Nursing at the University of Florida at Gainesville.

A few months later, in December, they all received a letter from Emory Morris notifying them that the Foundation planned to reorganize. "The change reflects . . . a consensus that Foundation activities are becoming . . . more and more focused on the problems of people rather than oriented to particular disciplines" (Emory Morris to Schlotfeldt, Smith, Scott, Belcher, and Harvey). As part of the change the Foundation abolished its seven divisions and re-designated the division directors as "program directors." Morris wrote that the Foundation would continue its interest in agriculture, health, education, and public affairs but that staff would no longer be specifically identified with their respective disciplines. In the future, he said, "advisory committees will be constituted on an 'ad hoc' basis . . . we will no longer utilize discipline-related committees as in the past" (W. K. Kellogg Foundation archives). Thus ended an era of exceptionally close cooperation between the Foundation and national nursing leaders.

The Foundation would continue its interest in agriculture, health, education, and public affairs but staff would no longer be specifically identified with their respective disciplines... thus ended an era of exceptionally close co-operation between the Foundation and national nursing leaders.

Nurses who knew Tuttle clearly recall her influence at the Foundation. According to Mary Kelly Mullane, "Mildred Tuttle was very well known, very well-liked. The Kellogg Foundation's contribution to nursing was Mildred Tuttle" (Personal communication, November 1988). Eleanor Lambertsen thought of Tuttle as

> a perfect match for a foundation at that time. She was highly respected by the people there. [Her] laid-back [style] was important. We were getting aggressive, losing some of the [deference] we had. That wouldn't have worked at Kellogg. She used tremendous skill in getting information. Ever a lady. She didn't ruffle feathers—it wasn't that she was innocuous—she was skilled. She did a great deal in her association with the leaders in nursing to bring them into the Foundation. Once you got on the Advisory Committee she used you. You felt a kinship with her.

Speaking of both Morris and Tuttle, Lambertsen went on, "I think they were terrific. They were out in the open. [Morris] had an astute way of questioning and learning" (Personal communication, December

1989). Lulu Hassenplug, who got support from the Foundation for initiatives at her School of Nursing, University of California at Los Angeles, remembered Tuttle as a "very forward-looking person" who took leads well—listened and acted. She remembered that Tuttle was shocked at the problems faced by nursing in the West (Personal communication with Lulu Hassenplug, December 1991). But, of course, with Mildred Tuttle's resignation and the move of Emory Morris out of the president's role, the direct, personal Foundation contacts enjoyed by many nurse leaders ceased.

Mildred Tuttle died in 1976 at the age of 73. She was eulogized as "a leader in movements to prepare clinical nurse specialists and associate degree nurses in this country, for improvement of nursing leadership in Canada and for strengthening nursing education and the training of nursing auxiliaries in Latin America" (W. K. Kellogg Foundation, 1976). According to Andrew Patullo, who worked with Tuttle for many years, she was a warm and compassionate individual and earned the affection and esteem of all with whom she came in contact during her long and successful career (Personal communication, November 14, 1988). Her colleague Emory Morris had died two years earlier.

Mildred Tuttle's successor at the Foundation was Barbara J. Lee, who was recruited from Yale University's program in hospital administration. Barbara Lee graduated from Skidmore College with a baccalaureate in nursing and then earned a master's in Public Health and Hospital Administration at Yale. Her career before joining Kellogg included administration posts in a number of hospitals (W. K. Kellogg Foundation, 1968). Appointed as associate program director, Lee reported to Andrew Patullo, who had recruited her believing she would "bring a different perspective" (Personal communication, November 14, 1988).

Russell G. Mawby took over as the new president of the W. K. Kellogg Foundation in 1970, succeeding Phillip Lackerby, who served from 1967 to 1970. President Mawby sought to use Foundation resources to "play a distinctive role . . . critical and catalytic, in providing for experimentation, redirection, [and] exploration" (Mawby, 1974). Robert E.

Kinsinger, who was much involved in the community college movement, became vice president. The Foundation, by this point, received most of its income from the trust created by Will Kellogg which, by 1970, held 18,088,240 shares of the Kellogg Company's common stock. The Foundation's income for 1970 was $16,814,889. New funds committed to nursing, in addition to ongoing projects, totaled $837,823 (W. K. Kellogg Foundation, 1970, p. 25).

According to both Barbara Lee and Andrew Patullo, Mawby introduced a more group-oriented and staff-driven style of program development and funding. Every project had to be endorsed by the staff leaders in agriculture, health, and education before it went to the board. Patullo remembered "the group came together and discussed objectives and priorities—where we wanted to go. [We had] more congeniality [and felt] more a part of the team" (Personal communication, November 14, 1988). Lee remembered the process of staff decision making as beginning with consensus about where the Foundation should be putting its money in the next year or two. The group process became the framework for programming, so it "was difficult if a given proposal didn't fit—I had to show Andrew Patullo and Russell Mawby where a particular proposal fit into my program plans" (Personal communication with Barbara Lee, November 1, 1988). Advisory committees from the various fields were eliminated; staff consulted but, according to Lee, "we did not have deans influencing [the Foundation] from outside" (Personal communication, November 1, 1988). In spite of this more staff-driven style, advisory committees and consultants were still invited on an ad hoc basis to the Foundation to give advice on specific projects or new agendas (see annual reports of the W. K. Kellogg Foundation, 1972, 1976, 1977, and 1979).

In the case of nursing, the recommendations laid out by the National Commission on Nursing guided funding decisions. Barbara Lee remembers spending a lot of time educating the board and the officers of the Foundation about nursing. "There was not a good understanding," particularly when it came to the unusual three-pronged educational system leading to licensure.

The 1970 Nursing Agenda at Kellogg

By 1977, after nearly a decade at the Foundation, Barbara Lee was a well-known figure on the national nursing scene and a program director. Invited to address an audience of nurse educators gathered together by the Southern Regional Education Board (SREB), she reviewed the health, education, and agriculture interest of the Foundation. She again encouraged the educators to propose creative programs demonstrating the application of new knowledge. And she reminded listeners that it was not the Foundation's practice to fund buildings and equipment, nor was it primarily interested in basic research. Rather, it preferred to provide seed money for experimental, pilot projects that, if successful, could be emulated and replicated in other communities, institutions, and organizations. She explained that Foundation staff decided on the issues and concerns in nursing (for instance) by taking stock of supply and demand of personnel, trends in education and employment, and evaluating strengths and weaknesses. The Foundation chose to fund "nursing education and service projects [focused] on such issues as . . . career mobility, . . . new roles and practice patterns, promoting interdisciplinary education, improving management skills, implementing curriculum innovations and finding ways to narrow the gap between nursing education and service." In the late 1970s, Lee reported that the Foundation took a "[positive] position on health promotion or health education because . . . prevention of illness, health maintenance, and educating people to take more responsibility for their own health is essential" (Lee, 1977).

The Foundation funded and encouraged three of the most interesting and controversial movements affecting nursing in the 1970s.

The Foundation funded and encouraged three of the most interesting and controversial movements affecting nursing in the 1970s. Of course, these movements—linking university nursing programs with academic health centers, encouraging collaboration between medicine and nursing, and expanding the scope of nursing practice—did not spring to life in 1970. All had roots in the Tuttle era at the Foundation, and all were the subject of the various studies of nursing done in the years since World War II.

University Nursing in Academic Health Centers

Improving patient care and nursing education by forging a clinical, educational, and research link between university schools of nursing and their respective university hospitals (a concept originated by Dorothy Smith, dean of Nursing at the University of Florida in Gainesville) slowly attracted supporters. Rozella Schlotfeldt was the dean who first garnered Kellogg Foundation support for her initiative at (Case) Western Reserve University. The Foundation began a five-year commitment with Western Reserve in 1966. The funding was intended to develop and evaluate "a new inter-institutional scheme" between the School of Nursing at Western Reserve University and the nursing department of the university hospitals. The faculty and hospital nursing department planned to reorganize to "attain high-quality nursing care of patients and [promote] excellence in education" (Tuttle, 1967-1968, p. 34, 35). The school and hospitals were linked by joint committees that dealt with overall policy decisions, nursing practice, patient records, and ambulatory care, among other issues. To achieve its goals, the project relied on joint decision making, the introduction of clinical specialists with both patient care and teaching responsibilities, collaboration between medicine and nursing, and re-allocation of certain ward responsibilities to ward managers. Certain faculty held clinical appointments in the hospital as well as academic appointments in the university.

What came to be called the "unification" movement, bringing together nursing education and practice in academic health centers, was fueled by the deep concern felt by many nurses about the separation of nursing education from its practice base. Mary Kelly Mullane, dean of the College of Nursing at the University of Illinois, argued that professors had to get involved in nursing practice in the hospitals, clinics, and homes where they taught. She supported the earlier campaign to separate nursing education and nursing service as necessary for each to "develop for its unique purposes without distraction or hindrance from the other." But, she argued, "the time has now come for reunification of

What came to be called the "unification" movement, bringing together nursing education and practice in academic health centers, was fueled by the deep concern felt by many nurses about the separation of nursing education from its practice base.

nursing *practice* (not nursing service) and nursing education in the teaching hospitals and agencies that constitute the practice and research laboratories for nursing students and professors" (Mullane, 1969). Mullane's remarks served to amplify her argument that faculty in schools of nursing must behave more like other faculty members in universities. They must, she insisted, base teaching on clinical investigation and submit for criticism results of their investigations to journals and in professional meetings. Barbara Lee included Mullane's 1969 paper in materials explaining the Rochester project, the second university "unification" effort she proposed for funding by Kellogg.

In May of 1967, before she retired, Mildred Tuttle made a field visit to the University of Rochester in New York at the invitation of Eleanor Hall, chair of the Department of Nursing of the School of Medicine and Dentistry. Hall was seeking funding to create a new College of Nursing. The president of the University of Rochester, economist W. Allen Wallis, was about to take up his task as president of the National Commission for the Study of Nursing and Nursing Education. Tuttle's notes on the visit concluded, "I agree with Dr. Morris that this is a situation with excellent leadership and administrative support. With financial assistance this program should rapidly develop into a 'prestige' College of Nursing where the emphasis is in the development of skilled practitioners and excellent teachers of nursing" (W. K. Kellogg Foundation, 1967). In September she prepared a draft memo for the Foundation board saying that Rochester had all the essential characteristics for funding consideration; she recommended $725,000 over a five-year period.

But then came a long pause, which probably was at least partially due to the transition of leadership at Kellogg. Years later, Barbara Lee recalled that, at some time in 1968, when they were on a plane trip together, Emory Morris asked W. Allen Wallis what he needed in terms of financial support to start a school of nursing (Personal communication with Barbara Lee, November 1, 1988). We don't know Wallis's exact answer but can assume he came up with a figure. The university and the Foundation were in constant communication in view of the fact that the National Commission, then at the height of its activity, was based at the University of Rochester.

By 1969 the correspondence about the Kellogg-Rochester project connected Foundation support for the new school of nursing with the unification idea then being worked on in Cleveland at Western Reserve. Kellogg staff felt that a strong commitment on the part of the school to the nursing service of the university hospital (Strong Memorial Hospital) should be part of the proposal. In 1968, the dean of the medical school at Rochester, Lowell Orbison, appointed a committee on nursing, which, among other things, recommended creating a new school and unifying that school and the nursing service of the university hospital under one nursing leader (Report of the Committee on Nursing, 1969).

The university followed up with a "Master Plan for Nursing at the University of Rochester," which proposed an organizational design for the new school encompassing the nursing service of Strong Memorial Hospital. The design established the position of dean and director of the College of Nursing, who would have authority for education, practice, and research in nursing. Dr. Virginia Brandtl, recruited from the University of Chicago in 1969 to head the expansion of the graduate program, authored the "Master Plan" (Joint Committee on Nursing and Kellogg Proposal Sub-Committee, November 4, 1970).

In 1971, Barbara Lee prepared a request to the Kellogg board for $930,000 over a five-year period to "develop a College of Nursing." Lee reviewed the frustrating history of nursing's inability to achieve parity in academia and characterized the Rochester proposal as a new approach:

> [Rochester] is committed to restructuring its Department of Nursing to compare organizationally with the School of Medicine in terms of autonomy, financing, qualifications of faculty, and facilities. The proposed College of Nursing could provide a model for every school of nursing located in a university which also supports a school of medicine. The uniqueness of the Rochester proposal is exemplified best by knowing that the leadership of the Medical School was instrumental in recognizing the plight of its Department of Nursing and organizing the machinery necessary to develop the proposal. The results of this long-overdue

> alliance cannot but help to contribute significantly to the troubled profession of nursing (Lee, 1971, p. 13).

The board funded the project and the University of Rochester committed $2 million. Loretta Ford, one of the originators of the nurse-practitioner movement, was selected to become the first dean of the new school and took office in fall 1972.

That same year Rush University in Chicago established its school of nursing using similar principles. Founded under the direction of Luther Christman, Rush joined with Western Reserve and Rochester to demonstrate and influence patterns of faculty practice and research as well as teaching in university nursing across the country.

Bringing Nursing and Medicine toward Collaboration

The National Commission specifically recommended creating a forum where nursing and medical leaders could work on issues important to both disciplines. In 1973, the American Medical Association Education and Research Foundation requested about $327,000 from the Foundation to assist the American Nurses' Association and the American Medical Association in supporting the National Joint Practice Commission (NJPC). Arguing that the nation's 700,000 nurses constituted an under-used resource and that joint planning by medicine and nursing would be essential to effective collaboration, the Kellogg staff urged the board to fund the NJPC proposal. The NJPC mission was intended to make recommendations on the roles of the physician and nurse in providing health care, suggest changes in education for both physicians and nurses, alleviate differences between the two disciplines, and help develop state-level joint practice committees.

Eight representatives each from nursing and medicine served on the Commission, which was funded, at first, via the National Commission's implementation budget. In April 1972, Robert A. Hoekleman, M.D., of the University of Rochester, became the first chair and Shirley Smoyak, RN, Ph.D, of Rutgers University, accepted the first vice-chair post. Later in 1972 the NJPC organized its first conference to stimulate the formation of joint practice committees around the country.

Over the next few years, the NJPC issued statements on medical and nurse practice acts, the certification of nurses and physicians, the definition of joint practice in hospitals, and the definition of joint practice in primary care. The Foundation continued its support, adding demonstrations and consultation to the repertoire of the NJPC (W. K. Kellogg Foundation, Quality of Health Care—National Joint Practice Commission). The demonstrations were intended to improve patient-care records by integrating nurse and physician input, to improve planning for care, and to increase participation of nurses in decision making. Ultimately, in 1981, the NJPC disbanded when the American Medical Association withdrew its support.

Expanding the Scope of Nursing Practice

In 1970, the Assistant Secretary for Health and Scientific Affairs of the federal government, Roger Egeberg, called together a task force to examine the scope of practice issues in nursing. In 1971 the panel's report, "Extending the Scope of Nursing Practice," defined patients' needs for care into three categories called primary, acute, and long-term and spelled out specified areas of function for nurses (The Secretary's Committee to Study Extended Roles for Nurses, 1971). The Committee "chose to view the subject from the perspective of the consumers of health services." A broad concept of health care, based on the comprehensive care ideas of the 1950s and 1960s, pervaded their brief report:

> Health Care in its entirety, from the point of view of providers and consumers alike, is the sum total of care rendered by all disciplines. It comprises more than diagnosis, treatment and rehabilitation associated with acute and chronic illness; it includes health education, health maintenance, prevention, and early case finding. It involves giving the public a voice in the design and operation of

> health systems and the allocation of health resources to meet changing demand. As such, health care is not the province of any one profession, nor does it lend itself to delivery through a rigid professional hierarchy (Preface, *Extending the Scope of Nursing Practice*, p. 3).

The Committee was influenced by nursing and public health demonstrations carried out in the late 1960s. At the University of Colorado, in the first of these experiments to be published, nurse Loretta Ford and physician Henry Silver launched a program "to prepare nurses with bachelor's or master's degrees to assume an expanded role in providing total health care for children" (Silver Ford, & Day, 1968, p. 88-92). Nurses were expected to provide comprehensive care to well children and identify, appraise, and temporarily manage certain acute and chronic conditions of those who were ill. The program prepared nurses for practice in private physicians' offices, clinics, and other areas without adequate health services for children. Subsequent demonstrations at the University of Kansas by physician Charles Lewis and nurse Barbara Resnik proved that nurses could, indeed, satisfactorily manage a broad spectrum of problems presented by adults as well as children seeking primary care.

Also influential, especially given the strained history between organized medicine and nursing, was the 1970 recommendation of the AMA Committee on Nursing, which was subsequently adopted by the AMA. Among other objectives intended to "[increase] the significance of nursing as a primary component in the delivery of medical services," the AMA supported the idea "of expanding the role of the nurse in providing patient care" (Medicine and Nursing in the 1970s, 1970, p. 1881-1883). Edmund Pellegrino, M.D., medical educator and philosopher chaired the AMA committee and served as a member of the 1971 Secretary's committee as well.

In general, practicing in this new nursing role was believed to require special training and experience to be able to provide a sophisticated level of care, comfort, and guidance based on the patient's health care problems. Hence the terms "extended" or "expanded"; the term "nurse practitioner," although a bit redundant, became attached to this concept by

1972. The idea of the nurse practitioner, to some extent, was built on previous positive experience with nurse midwives, nurse anesthetists, and public health nurses, who accepted broad and complex responsibilities above and beyond those accepted by the "average" nurse. Faye Abdellah linked the popularity of the nurse practitioner movement with the earlier progressive patient care experience, especially in intensive care units, where nurses assumed much more clinical responsibility and worked closely with physicians (Personal communication).

Nurse practitioners were expected to employ skills in history-taking, interviewing and physical diagnosis, and knowledge of physiological and behavioral processes in health and illness to evaluate and respond appropriately to patient needs. The nurse practitioner would then implement the appropriate action or refer the matter to a physician colleague. The nurse practitioner was also responsible for creating and maintaining records of care rendered. At first, nurses were prepared to become nurse practitioners through continuing education programs. Graduate nurses would add the new skills of diagnosis and treatment to their existing nursing knowledge and skills. In the mid- and late-seventies, the concept began to be integrated into the formal, degree-granting nursing education system.

The Kellogg Foundation found the concept of the nurse practitioner to be very consistent with its goals for improving quality and access to health service. In certain areas of the country, access to physicians or primary-care services was a problem. Nurses with enhanced clinical ability could, it was believed, bring health care to the people in those areas. The nurse could do health assessments, supervise care for people with chronic conditions, provide midwifery care, and care for children. The Foundation sponsored six state projects to both expand health service delivery and prepare nurses for nurse practitioner roles. The staff at the Foundation also encouraged the development of family nurse practitioner programs sponsored by universities in the West. The Mountain States Health Corporation (Boise, Idaho) studied the efficiency of geriatric nurse practitioners in nursing homes. This large and influential project both prepared nursing-home nurses with additional skills and tried to integrate these nurses back into the nursing home in new roles.

The Kellogg Foundation found the concept of the nurse practitioner to be very consistent with its goals for improving quality and access to health service.

Barbara Lee thought it imperative that nurses seize the opportunity presented by the nurse practitioner movement to become full participants in health care. She believed that nursing was entering a new and exciting era and that the momentum created by the practitioner idea had to be maintained by thoughtful evaluation, documentation of practice, and collaboration with physicians. The Foundation expended more than $6 million on various projects and demonstrations related to the nurse practitioner movement during her tenure (Lee, 1977; and Mawby & Sparks, 1983).

Although these new roles for nurses and changes in the practice relationships of nurses and physicians would remain continuously controversial, expanded practice and nurses' need for more education became the norm for the seventies. In 1973, citing the growing trend of nurses practicing in areas of care traditionally thought the province of physicians, the Division of Nursing of the United States Public Health Service funded a major study of the nurse practitioner movement. The Division of Nursing as well as Kellogg and other Foundations had invested heavily in educating nurse practitioners. Their *Longitudinal Study of Nurse Practitioners* examined educational programs, students in the programs, and graduates subsequent employment in "expanded roles" (results are in Sultz, et al., 1976, 1978, and 1980). The study reported favorably on the nurse practitioner movement and became a basis for continued, relatively generous support of the nurse practitioner idea.

The postwar redefinition of nursing revamped the limits of the 1940s, in effect re-conceptualizing nurses' work. But the struggle to find better and reliable ways to administer nursing services continued. Improving administration in nursing, to ensure that higher quality services actually reached the public, remained on the Foundation agenda from 1930 well into the 1980s.

Notes and References

Abdellah, Faye G. (1955). Methods of Determining Covert Aspects of Nursing Problems as a Basis for Improved Clinical Teaching (Doctoral dissertation, Teacher's College, Columbia University, 1955).

American Nurses' Association. (1965). Educational Preparation for Nurse Practitioners and Assistants to Nurses: A Position Paper. American Nurses' Association.

American Nurses' Association. (1969). Statement on Graduate Education in Nursing. New York: American Nurses Association.

Anderson, Odin. (1968). *Toward an Unambiguous Profession?: A Review of Nursing.* University of Chicago: Center for Health Administration Studies.

The Challenging Role of the Professional Nurse in the Hospital. (December 16, 1961). *Hospitals, 35,* p. 56-62.

DeVries, Robert A. (June 16, 1970). Progressive Patient Care. *Hospitals, Journal of the American Hospital Association.* DeVries was supported by the Kellogg Foundation to conduct an evaluation of the progressive patient care concept at one hospital in Michigan; he reported his review of five years' experience.

Emory Morris to Schlotfeldt, Smith, Scott, Belcher, and Harvey. Battle Creek, Michigan: W. K. Kellogg Foundation archives, administrative files.

The Ethical Implication in Changing Practice. (September 1964). *American Journal of Nursing, 64,* p. 110-112.

Fairman, Julie. (1992). New Hospitals, New Nurses, New Spaces: The Development of Intensive Care Units, 1950-1965 (Doctoral dissertation, University of Pennsylvania, 1992).

Frances Reiter, 1904-1977. (1988). In V. L. Bullough, O. M. Church, & A. P. Stein (Eds.), *American Nursing—A Biographical Dictionary.* New York and London: Garland Publishing, Inc.

Halderman, Jack C., & Abdellah, Faye G. Concepts of Progressive Patient Care. (May 16, 1959). *Hospitals.*

Henderson, Virginia. (1966). *The Nature of Nursing.* New York: The Macmillan Company.

Holt, Eric. (1970). *Coronary Units in Small Hospitals.* Battle Creek, Michigan: W. K. Kellogg Foundation.

Joint Committee on Nursing and Kellogg Proposal Sub-Committee. (November 4, 1970). A Master Plan for Nursing at the University of Rochester. Author's personal files. The committees included leadership from Strong Memorial Hospital's nursing service, the Department of Nursing of the School of Medicine and Dentistry, the School of Medicine, and the Strong Memorial Hospital administration.

Lee, Barbara J. (April 1, 1977). Meeting New Challenges in Health Care: Nursing Education and Practice. Talk presented at the 27th meeting of the Council on Collegiate Education for Nursing, Southern Regional Education Board.

Lee, Barbara J. (April 1977). Meeting New Challenges. And Mawby, Russell G., & Sparks, Robert D. (August 12, 1983). WKKF Contributions to Nursing. Memorandum to the W. K. Kellogg Foundation Board of Trustees.

Lee, Barbara. (October 18, 1971). University of Rochester, College of Nursing. *Report to the Board of Trustees.* W. K. Kellogg Foundation archives.

Lynaugh, Joan. (February 1980). The Entry into Practice Conflict—How We Got Where We Are and What Will Happen Next. *American Journal of Nursing, 80.*

Lysaught, Jerome P. (1970). *An Abstract for Action.* New York: McGraw-Hill Book Company.

Mawby, Russell G. (1974). A Man Stands Up. *Annual Report.* Battle Creek, Michigan: W. K. Kellogg Foundation archives.

Medicine and Nursing in the 1970s: Position Statement. (September 14, 1970). *Journal of the American Medical Association, 213.*

Meltzer, Lawrence; Pinneo, Rose; & Kitchell, J. R. (1965). Preface. *Intensive Coronary Care.* Bowie, Maryland: The Charles Press.

Mullane, Mary Kelly. (February 1969). Nursing Faculty Roles and Function in the Large University Setting. Memo to Members—Council of Baccalaureate and Higher Degree Programs. New York: National League for Nursing.

The Nurse Must Know—the Nurse Must Speak. (March 1960). *American Journal of Nursing,* 60.

Peplau, Hildegard. (1952). *Interpersonal Relations in Nursing—A Conceptual Frame of Reference for Psychodynamic Nursing.* New York: G. P. Putnam's Sons.

Pinneo, Rose. (February 1965). Nursing in a Coronary Care Unit. *American Journal of Nursing, 65.*

Report of the Committee on Nursing, University of Rochester Medical Center (January 31, 1969). Author's personal files. Chaired by Barbara Bates, MD, the committee consisted of four nurses, five physicians, and three university representatives.

The Secretary's Committee to Study Extended Roles for Nurses. (1971). *Extending the Scope of Nursing Practice.* Washington, D.C.: U.S. Government Printing Office.

Silver, Henry; Ford, Loretta; & Day, Lewis. (April 22, 1968). The Pediatric Nurse Practitioner Program. *Journal of the American Medical Association, 204.*

Sultz, Harry A., et al. (1976, 1978, and 1980). *Longitudinal Study of Nurse Practitioners:* Phases I, II, and III. Washington, D.C.: U.S. Government Printing Office.

Tuttle, Mildred. (1967-68). Progress Report to the Board of Trustees. W. K. Kellogg Foundation archives.

W. K. Kellogg Foundation. (1970 and 1972). Appropriation Requests. Barbara Lee personal files.

W. K. Kellogg Foundation. (1970). *Annual Report.* Battle Creek, Michigan: W. K. Kellogg Foundation library.

W. K. Kellogg Foundation. (1980). *W. K. Kellogg Foundation: The First Half-Century, 1930-1980: Private Approaches to Public Needs.* Battle Creek, Michigan: The W. K. Kellogg Foundation.

W. K. Kellogg Foundation. (April 2, 1976). Press announcement. Battle Creek, Michigan: WKKF archives.

W. K. Kellogg Foundation. (January 1968). Announcement. Battle Creek, Michigan: WKKF archives.

W. K. Kellogg Foundation. (May 26, 1967). Field visit to University of Rochester. Battle Creek, Michigan: W. K. Kellogg Foundation archives.

W. K. Kellogg Foundation. Quality of Health Care—National Joint Practice Commission. Battle Creek, Michigan: W. K. Kellogg Foundation archives. The Kellogg board approved a request for $1,155,765 in July of 1977.

Weeks, Lewis E. (Ed.). (1983). *Faye G. Abdellah, In the First Person: An Oral History.* Chicago: American Hospital Association and Hospital Research and Educational Trust.

6 Delivering Care on Demand: Administering and Regulating Nursing Services

When substantial numbers of people who cannot care for themselves are gathered together, even temporarily, in a circumscribed space, organization, control, and reliable performance by competent people are essential to avoid chaos. Any news photograph of natural disaster, war casualties, or victims of famine or illness vividly makes this clear. We don't think of health care institutions as necessarily chaotic but, of course, the potential for disaster is inherent. Hospitals and other institutions exist to shelter dependent people whose conditions cause them to need help more or less continuously.

To the extent that institutions admit only those who are acutely ill or whose chronic problems are quite severe, the tempo of patient-care needs speeds up and the potential for chaos accelerates. By 1950 most non-psychiatric, community hospitals restricted their services to acutely ill or surgical patients, people with debilitating injury or severe chronic illness, and maternity patients. The continuing fall in the average length of patient stay in hospitals measured the disappearance from the nation's hospitals of the convalescent, the person admitted for medical observation, or the person with stable chronic illness. In 1946, nursing homes admitted relatively less-sick old people. However, as time went on, even nursing homes began to restrict admissions to only those sick persons who needed substantial help with daily living. Thus, in all the nation's

> *Thus, in all the nation's health care institutions, the potential for chaos rose along with the proportion of patients who were acutely ill or very dependent.*

health care institutions, the potential for chaos rose along with the proportion of patients who were acutely ill or very dependent.

Almost from the time of its founding, the Foundation acted on its interest in the problems of clustering the sick, injured, elderly, or very young in institutions. Of course, nursing was invented in the 19th century because of the problems of managing care in hospitals. Most of the impetus to found nursing as a distinct occupation came from reformers' need to find able and acceptable caretakers to keep order in institutions (Rosenbert, 1987; or Lynaugh, 1989). America's post-World War II decision to expand and improve its hospitals forced reexamination of their management and, in particular, staffers at Kellogg questioned whether hospitals could deliver on their promise of safe and effective nursing care for those facing acute illness, severe injury, or major surgery.

Hospital expansion meant reforming a system that, at the beginning of the Great Depression, totaled about 7,500 hospitals. Thirty percent of these institutions were very small, with fewer than 25 beds; only 23% could accommodate more than 100 patients, while the majority held between 50 and 100 beds. Although many marginal hospitals failed during the Depression, those that survived were still small, impoverished, and isolated from one another.

Nurses as Administrators

Before World War II the administration of these small, local hospitals was in the hands of physicians, nurses, nuns, or experienced lay men or women. In 1927, Michael Davis estimated that about 50% of hospital superintendents were women (Davis, 1929). Noting that hospitals had the "responsibilities of a business but the characteristics of a philanthropy," Davis cited lack of training and the low salary paid nurse administrators. Male administrators, whether physicians or laymen, earned between $4,000 and $6,000 per year while the most frequent salary paid nurse superintendents was $1,800 plus room and board (Davis, 1929, p. 6, 14). Some nurse superintendents earned more, of course, and hospital administration was a popular and secure career choice for women who remained in nursing for a lifetime. When the hospital administrator was a nurse, he or she usually was also directly responsible for the nursing service. This was a common merged role, usually learned on the job.

Early hospital administrators formed professional associations (the American Hospital Association, for example) and courses in hospital management, journals and textbooks abounded. But hospital administration as a distinct field of university study did not emerge until the mid-1930s; it took another 10 years before the field developed. Then, with help from the private sector, notably the Kellogg Foundation, and also from the federal government, programs in hospital administration began to multiply in universities across the country.

Hospital administration as a distinct field of university study did not emerge until the mid-1930s; it took another 10 years before the field developed. Then, with help from the private sector, notably the Kellogg Foundation, and also from the federal government, programs in hospital administration began to multiply in universities across the country.

According to Eleanor Lambertsen, who taught in the hospital administration program created at Columbia University, it took time before the new schools could gather prestige. She pointed out that the student group was a mix of young college graduates, more seasoned nurses, and an ever-shrinking group of physicians; that women who were not nurses did not choose this career; that the future for nurses in hospital administration was bleak. "They don't end up at the top," she said (Personal communication).

The traditional management structure of hospitals gradually changed after the war. The usual triangular relationship involving the nurse or other superintendent, the board of trustees or other owner, and the medical staff, was replaced by an organization where the chief nurse slid down the hierarchy to report to a professional manager instead of directly to the hospital board or owner.

This demotion of nursing in hospital management was a very gradual and erratic process. Kellogg's Andrew Patullo remembered beginning his hospital management experience in a medium-sized hospital under the tutelage of a respected nurse superintendent (Weeks, 1983). Occasionally, even a large teaching hospital retained a nurse as superintendent; Betty Barrang, RN, was superintendent of the Hospital of the University of Pennsylvania until the 1960s. In the 20 years after 1940, however, physicians, nurses, and laypeople steadily gave way to professional hospital administrators. Well-prepared hospital administrators became particularly important as hospitals and their budgets grew larger. Financial control became ever more critical to hospital survival.

Well-prepared hospital administrators became particularly important as hospitals and their budgets grew larger. Financial control became ever more critical to hospital survival.

Paralleling the expansion of the new field of hospital administration was a sharper emphasis on specialized nursing administration. The nagging nurse-supply problem, added to other burgeoning hospital demands, served as an impetus for the development of nurse managers. "The next creative step in the improvement of nursing services," wrote Herman Finer, director of the Kellogg Foundation Nursing Service Administration Research Project, "is better administration" (Finer, 1952, p. vii). The director of nursing service was responsible for "organizing, directing and supervising the nursing service to insure sufficient and competent nursing care . . . setting up the budgeting request and administering the budget appropriation for the nursing service" (Finer, 1952, p. 145). Planners assumed that in order to perform well, nursing administrators would have to know the full range of nursing activities in the hospital. The nurse administrator should "know the functions she is administering, who ought to be doing them, in what relationship of teamwork, with what degree of quality, and at what comparative compensation . . . an intimate, inward acquaintance with each function is indispensable" (Finer, 1952, p. 145).

Adding to this daunting list of obligations, the accounting firm of Harris, Kerr, Forester & Company emphasized the Director of Nurses' financial responsibilities to the hospital in a 1953 survey report. They believed that nursing administrators should propose their own nursing budgets (not a common occurrence) and hold economy of nursing services as their primary goal (Maschal, 1953, p. 64-71, 76, 77, 94). Agreeing with a popular idea of the time, the firm stressed the use of time and efficiency studies as ways to measure and link nursing functions with cost saving. Herman Finer, while concurring that time and motion studies were fundamentally worthwhile, warned that "[they] are liable to become facile diversions from more difficult, but more important researches in nursing services; there is a danger of being fascinated by the obvious" (Finer, 1952, p. 317).

On the other hand, neither financial management nor institutional planning was generally believed to be "women's work": directors of nursing who sought control of the nursing budget and/or advancement to the top of the revised hospital hierarchy found little encouragement. Kellogg's Barbara Lee, a nurse prepared and experienced as a hospital administrator, believed that hospital management required someone with a lot more training . . . [than a nurse]" (Personal communication). Hospital administrators and physicians, though, were usually active participants in planning and decision-making for nursing. Historian Susan Reverby suggests that nursing was so essential to the hospital and medical postwar agenda that it made administrators and physicians "fearful of the competition from nurses and unwilling to let a group of women determine their own destiny" (Reverby, 1987, p. 121).

Neither financial management nor institutional planning was generally believed to be "women's work": directors of nursing who sought control of the nursing budget and/or advancement to the top of the revised hospital hierarchy found little encouragement.

Nursing administration during the 1950s emphasized and focused on the internal domestic management of nurses' work. Final decisions about appropriate nurse staffing or the numbers and types of patients admitted to the hospital were usually outside of the responsibility of nurse managers. These crucial decisions increasingly became the province of senior hospital administrators who were neither physicians nor nurses.

Creating a Staff

No matter who was in charge of the 1950s hospital, the task of managing patient care called for one overarching objective: The hospital had to build a nursing staff capable of predictable, competent, and continuous care in an increasingly technological and unpredictable environment.

Traditionally, many hospitals used the supervision of head nurses. This kind of staffing meant that patients were cared for by constantly rotating and under-skilled nurses-to-be. To supplement the few graduate nurses and the students, hospitals helped patients find private duty nurses to give them individual care. The feasibility of this approach was undermined by patients' lack of funds to pay their own nurses and a growing shortage of qualified private duty nurses.

The nursing work on the typical hospital unit was difficult and demanding. Before the ideas of progressive patient care (see Chapter 5) began to take hold, patients usually were assigned to beds classified by medical, surgical, maternity, or pediatric categories, by the identity of the admitting physician, by the patient's ability to pay, or by some combination of these. So, on any unit, there could be a mix of patients capable of self-care, others needing intermediate care, some under post-operative observation, and others who were critically ill. If a patient was critically ill, the wise family hired a private duty nurse. That, of course, depended on whether they could find and afford one.[1] Often requests for private duty nurses went unfilled or were delayed.

Thus, patient care in the hospital could be erratic, and conditions on any single unit varied widely and changed rapidly. Often inexperienced nurses or student nurses faced new emergency situations without consultation or appropriate knowledge of their own. As one physician lamented, "many nurses in their training and immediately afterward have been in contact with so few cases requiring intensive therapy . . . that they know relatively little about their management" (Sadov, Cross, & Higgens, 1954, p. 65-70). As historian Julie Fairman has noted elsewhere, "putting out brush fires" characterized care rather than the systematic approach to each patient the nurses were supposed to learn in their training (Fairman, 1990). In effect, critically ill patients were at risk of neglect except as nurses responded to each immediate crisis. Moreover, most nurses expected their care responsibilities to be technical and procedural, the sort of work reasonable to expect from partially trained and inexperienced staff.

[1] *Some estimates suggest the number of active private duty nurses fell by more than 40% during World War II. See Roberts, Mary. (1954). American Nursing. New York: The MacMillan Company, p. 363.*

An Experiment in Staffing the Hospital

In 1950, the Foundation funded a study of nursing team organization and functioning conducted by Eleanor Lambertsen of Teacher's College, Columbia University. R. Louise McManus, who instigated the study and attracted Kellogg's interest, acknowledged the complex causes of the hospital nursing crisis. She argued for the team concept on the basis that there could be no "justification for postponing an all-out attack on the problem" (McManus, 1953, p. iii, iv).

The premise of "team nursing" assumed that the professional nurse could extend her or his capabilities by leading a care team made up of practical nurses and nursing assistants. Groups of patients were assigned to each team by the head nurse. Patient needs were evaluated in relation to the personnel available and the various team members assigned to patients by the team leader. The team leader retained responsibility for all the patients and for continuous personal contact with the team's patients as "their nurse." Written plans and conferences led by the team leader coordinated the efforts of the group.

The premise of "team nursing" assumed that the professional nurse could extend her or his capabilities by leading a care team made up of practical nurses and nursing assistants. Groups of patients were assigned to each team by the head nurse. Patient needs were evaluated in relation to the personnel available and the various team members assigned to patients by the team leader.

Those who urged the team method stressed the crucial role of the team leader. The team leader, one definition went,

> must be a good nurse, capable of giving good bedside care. She must know team work and believe in the team plan and have enthusiasm for it. She should have mature judgment and interest in the broadening field of nursing and insight into the need to depend on the team rather than upon herself (Schafer, 1953, p. 58; and Lamberts, p. 17-20).

But most nurses also understood that the ratio of one professional nurse to two assistants was ideal; more assistants per professional nurse meant that the nurse spent all her time supervising and had little time to care for patients needing professional services (Shafer, 1953). The team nursing idea relied heavily on planned differentiation of work and placed a high value on efficiency.

The problem was, however, that hospitals did not employ enough professional nurses to act as team leaders—at least in the ratios recommended by its proponents. Most of the time the nursing staff "made do" with the people available. The work was divided into specific tasks by the team leader, who was, usually but not always, a registered nurse; a student or graduate nurse gave medications, and various nursing assistants and licensed practical nurses (LPNs) did the rest. On a single unit there might be 30 to 60 patients; they were grouped into two or three clusters of 15 to 20 patients per team. Depending on whether the hospital maintained a nurse training school, each patient group might be tended by one or two graduate nurses, three or four students, and/or one or two practical nurses or nursing assistants. Team nursing spread through the American hospital system during the 1950s but complaints about the quality of care given by hospital nursing staffs continued.

Criticisms of care and team nursing focused on the nurse's absence from the bedside, which, in turn, was blamed on hospital administrators' over-emphasis on "technique and efficiency" (Rogge Meyer, 1960). According to these critics, adding variously skilled layers of nursing service personnel meant that administration and supervising all these workers absorbed more and more of the professional nurses' time, leaving patient care to less-skilled people. Both Faye Abdellah and Frances Reiter participated in this critique; both also advised the Kellogg Foundation on a regular basis (Abdellah & Levine, 1954, p. 15; and Reiter, 1966, p. 274-280).

Could Something Better Be Feasible?

The invention and spread of team nursing and the criticism it attracted was one more episode in the historical conflict in administering nursing care. The basic question was, who would actually do the work? Nurse administrators were responsible for groups of sick people and needed to supply continuous and reliable coverage. The nursing promise to provide care differed from the physician's obligation in that the nurse had to be continuously physically present while the physician could come and go. The physician's promise to diagnose and treat illness usually depended on the assumption that someone else, either family, servant, or nurse, would be there to assist the patient and carry out the therapeutic plan.

On one side of the question of who would do the work was Isabel Stewart, an influential nurse educator during the interwar period, who believed firmly in the concept of scientific management. She thought differentiating work on the wards and assigning tasks to workers according to their abilities was essential to creating an affordable and efficient nursing staff. It is "uneconomical to use a skilled person, such as the nurse, to perform unskilled work," she argued (Hanson, 1991, p. 343). Stewart correctly foresaw the pressures on nursing which would be created by an aging society, and the increase in chronic illness and shifting boundaries of responsibility and practice among professionals and lay persons.

Stewart correctly foresaw the pressures on nursing which would be created by an aging society, and the increase in chronic illness and shifting boundaries of responsibility and practice among professionals and lay persons.

Taking the opposite perspective was Virginia Henderson, who assailed the notion that minimally trained people could safely nurse. "The danger of turning over the physical care of the patient to relatively unqualified nurses is two-fold," she argued. The unqualified nurse might fail to assess patient needs adequately; and the qualified nurse, "being deprived of the opportunity while giving care to assess his needs, may not find any other chance to do so" (Henderson, 1960, p. 13). Thus, for Henderson, the suggestion that rising demands on nursing service departments be met by routinization and assembly-line methods was anathema.

Historian Susan Reverby believes that nurse leaders' enthusiasm for scientific management and their reliance on time and motion research as a way of understanding nursing work did encourage the application of assembly-line, industrial engineering concepts to hospital work. In fact, to hospitals, efficiency was almost always defined as spending less, while at the same time having more workers on the wards. They kept trying to find ways to use inexpensive caregivers to the on-going detriment of staff nurses' satisfaction with practice (Reverby, 1979, p. 206-225).

Complicating the 1950s labor situation were the simultaneous changes in hospital training schools, which led to higher costs for education. The tradition of relatively inexpensive care-giving by always-available student nurses slowly gave way to more class time for students, reduced working hours, standard clinical experiences, and better living conditions. Thus, the subsidy formerly provided by student labor to hospitals gradually became a liability.

Mildred Tuttle thought the problem of nursing service was "the most difficult one ever considered by [her] Divison. . . . The number of unprepared graduate nurses in administrative and supervisory nursing service positions is so large that to undertake the preparation of the present group through a regular academic program of study would be ridiculous" (Tuttle, p. 3). As was her custom, she convened an advisory group of "carefully selected directors of nursing" at the Foundation. These advisors represented the Veteran's Administration, university hospitals, and a large general hospital; a few educators attended.

Lucille Petry, chief nurse officer and assistant surgeon general, estimated that the number of nursing administration positions requiring advanced preparation totaled 67,500. According to Petry, only about 20% of the incumbents in those positions were qualified for their jobs. Meeting in April 1950, Tuttle's advisors listed a series of problems facing hospital nursing service administrators.

First, they pointed out the common expectation that, in addition to their own responsibilities, nursing administrators would take over hospital administration from 5 p.m. to 7 a.m. Making the nurse handle calls for the pharmacy, operating room, and admissions office, they argued, distracted from the main task of meeting patient needs. At the same time they noted that nurses in charge in the evening and during the night were often less prepared than the daytime supervisors. Second, they criticized the deficiencies in the current nursing administrators' ability to evaluate patient-care effectiveness, analyze data and present statistics in a meaningful way, write reports to interpret nursing needs to hospital boards or community leaders, and handle interpersonal relationships. The advisors urged the Foundation to develop educational programs to "enlarge the horizons of . . . nurses now employed to direct the work of others" and to create professional degree programs to attract young graduates who "have the . . . qualities to direct others" (Advisory Committee of Nursing Service Administration, 1950, p. 4-5). They recommended a dual approach that would upgrade current nursing practice and build university programs at the same time.

The program Kellogg launched in response to the aura of crisis surrounding administration of nursing service proved to be both an ambitious and somewhat troubled undertaking. Following a pattern tested in her "Post War Plan for Nursing Education" for improving clinical education in university schools of nursing, Miss Tuttle and Emory Morris invited 14 universities around the country to participate.

The program Kellogg launched in response to the aura of crisis surrounding administration of nursing service proved to be both an ambitious and somewhat troubled undertaking.

As a centerpiece of the new project, each university sent two representatives to a five-month-long seminar held at the University of Chicago. The two participants were supposed to represent education and nursing practice and expected to help launch programs in nursing administration at their respective universities when the Chicago seminar was finished.

For a variety of reasons, the seminar proved unsatisfying to both the Foundation and several of the participants. The seminar leader, Herman Finer, was a professor of political science at the University of Chicago. Seminar participants reacted negatively to the philosophical, theoretical and perhaps pedantic quality of his presentations. The other seminar leaders were not strong, no nurse leader emerged, and some critics felt more emphasis should have been placed on the specifics of management and supervision. At the heart of the problem, however, was uncertainty about short- and long-term goals (Myrtle E. Adelotte to Amy Viglione, 1951). As Mildred Tuttle put it in the annual report, "the objectives of the Seminar were not clearly understood," and the group was "non-homogeneous" (W. K. Kellogg Foundation, 1950-1951, p. 170). Mary Kelly Mullane, who helped lead the follow-up, program-development phase of the project, recalled that the seminar leaders and participants wanted different things from the experience. She and some of her colleagues wanted to work out a research strategy to take advantage of hospitals' need for new solutions to problems. The seminar group, however, was not only heterogeneous but also not ready or able to take advantage of the opportunity.

Finer's 1952 report, based, as he put it, "on my research experiences, as assisted by funds from the Kellogg Foundation," hints at the tensions in the seminar's merger of people in administration, education, and nursing. Disclaiming any Foundation responsibility for his report, Finer insisted his report was his version of the "intellectual operations in the seminar" (Finer, 1952, p. vii). It is clear from his text that the advice of hospital administrators, representatives of the American Hospital Association, and the medical men, as well as the support of the Foundation's leaders, meant much to Finer. However, his condescending attitude toward the seminar participants and nurse advisors is equally clear.

Nevertheless, Finer made crucial observations and raised fundamental questions for the hospital and nursing administrators of the time. Asserting as a principle that competent administration could only be built on clinical competence, he identified the dual and often conflicting expectations of nurse managers, which would create so much difficulty in the future. And, he reminded them, "all the administration in the world will not entirely wipe out the effects of serious underpayment [of nurses]." Finer wondered how nurse managers planned to handle the transition from the hospital supervisor tradition of strict hierarchy and layered authority. He criticized the switch to "easy democratic methods generally characteristic of American collective effort and progress by persuasion" (Finer, 1952, p. 3-32). At the time, interpersonal relationships and group dynamics were stressed as essential leadership tools; these ideas particularly pervaded the literature about team nursing.

Finer also challenged the seminar on the quality issue. "How high is the nurse to aim?," he asked. "Is she to put herself, *in her own field*, on the same level on which the finest doctors have put themselves?" (Finer, 1952, p. 126). "What is to be the standard?" The seminar participants wavered on this point. Finer was irritated: "Such uncertainty makes it extremely difficult to gather whether the profession is satisfied with the quality of care now given or not. . . . Practitioners of any profession often err not out of laziness, and certainly not out of ill will, but out of a lack of awareness that something better is feasible" (Finer, 1952, p. 127).

Seeking New Nurse Administrators

Troubled as the seminar may have been, Mildred Tuttle put the best possible face on it, saying, "Many of the problems, which they as individuals and as a group experienced, would probably appear in their own programs." She hoped "the seminar experience would help them recognize [and manage] the problems early" (W. K. Kellogg Foundation, 1950-1951, p. 171).[2] In June of 1951 the seminar participants went home, where they immediately began to develop nursing service education programs.

This was a major initiative for the Foundation, involving universities from coast to coast and extending over a five-year period. May Kelly Mullane's 1959 summary shows that 751 nurses earned baccalaureate or master's degrees, which included majors in nursing administration. Just over half of the graduates who could be traced "went directly from school to a job as director or assistant director of a hospital nursing service and two out of every three took an administrative post" (Mullane, 1959, p. 95). Probably most significant was establishing or reinforcing 13 nursing master's programs in universities. Mullane summed up needs still to be implemented: better clinical training to reinforce classroom study, research on nursing administration issues, readily available continuing education, funding for educating and supporting faculty, and encouragement to nurses seeking higher education in the form of peer approval and assistance with costs (Mullane, 1959, p. 199-203).

[2] *Universities funded were: Boston University, University of Chicago, Columbia University, University of Iowa, University of Minnesota, University of Mississippi, University of Pittsburgh, St. Louis University, Syracuse University, University of Texas, University of Washington, Wayne University (Detroit), and Western Reserve in Cleveland.*

In 1957, Mildred Tuttle prepared a detailed, final report on the Nursing Service Administration project. She clearly wanted to call attention to influential innovations in teaching and practice, which she felt were a direct result of the project. Drawing praise were the "case method" of teaching, borrowed and modified by Boston University School of Nursing from the Harvard Business School; development and testing of the "team nursing" idea at Columbia University; nursing application of scientific management techniques by the University of Pittsburgh; and the University of Washington's introduction of "human relations" content into curriculum. Tuttle asserted that "nursing service administration" was an almost unknown phrase in 1951; "now," she wrote, "nursing *service* is receiving as great, if not greater, attention than nursing *education* (Tuttle, Nursing Service Administration, p. 18-22).[3]

Although Kellogg supported programs for hospital administrators at Columbia and Minnesota, the same schools where nursing service administration programs were founded, there did not seem to be extensive communication between hospital administration and nursing service administration programs.

Although Kellogg supported programs for hospital administrators at Columbia and Minnesota, the same schools where nursing service administration programs were founded, there did not seem to be extensive communication between hospital administration and nursing service administration programs. Kellogg also supported hospital administration programs at Yale, Washington University in St. Louis, Johns Hopkins, and the University of Toronto. All of the early hospital administration programs (except Washington) were in Schools of Public Health. The tone of the minutes of the Foundation's Hospital Advisory Committee from 1944 to 1965 suggests leaders in the new field were very careful to keep clear boundaries with other professions. In the early years educators of hospital administrators didn't want either nurses or physicians to invade their territory (W. K. Kellogg Foundation, 1944-1965).[4]

[3] *Tuttle had some disappointments with this project though. In an earlier note (1956) she decided that experience with the University of Chicago in both the Post War Nursing Program and the Nursing Service Program "has been such that it might prevent the Foundation from entering into further negotiations on nursing education projects." In her usual elliptical style she indicated loss of patience with lack of cooperation and follow-through on funded projects. [See Tuttle, Mildred. (July 19, 1956). Field Visit Report. Battle Creek, Michigan: WKKF archives microfilm collection, reel 616, 2.] Some interpersonal problems with one of the project staff, which were external to the project, probably undermined both the success of the project and the Foundation's satisfaction with it.*

[4] *The first graduate program in hospital administration began at the University of Chicago in 1934. Prior to that, non-degree programs were offered at Duke University in Durham, North Carolina. By 1982 there were 60 accredited graduate health administration programs in the United States and Canada. Over that period the Kellogg Foundation invested about $28 million in hospital and health management. A total of about $7 million was invested in nursing services administration.*

Shifting the Emphasis

By the early 1970s, however, several new ideas were percolating through the Foundation's discussions about administering nursing. First, specialized or advanced nursing practice and integrating or re-linking practice and education gained strength and popularity throughout health care. Much more attention was being paid to administering nursing services in nursing homes and in-home care. And finally, the context for studying and teaching nursing service administrators began to shift to inter-disciplinary programs. That is, staff began to believe that graduate nursing service administration programs in a school of nursing might be less relevant or less sophisticated than inter-disciplinary programs with schools of business, public health or health services administration (Lee, 1977). In 1972, a new Advisory Committee on Management of Nursing Services was invited to the Foundation to take another look at the issues.

By the early 1970s, several new ideas were percolating through the Foundation's discussions about administering nursing.

Based on their suggestions and staying with its focus, the Foundation sponsored several conferences and developed new funding initiatives in nursing administration during the 1970s. Not only did nursing service administration have to now compete harder for qualified people who were increasingly drawn to specialty nursing practice, but new managers faced an array of problems; some were familiar, some new:

- Maintaining clinical practice in the field and the time demands on a nurse manager;
- Adopting an adversary role, as a member of the administrative team as organizations moved through collective bargaining with nurses;
- Adopting to the diversity of settings that require nurses and nursing service administrators;
- Leading the organization's nurses, who came from an extremely wide range of educational programs; and
- Being often the only female manager on a male-dominated administrative team (Slater, 1978, p. 7).

Improved Management of Nursing Services

The inter-disciplinary programs launched by the Foundation in 1972 and continued into the 1980s tried to blend content from the two fields. For instance, nursing content such as specialized clinical knowledge, decision making in nurse practice, and case-oriented nursing service problems, was put together with hospital administration content such as economics of health services, organizational behavior, operations research, and statistical decision-making. The goal was to retain the nurse administrator's identity and ability in nursing while introducing administrative knowledge and skills.

According to Barbara Lee, who was very committed to these programs, the great problem was getting the two (or more) schools in each university to really function jointly. Often they did not; one school or the other would dominate (Personal communication). Most of the leadership and enthusiasm came from the schools of nursing which got the funding under the Improved Management for Nursing Services program.

This second major investment totaled, by 1977, about $3.5 million. By then, eight universities agreed to bring together basic health administration programs and nursing in a collaborative curriculum. City University of New York (Baruch College), Columbia University, University of California at Los Angeles, and the Universities of Michigan, Alabama, Washington, Colorado, and Minnesota participated.[5] Several projects branched out from preparing hospital nursing managers to teach administration in home care, community health and long-term care.

After 30 years of effort, only about 20% of nurse administrators held a baccalaureate while another 15% held a master's degree in 1980.

Still, after 30 years of effort, only about 20% of nurse administrators held a baccalaureate while another 15% held a master's degree in 1980. Schools of nursing with master's programs gave up their so-called "functional majors" in teaching and administration in favor of clinical specialist, midwifery, and nurse practitioner emphases. And, the pool of baccalaureate-prepared candidates from which applicants for master's

[5] *Two more were added in 1980 and 1981, the Medical College of Virginia at Richmond and the University of Pennsylvania in Philadelphia. The program at Baruch College did not succeed and was terminated.*

study could be drawn was still small. A combination of minimal entry-level education, competition from clinical specialty practice, and the demands of the nurse-administration career kept the Foundation's goals elusive.

Administrative Problems outside the Hospital

Managerial gaps in nursing homes and in ambulatory care practices led the Foundation to direct some of its funding for administration toward upgrading nursing skills in nursing homes and in primary care in 1976. Forecasting its own increasing movement away from hospitals and back to community settings, the Foundation funded the Hospital Research and Educational Trust (of the AHA) to devise new models of primary care delivery. These projects focused on selected elements of management such as record-keeping, auditing, and development of services, which were thought vital to primary care success. The project also developed a guide for managers and practitioners working in primary care centers (Bisbee, 1982).

Throughout the seventies the staff at the Foundation supported a variety of innovative smaller projects trying to use nurse practitioners and better organizational designs to deliver primary care in rural areas and to improve access to care on a statewide basis.

Also, throughout the seventies the staff at the Foundation supported a variety of innovative smaller projects trying to use nurse practitioners and better organizational designs to deliver primary care in rural areas and to improve access to care on a statewide basis. Some of these focused on demonstrating joint medical and nursing practice; others intended to strengthen educational programs by building clinical practice sites. Applications from the Midwestern and mountain states received funding in line with the Foundation's interest in underserved rural populations.[6]

One of the Foundation's largest nursing-oriented investments during this period was its substantial support for the Mountain States Heath Corporation in Boise, Idaho. The demonstration tried to improve

[6] *Joint-practice demonstrations in California, North Dakota, Missouri, Montana, and Idaho were tried out in the years after 1972 when the federal government encouraged "expanding the scope of nursing practice."*

patient care in nursing homes by "updating administrative practices and introducing geriatric nurse practitioners (GNP)" (Improving the Quality of Long-Term Care, Mountain States Health Corporation, 1976). Specifically, here's what the project planned:

- Redefine existing health professionals' roles and introduce geriatric nurse practitioners and medical directors;
- Better utilize supporting patient care staff;
- Improve medical record systems and practices;
- Introduce shared services (among the nursing homes) in personnel, education, and purchasing;
- Educate administrative staff about federal regulations and Occupational Safety and Health Act standards.

The Mountain States project may be fairly said to have demonstrated a new kind of nursing practice in a way that attracted national attention.

Mountain States proposed to do this by means of a traveling team made up of a geriatric nurse practitioner, an administrator or accountant, and a physician, to visit and consult with participating nursing homes. A major project goal was training and placement of nurse practitioners in the nursing homes. In 1976, the Foundation granted more than $500,000 for a three-year project. This was followed up in 1981 with an expanded training program for geriatric nurse practitioners; more elaborate involvement of opinion makers in the communities; and joint participation of schools of nursing at the Universities of Washington, California at San Francisco, Arizona, and Colorado. Funding for the second project totaled $2.75 million for three years.

The geriatric nurse practitioner was expected to act as a link to the nursing home patients' primary care physicians and to offer better clinical care to patients while also helping upgrade other staff. In the Mountain States project, the nursing homes employed the GNPs; thus they did not act as independent care providers nor compete with physicians. The Mountain States project may be fairly said to have demonstrated a new kind of nursing practice in a way that attracted national attention. Subsequent evaluation of the impact of GNPs on patient care in the Mountain State nursing homes seemed equivocal, however, and disappointed proponents of the concept (Kane & Gerrard, The Geriatric

Nurse Practitioner in the Nursing Home; and Kane, et al., 1989, p. 1271-1277).

Barbara Lee was enthusiastic about the potential of nurse practitioners in nursing homes, schools, and other places where basic care needed improvement. She thought of the Mountain States project as a logical consequence of the recommendations of the National Commission on Nursing. She thought part of the problem of implementing the GNP role in nursing homes was that the homes were not ready to employ master's-prepared nurses (Personal communication).[7] Since most nurses employed by nursing homes held hospital diplomas or associate degrees, the idea of taking nurses out of the nursing homes, to train them and send them back, came into conflict with the emerging national standard of master's preparation for nurse practitioners. Notwithstanding this problem, the Mountain States project set in motion a durable trend that led to continuing demonstrations through the 1980s and ultimate acceptance of the GNP role (Shaughnessy, et al., 1995, p. 55-83).[8]

The Context of Delivering Nursing Care

During the 1970s, the nagging problem of regulation and control of practice for health care workers began to assume worrisome significance. First, the U.S. Department of Health, Education, and Welfare produced a critical study of licensure and certification of health personnel" (U.S. Department of Health, Education, and Welfare, 1971). The study labeled as major problems excessive professional control over practice, rigid state boards, barriers to practice, and fragmentation among services. New pressures on the traditional regulatory systems were caused by rap-

[7] *Barbara Lee and Jack Gerdes, who was the founder and leader of the Mountain States Project, had worked together at the University of Arkansas, Little Rock, Medical Center from 1957 to 1961. They were lifelong friends. Gerdes first worked in the Regional Medical Program (RMP) in Idaho, Montana, Nevada, and Wyoming, then developed the Mountain States Project as a component part of RMP. The Regional Medical Programs, funded by the federal government during Lyndon Johnson's administration, were intended to speed dissemination from university medical centers of new medical and health care knowledge and skills throughout the regions they served.*

[8] *The idea of geriatric nurse practitioner and the "teaching nursing home" continued through the 1980s. A subsequent evaluation of the GNP concept indicated substantial clinical improvements related to the use of GNPs in nursing homes.*

idly changing roles in nursing, medicine, and other fields; the 1971 report stirred and stimulated re-examination of state regulation as well as certification for specialty practice. In addition, new documentation requirements from Medicare and from the Joint Commission on the Accreditation of Hospitals, as well as other interested bodies, accentuated calls for reform.

Until this time coordination of the state boards of nursing, such as it was, came under the auspices of the American Nurses Association. State boards of nursing are charged with licensing discipline, interpreting the scope of practice, and setting rules to determine competence. In nursing, state regulation dates from 1903; mandatory licensure requirements became widespread in the late 1930s. Criticism of too-close links between the professional organization and the regulatory boards (especially the 1971 report and subsequent legal opinions) led to a request to the Foundation to help underwrite a free-standing coordinating group for the state boards.

In 1978, the American Nurses Association asked for funding to create the National Council of State Boards of Nursing (NCSBN) as an independent body. The objective of the NCSBN was to "assist state boards of nursing in ensuring nursing competence" (Dvorak & Schowalter, 1990, p. 259-264; also see Association of State Boards of Nursing—Development, 1978). A crucial function of the NCSBN was ownership and administration of the National Council Licensure Examination, formerly called the State Board Test Pool Examination, the licensing examination for entry into nursing. The national examination facilitates movement of nurses across states and provides a uniform standard for entry into practice.

The public position taken by the NCSBN became that licensees (i.e., nurses) benefit from licensure but so does the public. The obligation of the state board is to exercise its police power on behalf of the public. In effect, the NCSBN oversees vocational and professional licensure by coordinating, doing studies, and providing information to the various state boards. In the end, the decade-long negotiation between the public and the profession agreed that states should continue to regulate entry into practice with an oversight board comprised of mostly professionals.

Credentialing for specialty practice, however, continued to be split among an ever-increasing number of groups. The American Nurses Association sponsors a large credentialing program, the numerous specialty organizations certify in their own fields, and some states certify specialized practice. The goal of certification is, of course, to verify the truth of the claim that the nurse is in possession of knowledge for specialized practice. This debate continues; professionals generally believe that certification for specialty practice should remain in the hands of professional bodies, but issues of quality, monopoly, equity, and access to service sustain tension between public and private controls.[9]

The goal of certification is, of course, to verify the truth of the claim that the nurse is in possession of knowledge for specialized practice. This debate continues; professionals generally believe that certification for specialty practice should remain in the hands of professional bodies, but issues of quality, monopoly, equity, and access to service sustain tension between public and private controls.

Still another complexity in the regulatory context of nursing practice attracted the attention of the Kellogg Foundation in the 1970s. The Commission of Graduates of Foreign Nursing Schools (CGFNS) asked for support to develop a specific examination system for foreign nurses who planned to enter the United States and practice. Originated under the auspices of the American Nurses Association and the National League for Nursing, CGFNS gave the first such exams in 1978. CGFNS responded to the relaxation of quotas in the Immigration and Naturalization Act, which led to higher numbers of non-English-speaking nurses migrating and seeking employment in the states. According to some estimates, over 42,000 nurses entered the country between 1972 and 1976; many were attracted by hospital recruiters and word of a nursing shortage.[10] When immigrant nurses took the licensing exams in the states where they hoped to practice, a large proportion, as many as 80%, failed. The CGFNS proposal copied a 1950s strategy used by medicine, which

[9] *A recent analysis of these issues can be found in Finocchio, et al. & the Taskforce on Health Care Workforce Regulation, 1995.*

[10] *For the detailed examination of this phenomenon, see Brush, Barbara L. (1994). Sending for Nurses: Foreign Nurse Immigration to American Hospitals, 1945-1980 (Doctoral dissertation, University of Pennsylvania, 1994).*

set up a system to examine immigrating physicians before they migrated into the country.

In 1980, the Foundation again was asked to provide support for CGFNS because anticipated income from examination fees was less than expected. What had happened was that the Immigration and Naturalization Service (INS) delayed issuing needed regulations to implement the CGFNS system in spite of their earlier promise to do so. According to Barbara Lee's 1980 appropriation request to the Kellogg board, the INS delay was due to pressure from hospital associations, commercial recruiting agencies, and individual nurses who feared the CGFNS exam would seriously reduce movement of nurses across national borders (Credentialing Foreign Nurse Graduates, Commission of Graduates of Foreign Nursing Schools, 1980). CGFNS finally prevailed and the regulations were finalized in May 1980. During the first four years of testing, only 17% of foreign nurses eligible to take the exam were able to pass.

The Everlasting Task

Finding means to deliver nursing services to the public in a reliable, effective, and safe way preoccupied nurses and the Foundation staff from the 1930s into the 1980s.

Finding means to deliver nursing services to the public in a reliable, effective, and safe way preoccupied nurses and the Foundation staff from the 1930s into the 1980s. Adapting nursing practice and education for nursing administration to a changing institutional, economic, and political environment proved to be an erratic and halting endeavor.

Most nurses—65-70%—still practiced in hospitals in 1980. Hospital nurses specialized and took on more clinical responsibility. Some nurses transferred their practice to home care or ambulatory care. But in spite of the Mountain States' demonstration and other efforts, nurses did not move to nursing homes in large numbers.

As Herman Finer remarked in the 1950s, the best administration in the world could not overcome unwillingness to pay for nursing services. Money to pay nurses in nursing homes did not materialize in spite of exponential growth in nursing home beds during the period after 1965.[11]

[11] By the mid-1980s the number of patient days spent in nursing homes far exceeded (nearly doubled) the number of patient days spent in hospitals. See National Center for Health Statistics, 1987.

About half the care in nursing homes was purchased privately; the other half was paid by Medicaid or other third party sources. Sicker patients in nursing homers made their wards resemble the 1950s hospitals, but with one major difference. Far fewer professional nurses could be found in the nursing homes of 1980 than were practicing in the troubled hospitals of 1950.

The expanded scope of nursing practice and redefinition of the nurse-physician working relationship combined with a vastly enlarged nursing workforce to change the face of hospital nursing in the years after 1965. The distribution of nurses across the care system, that is, in hospital, home care and long-term care, and how to best administer their services, remained a major problem in search of solutions. And as in its beginning, Kellogg figured into the problem-solving.

Notes and References

Abdellah, Faye, & Levine, Eugene. (June 1954). *Nursing Outlook, 3.*

Advisory Committee of Nursing Service Administration. (April 6 and 7, 1950).

Association of State Boards of Nursing—Development. (1978). Appropriation Request. Barbara Lee files.

Bisbee, Gerald E. (Ed.). (1982). *Management of Rural Primary Care-Concepts and Cases.* Chicago: The Hospital Research and Educational Trust.

Brush, Barbara L. (1994). Sending for Nurses: Foreign Nurse Immigration to American Hospitals, 1945-1980 (Doctoral dissertation, University of Pennsylvania, 1994).

Credentialing Foreign Nurse Graduates, Commission of Graduates of Foreign Nursing Schools. (1980). Appropriation Request. Barbara Lee files.

Davis, Michael. (1929). *Hospital Administration: A Career.* New York: NP. Davis's study of hospital administration was supported by the Rockefeller Foundation. Davis, who became an officer of the Julius Rosenwald Fund, founded the first hospital administration program at the University of Chicago in 1934.

Dvorak, Eileen McQuaid, & Schowalter, Joyce M. (1990). The Role of the National Council of State Boards of Nursing in Consumer Protection. In J. C. McCloskey & H. K. Grace (Eds.), *Current Issues in Nursing,* third edition. St. Louis: The C. V. Mosby Company.

Fairman, Julie. (September 1990). Do We Need More Nurses? Presented at the American Association for the History of Nursing annual meeting, Baltimore, Maryland.

Finer, Herman. (1952). *Administration and the Nursing Services.* New York: The MacMillan Company.

Finocchio, L. J., et al., & the Taskforce on Health Care Workforce Regulation. (1995). *Reforming Health Care Workforce Regulation: Policy Considerations fort the 21st Century.* San Francisco: Pew Health Professions Commission.

Hanson, Kathleen S. (November-December 1991). An Analysis of the Historical Context of Liberal Education in Nursing Education from 1924 to 1939. *Journal of Professional Nursing.*

Henderson, Virginia. (1960). *Basic Principles of Nursing Care.* Geneva, Switzerland: International Council of Nurses.

Improving the Quality of Long-Term Care, Mountain States Health Corporation. (1976). Appropriation Request. Barbara Lee files.

Kane, Robert L., & Gerrard, Judith. (No date). The Geriatric Nurse Practitioner in the Nursing Home: Methodological Design for Assessing Quality of Care and Cost Effectiveness. Funded by the Robert Wood Johnson Foundation and the Health Care Financing Administration, NIH.

Kane, Robert L., et al. (September 1989). Effects of a Geriatric Nurse Practitioner on Process and Outcome of Nursing Home Care. *American Journal of Public Health, 79.*

Lamberts, Eleanor. *Nursing Team Organization and Functioning.* New York: Teachers College-Columbia University.

Lee, Barbara J. (April 1977). Meeting New Challenges in Health Care—Nursing Education and Practice. Talk presented at Southern Regional Education Board.

Lynaugh, Joan. (1989). *The Community Hospitals of Kansas City, Missouri, 1870-1915.* New York and London: The Garland Press.

Maschal, Henry T. (January 1953). Developing the Nursing Service Budget. *Hospital Management, 75.*

McManus, R. Louise. (1953). Foreword. In E. Lamberts, *Nursing Team Organization and Functioning.* New York: Teachers College-Columbia University.

Mullane, Mary Kelly. (1959). *Education for Nursing Service Administration: An Experience in Program Development by Fourteen Universities.* Battle Creek, Michigan: W. K. Kellogg Foundation.

Myrtle E. Adelotte to Amy Viglione. (March 21, 1951). Battle Creek, Michigan: WKKF archives microfilm collection, reel 153. Adelotte was one of the seminar participants.

National Center for Health Statistics. (1987). *Advance Data from Vital and Health Statistics, 42.* Hyattsville, Maryland: Public Health Service.

Reiter, Frances. (February 1966). The Nurse Clinician. *The American Journal of Nursing, 66.*

Reverby, Susan. (1979). The Search for the Hospital Yardstick: Nursing and the Rationalization of Hospital Work. In S. Reverby & D. Rosner (Eds.), *Health Care in America.* Philadelphia: Temple University Press.

Reverby, Susan. (1987). *Ordered to Care: The Dilemma of American Nursing, 1850–1945.* Cambridge and London: Cambridge University Press.

Roberts, Mary. (1954). *American Nursing.* New York: The MacMillan Company.

Rogge Meyer, Genevieve. (1960). *Tenderness and Technique: Nursing Values in Transition.* University of California at Los Angeles: Institute of Industrial Relations.

Rosenbert, Charles. (1987). *The Care of Strangers.* New York: Basic Books.

Sadov, M.; Cross, J.; & Higgens, H. (January 1954). The Recovery Room Expands Its Services. *The Modern Hospital, 83.*

Schafer, Margaret K. (August 1953). Efficient Nursing Care Results for Making the Best Use of the Nursing Team. *Modern Hospital, 790.*

Shaughnessy, Peter W., et al. (Summer 1995). Quality of Care in Teaching Nursing Homes: Findings and Implications. *Health Care Financing Review, 16.*

Slater, Carl H. (Ed.). (1978). *The Education and Roles of Nursing Service Administrators.* Battle Creek, Michigan: The W. K. Kellogg Foundation.

Tuttle, Mildred. (July 19, 1956). Field Visit Report. Battle Creek, Michigan: WKKF archives microfilm collection, reel 616.

Tuttle, Mildred. Administration of Nursing Services. Battle Creek, Michigan: WKKF archives microfilm collection, reel 154.

Tuttle, Mildred. Nursing Service Administration. Battle Creek, Michigan: WKKF archives microfilm collection, reel 153.

U.S. Department of Health, Education, and Welfare. (1971). Report on Licensure and Related Health Personnel Credentialing. Washington, D.C.: U.S. Government Printing Office.

W. K. Kellogg Foundation. (1944-1965). Hospital Advisory Committee Minutes. Battle Creek, Michigan: WKKF archives.

W. K. Kellogg Foundation. (1950-951). *Annual Report, 1950-1951.* Battle Creek, Michigan: WKKF archives.

Weeks, Lewis E. (Ed.). (1983). *Andrew Patullo, In the First Person: An Oral History.* Chicago: Hospital Research and Educational Trust. The interview was done in 1978.

7 Nursing and the Foundation: New Priorities

By 1980, nursing and the Foundation shared a 50-year, many-faceted history. This history encompassed a time of rapid reconfiguration of virtually every aspect of health care and nursing. In 1982 and 1983, the Foundation, as it did in almost every previous decade, reviewed and debated its health programming priorities and positions.[1]

The early 1980s also was a time for several important staff changes at the Foundation. Both Barbara Lee and Robert Kinsinger retired in 1983. Robert D. Sparks took over as president of the Foundation in 1982, while Russell Mawby moved up to become the Foundation's chairman and chief executive officer.[2] Helen Kennedy Grace became program director and coordinator of health programs in 1983. A well-known academic leader in nursing, Grace moved to the Foundation from a deanship at the University of Illinois College of Nursing.

Foundation priorities for the 1980s, as named in 1982 by then-Foundation president Russell Mawby, included availability and access to health care, comprehensiveness and continuity of care, cost containment and productivity, and greater emphasis on health

---

[1] *The process extended over several years. In 1984 the Foundation told a funding newsletter it was still "reassessing its priorities and making adjustments." In 1983 the Foundation funded 740 projects totaling $54.6 million. See Health Grants and Contracts Weekly (September 25, 1984), p. 7. This level of grant-making in health placed the Foundation in the top echelon of private funding sources with the Robert Wood Johnson Foundation, the Ford Foundation, and the Rockefeller Foundation.*

[2] *Sparks served as president until 1987, when Norman A. Brown took over as president and chief programming officer.*

In 1982 and 1983, the Foundation, as it did in almost every previous decade, reviewed and debated its health programming priorities and positions.

promotion, disease prevention and public health (Mawby, 1982). These emphases reflected the Foundation's shift away from its earlier focuses on education for professionals, innovation in care delivery, and expansion of a hospital-based health care delivery system. Now Kellogg's nursing-related interests moved toward multidisciplinary, system-wide demonstrations, and an even stronger emphasis on community, as opposed to specific institutional projects.

After 35 years, Foundation staffers, along with other leaders in American health care, seriously began to reconsider their definitions of success in health. "Doing Better and Feeling Worse," the title of a widely read issue of *Daedalus* in 1977, in some sense summed up the dissatisfactions evoked by American health care (Doing Better and Feeling Worse: Health Care in the United States, 1977).[3] While Americans, as Russell Mawby phrased it, continued to believe that "parts of the health care system were the finest in the world," worries persisted about inequality, lack of access, lack of health promotion, and, perhaps most of all, escalating costs. Themes that began to resonate more strongly than at any time since World War II ended were slowing down growth in the system and decentralizing responsibility away from government and professionals and back to communities and individuals.

When it came to nursing, however, the discussion of these latter themes took on a rather eccentric character. More, rather than fewer, nurses seemed necessary if the nation was to go on a health care diet, somewhat the way that substituting more pasta and beans improve a diet too laden with red meat. The professional literature was permeated with questions like the following: "If the nation needs better access to primary-care services, should we not educate more advanced-practice nurses who will ensure access at lower cost?" "If we even further reduce the length of stay in our hospitals, thus driving up their acuity level of the patients who are still there, will we not need more nurses per patient rather than less?" (Gortner, 1982, p. 495-502).

[3] *The issue included articles by distinguished academic physicians and others with concerns ranging from the "medicalization" of American life, technology worries, and the financing of health care, to problems of children and those with mental illnesses.*

In the context of calls to slow down growth in the health care system, hospitals insisted that they faced 100,000 vacant nursing positions; indeed, cries of nursing shortages permeated the health care rhetoric for much of the 1980s.[4] For context, though, while there were 249 nurses per 100,000 in 1950, by 1980 the number soared to 506 nurses per 100,000.[5] In 30 years the number of nurses per capita doubled, but that was still not enough to allay fears of nursing shortages.

In 1983 a memorandum to the Kellogg board of trustees summarized the Foundation's direct investments in nursing from 1930 to 1983; the total came to $58,336,202 (Mawby & Sparks, 1983).[6] Defining nurses as "responsible partners in the provision of quality, cost-effective health services," the document reflected the Foundation's judgment that, in spite of its problems, a newly configured health care system had emerged from the developmental period of earlier decades (Mawby & Sparks, 1983, p. 22). This moment of change in Foundation priorities and staff provides a useful and historically convenient vantage point to review and sum up 50 years of interaction between nursing and the Foundation.

Continued investment of the Kellogg Foundation in social goals consistent with nursing's growing role in American health care seemed likely to sustain their historical relationship, at least up to this writing. The elements of that relationship, that is, how philanthropy, health care and nursing came together in a time of rapid change, and the impact of their professional and philanthropic goals, show how the case of nursing and the W. K. Kellogg Foundation contributes to the historical record.

Philanthropy and Nursing

For a significant part of its first 50 years the Foundation tended to work through professionals from various fields to achieve social and health

---

[4] *See, for discussion of this paradox, Nurses for the Future, 1987, p. 87; or Lynaugh, 1990, p. 169-175.*

[5] *By 1982 there were 1.33 million registered nurses in active practice, plus about 500,000 practical nurses. Nurses were by far the largest group of health professionals.*

[6] *Foundation support of nursing was sometimes embedded in other divisions' funding, particularly administration.*

goals set by the board and officers. From its early days as an operating foundation, when the staff of the MCHP worked directly to create new public health organizations in several counties of southeastern Michigan, the habit of turning to professionals for advice, counsel, or to actually implement programs was strong. The first Foundation leaders, Morris, Tuttle, Davis, and others were themselves professionals trained in various health-related fields. They had confidence in professional expertise and sought and obtained cooperation from people around the country who were, in effect, their colleagues.

In a way, this idea of doing "good works" through professionals or managers is reminiscent of the Visiting Nurse Association (VNA) Boards of Lady Managers' relations with "their nurses." The ladies of VNA boards however, were elite community members who did not see the nurses, professional though they might be, as their equals.

In the 1940s and 1950s at the Kellogg Foundation, a different dynamic was at work. Priding itself on its Midwestern heritage, the Foundation kept a low profile and adopted an egalitarian stance toward its constituents (Personal communication).[7] Confidence in local decision making, a habit of targeting and getting help from respected consultants, and staying focused on their relatively narrow set of social goals smoothed the reciprocal relationship between professionals and the Foundation. Nurses and hospital administrators were pleased to be asked for advice and ready to be informed and influenced by the Foundation's preferences.

Foundation staffers, particularly in the Morris-Tuttle era, needed ideas from professionals and also needed a means for carrying out their agenda. As pointed out by hospital division head and later vice president, Andrew Patullo, the Foundation staff faced an obligation to distribute a certain amount of money each year. Their problem was to find a way to meet that obligation as responsibly and effectively as possible.

In nursing, the Foundation found a valuable social resource needy for assistance. The "special relationship" so widely recognized among nurses

[7] *Andrew Patullo particularly remembered that Emory Morris did not want much publicity attracted to the Foundation and chided Patullo when a speech he gave got on the Associated Press wire service.*

and tacitly acknowledged at the Foundation fit in very well with the Foundation's image of itself. "Helping people to help themselves," a theme and slogan originated by Will Kellogg himself, was also part of the definition of nursing. The idea of philanthropy as an ameliorating link between individual and corporate self-interest (the engine of capitalism), and the cooperative and shared interests of civilized society could be, and was, realized by encouraging the growth and development of nursing.

In nursing, the Foundation found a valuable social resource needy for assistance. The "special relationship" so widely recognized among nurses and tacitly acknowledged at the Foundation fit in very well with the Foundation's image of itself. "Helping people to help themselves," a theme and slogan originated by Will Kellogg himself, was also part of the definition of nursing.

On the other hand, the poverty of nursing, its very small and relatively uninfluential group of leaders, its ambiguous educational status, and its inexperience in the public arena, made the field vulnerable to pressure. To a considerable extent, nurse leaders behaved as though they had to wait until asked for advice before they could hope to influence decisions at the Foundation. When Mildred Tuttle and Emory Morris left the Foundation many nurse leaders felt shut out of a consulting role they had come to value. When the Foundation staff of the 1970s took more controversial decisions affecting nursing, particularly in the area of associate degree nursing education, many nurses were critical but apparently felt powerless. The Foundation was still a major financial resource for ambitious but struggling academic and clinical leaders in nursing. They might disapprove of some Foundation funding choices but found it hard to directly or publicly criticize its agenda.

Another part of the special relationship between the Kellogg Foundation and nursing resides in the "shirt-sleeve, practical" image cultivated by the Foundation during this era. Nursing was by no means an elitist field, whatever its leaders' academic ambitions might have been. Nursing's goals to get its educational plant transferred to colleges, to improve care in community hospitals, and to find better ways to deliver nursing services, were all imminently practical and down to earth. The

Foundation stressed the application of knowledge, not research per se. For several decades the two entities remained a good fit. Even at the end of this part of the story in the 1980s, as Foundation officers increasingly began to think of institutions and professionals as part of the problem rather than part of the solution, nursing held on to its service-oriented, utilitarian image (Personal communication, August 3, 1989). Of course, the operating origins of the Foundation in the MCHP, a demonstration of health care delivery where nurses played important parts, and the subsequent rise to Foundation leadership of Mildred Tuttle and Emory Morris, helped establish the pattern of the relationship in the first place.

Reciprocal Impact

Every study of nursing, beginning in 1948 and continuing through the 1980s, recommended standardization of professional nursing education. This goal was shared by the Foundation; it became a keystone of funding for nursing. Against a background of constantly changing practice, medical and scientific knowledge, and public expectation, nursing and its benefactor, the Kellogg Foundation, sought an acceptable solution. Viewed from a bit of distance, at least four factors interfered with finding that elusive, acceptable solution to the problem of nursing education. First, planners could not accurately anticipate the overall demand for nurses. They found the task of differentiating among nurses and then projecting the number of nurses needed at different levels to be undoable. Thus, in the decades after World War II, policy was repeatedly based on a combination of expert opinion and projections from past experience.

Every study of nursing, beginning in 1948 and continuing through the 1980s, recommended standardization of professional nursing education. This goal was shared by the Foundation; it became a keystone of funding for nursing. Against a background of constantly changing practice, medical and scientific knowledge, and public expectation, nursing and its benefactor, the Kellogg Foundation, sought an acceptable solution.

Second, no one was willing or able to let "market forces" operate to determine the supply of nurses. Hospital employers of nurses, the fastest growing segment of demand, did not respond to shortage by raising wages or improving conditions to attract workers in the usual labor market sense. So, nurses and others, including staffers at the Foundation, who were concerned with safe care in hospitals, felt it essential to "manage" the supply of nurses. This means they almost always added to the supply by educating more new nurses, perpetuating a vicious cycle of over-supply of entry-level nurses, which kept wages down and turnover high.

Third, the unplanned, ad hoc, and sweeping changes in the clinical responsibility of nurses continued to make educational reforms obsolete. Beginning in the mid-1950s, nurses began to assume more and more technical and substantive clinical responsibility either from physician delegation or because of clinical innovations. Training nurses for this new practice began outside the educational mainstream; it was at least a decade before formal nursing programs could respond effectively.[8] This is illustrated by the new knowledge that critical care nurses needed when intensive care units proved successful. Similarly, nurse practitioners could solve problems of access but required additional education to do so.

Meantime, the major educational reform, the associate degree program, focused on preparing nurses with minimal science but sufficient skills to work under direct supervision. Moreover, educating graduate nurses with content drawn from business or education schools did little to respond to the demand for clinical expertise. The supply of clinically expert nurses, who were needed to guide hospital nursing care and whose practice roles were in the process of creation, lagged in spite of large infusions of money for education from the federal government.

Fourth, the social environment of the 1960s and 1970s, buffeted by demands for civil rights, women's rights, peace in Vietnam, and by assassinations and political scandal, proved a tumultuous background for systematic reform of any kind. The women's movement surely helped schools of nursing prosper on college campuses. But at the same time, the drive for egalitarian access to the professions made maintaining or improving standards of entry to the profession impossible.

[8] *For a discussion of this phenomenon as it relates to critical care, see Fairman & Lynaugh (in press),* Intensifying Care: A History of the American Critical Care Movement.

The social upheavals of the 1960s nourished and helped sustain a certain logic of minimalist preparation for entry into nursing. That is, easy access to nursing as a field of work became a social goal. Improving access to health services was another social goal which, some believed, could be met by expanding the scope of nurses' responsibilities. The temptation to link them, to create a career ladder, proved irresistible.

Much of the Foundation's 1970s nurse education grant-making was absorbed by unique programs intended to repair the educational deficits of nurses with less than a baccalaureate.

At the Foundation in the late 1960s and throughout the 1970s, ideas of easy access to nursing and mobility within nursing held sway. The ADN concept was converted at the time of the 1970 National Commission on Nursing study from a terminal degree to the first step on a career ladder. The "flexible baccalaureate" initiatives, which included the external degree programs in New York State and California, as well as a wide array of innovative "ladder" programs all promoted upward mobility in nursing for the minimally prepared entrant to the field.[9]

But the logic of minimalist preparation also meant that ADN and hospital-trained diploma nurses were under immediate pressure to upgrade their abilities and qualifications if they hoped to have any career mobility. The growing complexity of the nurse education system was due, in no small measure, to a general unwillingness to commit resources to complete generalist, entry-level preparation for an adequate proportion of the nation's nurses. Much of the Foundation's 1970s nurse education grant-making was absorbed by unique programs intended to repair the educational deficits of nurses with less than a baccalaureate. Similarly, federal government investments in undergraduate nursing education were divided between various RN completion programs and generic baccalaureate education (Personal communication with Jesse Scott, November 30, 1989).[10]

[9] *The idea behind the external degree concept was that "nurses would be able to obtain a college degree in nursing entirely or in large measure, by examination." The programs depended heavily on proficiency examinations; they were intended to help practical nurses move to the associate degree level, and ADN or diploma nurses move to the baccalaureate. For Foundation involvement, see the following appropriation requests: External Degree in Nursing, New York State Education Department, 1973; Nontraditional Education-Nurses, The University of the State of New York, 1979; and Access to Health Care Services, The University of the State of New York, 1976.*

[10] *Scott, who was chief of the Division of Nursing of the United States Public Health Service during 1960s and 1970s, felt that the Foundation's support for the ADN program aggravated instead of solved a problem for nursing by forcing millions of dollars of public money to be used to move ADN plus diploma graduates into the baccalaureate. The Division of Nursing administered most federal funding for nursing education.*

Faye Abdellah characterized nursing as "residing in many houses" (Weeks, 1983, p. 40). She meant that nursing draws its concepts and science from many fields and cuts across disciplinary categories in its practice. This eclectic character of nursing tended to fit the pragmatism of the Kellogg Foundation. In the highly specialized world of the late 20th century, however, finding a language and standard for nursing and nursing practice proved to be difficult. The social obligation of the professional, that is to serve a specific ordained social role, was easily understood, but the theoretical and research basis for nursing practice, essential to 20th century credibility, simply eluded most observers. Particularly at the Kellogg Foundation, where basic research was not part of the mission, interest in nursing's need for foundational scholarship did not exist. In that sense, nursing, as was true of other professional and occupational groups, remained essentially an instrument useful to the goals and mission of the Foundation rather than a source of ideas.

Gradually, however, the expanding scope of nursing practice and growing sophistication of nursing leadership produced a new vision of nursing. The concept of primary nursing in hospitals, for instance where the fully prepared care group is led by a professional able "to assemble various wills into one unified, driving force" proposed quite a different idea of nursing from the authoritarian supervisor-heavy hierarchy of functional or team nursing (Clifford, 1982, p. 103).

In the 1960s and 1970s many nurses voted with their feet and abandoned care-giving systems where nurses lacked accountability or faced a hopelessly broad scope of responsibility. The arenas that attracted and retained nurses were those that required and allowed them to practice to their fullest ability.

In the 1960s and 1970s many nurses voted with their feet and abandoned care-giving systems where nurses lacked accountability or faced a hopelessly broad scope of responsibility. The arenas that attracted and retained nurses were those that required and allowed them to practice to their fullest ability: primary systems in which all nursing responsibility for the hospitalized patient rested in the professional nurse, ambulatory care and other arrangements in which practitioners were responsible for groups of patients, and specialties such as midwifery, nurse anesthesia, and critical care.[11]

In the Scheme of Things

In the 1980s, the Foundation returned to a more community-oriented, care access-focused agenda—an agenda in a way reminiscent of the Depression-era Michigan Community Health Project.

Nursing and the Foundation faced a future sure to be as complicated and difficult to interpret as that faced by those coping with the Depression, the postwar era, or the social crisis of the late 1960s. This relationship of necessity and convenience, sustained by shared interests, individual effort, and a shared commitment to a civil society, would most likely endure. In the 1980s, the Foundation returned to a more community-oriented, care access-focused agenda—an agenda in a way reminiscent of the Depression-era Michigan Community Health Project. Nursing underwent radical philosophical, intellectual, clinical and social change, emerging as a new field. It ultimately became a different occupation from that which linked with the Foundation in the 1930s.

The Kellogg Foundation and nursing are both grounded in and profoundly influenced by changing American understandings and expectations of health and the health care system. For the first 50 years, as this book shows, many people expended intense effort to try to guarantee an adequate, high quality supply of nurses through improving education. Similarly strong efforts tried to improve health services by improving and re-thinking nursing practice. As the nature of all entities constantly changes, the nature of the interactions between the Foundation and nursing surely changed too. With their interests focused on broader conceptions of health and health care, they offered new possibilities based upon five crucial decades of investments, experiment, disappointment, and success.

[11] *For an analysis of practice factors that enhance nursing practice, see McClure, Poulin, Sovie, & Wandelt, 1983.*

Notes and References

Access to Health Care Services, The University of the State of New York. 1976. Appropriation Request. Barbara Lee files.

Clifford, Joyce. (1982). Professional Nursing Practice in a Hospital Setting. In L. Aiken (Ed.), *Nursing in the 1980s: Crisis, Opportunity, Challenges.* Philadelphia: J. B. Lippincott.

Doing Better and Feeling Worse: Health Care in the United States. (Winter 1977). *Journal of the American Academy of Arts and Sciences.*

External Degree in Nursing, New York State Education Department. 1973. Appropriation Request. Barbara Lee files.

Fairman, Julie, & Lynaugh, Joan (in press). *Intensifying Care: A History of the American Critical Care Movement.* Philadelphia: University of Pennsylvania Press.

Gortner, Susan. (1982). Commentary. In L. Aiken (Ed.), *Nursing in the Eighties: Crisis, Opportunities, Challenges.* Philadelphia: J. B. Lippincott Company.

Health Grants and Contracts Weekly (September 25, 1984).

Lynaugh, Joan. (1990). Is There Anything New about this Nursing Shortage? In J. C. McCloskey and H. K. Grace (Eds.), *Current Issues in Nursing.* St. Louis: C. V. Mosby Publishing Company.

Mawby, Russell G., & Sparks, Robert D. (August 12, 1983). WKKF Contributions to Nursing. Memorandum to the W. K. Kellogg board of trustees. Battle Creek, Michigan: WKKF archives.

Mawby, Russell. (1982). A Layman's Perspective. Talk delivered at the Health Profession's Education Conference, University of Illinois-Chicago.

McClure, Margaret; Poulin, Muriel; Sovie, Margaret; & Wandelt, Mable. (1983). *Magnet Hospitals: Attraction and Retention of Professional Nurses.* Kansas City, Missouri: American Academy of Nursing.

Nontraditional Education-Nurses, The University of the State of New York. 1979. Appropriation Request. Barbara Lee files.

Nurses for the Future. (December 1987). *American Journal of Nursing.*

Weeks, Lewis E. (Ed.). (1983). *Faye G. Abdellah, In the First Person: An Oral History.* Chicago: American Hospital Association and Hospital Research and Educational Trust.

Part II

An Insider's View

Helen K. Grace
Gloria R. Smith

8 Nursing Development in the Political Context of the Foundation: 1930–1980

Helen K. Grace

Dr. Lynaugh's analysis of the relationship between nursing and the Kellogg Foundation focused primarily on nursing and the broader health field as reflected in archival documents. For my part, I have spent the past eight years working to compile a comprehensive database for the entire Kellogg Foundation, reading and indexing all board of trustees meetings from 1930 to 1994 and codifying all of the grants made as a result of the board's decision making. I also served as a program director and vice president for a 15-year period from 1982 to 1997. While my views are certainly not as objective as those of Dr. Lynaugh, I bring to this discussion a broader view of the Kellogg Foundation as a whole, and an intimate knowledge of the political context in which decision making occurred. It is from this perspective that I will frame the interaction between nursing and the Kellogg Foundation over the first 50 years covered by Dr. Lynaugh's analysis. Many of the important points that Dr. Lynaugh has made become even clearer when placed in the political context of the Foundation.

The Early Years

Kellogg's personal philanthropic work was driven by a concern for youth in his community—particularly those whose development was impeded by social isolation.

Starting Points. It is important to note that in starting the W. K. Kellogg Foundation, Will Keith Kellogg did not set out to focus primarily on the health field, though that became the major focus for the first 50 years. Kellogg's personal philanthropic work was driven by a concern for youth in his community—particularly those whose development was impeded by social isolation. On the one hand he worried about youth in rural areas whose educational experiences were limited because of the small rural schools they attended. He also worried about youth with handicaps whose life experiences were limited because of their physical limitations. Before he set up the Kellogg Foundation, Mr. Kellogg funded the development of the Kellogg Agricultural School, which consolidated smaller schools into a larger one in order to enrich the educational experiences for rural youth. Growing out of his interest to help those with physical handicaps, his funding led to the building of the Ann J. Kellogg School in Battle Creek. This school was designed to bring together area children with and without handicaps in a comprehensive program. The Ann J. Kellogg School became a visible national model for "educational mainstreaming," which later became a national trend. The starting point for the Kellogg Foundation was the education of youth, and schools were the hub for the initial funding.

The Initial Focus. While Mr. Kellogg's early concerns were for the educational needs of children, participating in the White House Conference on Youth provided him a more comprehensive framework for shaping his philanthropic work. In identifying leadership for his newly formed foundation, Mr. Kellogg looked to the Kellogg Company physician, Dr. A. C. Selman, formerly a missionary to China, to become the first part-time director. Understandably, Dr. Selman's views rapidly expanded the Foundation's youth-focused concerns to include health. With the schools as the hub of activity, the Foundation developed a health model built upon the concept that all children should receive regular physical and dental examinations. If problems were detected, remedial work would be provided. This concern for the health of children was coupled with a concern for being supportive of the private practice of physicians and dentists. Nurses were not a part of the original design. In the first few years, doctors and dentists would go to the schools to provide physical and dental examinations, and children needing remedial work were then referred to their offices for follow-up. Great care was given to preserve the private-practice medical and dental care system. However, it soon became evident that going out to the schools was time-consuming for the dentists and doctors. As an alternative, children were to come to the private practitioners' offices for the routine physical examinations. This required a great deal of coordination with the schools. While nurses had not been part of the original plan, the Foundation hired them to coordinate the rapidly expanding work between the schools and the private practitioner. In addition to providing the link between the children in the schools and the medical and dental practitioners, nurses also helped teachers add a health education component to the basic curriculum. Additionally, they connected with parents and were involved in programs to address the needs of families such as the Home Maternity Nursing Service, led by Dr. Mary Selman, the wife of Dr. A. C. Selman. Even though the nurses did a tremendous amount of work, it was repeatedly emphasized that they worked under the supervision of the doctors. In the home maternity nursing program, physicians requested that the nurses provide prenatal visits in the home. When a woman was in labor, they called the maternity nurse who came to the home and remained with the patient until she was about to deliver. Then the nurse called the doctor, who would come for the delivery. Nurses also provided follow-up care in the home. Again, this was all under the supervision of the physician. The Foundation did not want to do anything that would be in conflict with the practices of organized medicine. But despite this limitation, the model of community-health nursing developing in rural Michigan began to gain national visibility.

This concern for the health of children was coupled with a concern for being supportive of the private practice of physicians and dentists. Nurses were not a part of the original design.

The Michigan Community Health Project (MCHP)

While Rockefeller was primarily involved with health issues in urban settings, the Kellogg Foundation's location in rural Michigan provided opportunity to extend the work, particularly in the public health field, into a new frontier.

As Lynaugh noted, established foundations such as Rockefeller and the Commonwealth Fund had focused primarily on the charitable cause of caring for the poor in hospital settings and in the community, and on educational reform. The Rockefeller Foundation's earlier work profoundly influenced the direction of the Kellogg Foundation. While Rockefeller was primarily involved with health issues in urban settings, the Kellogg Foundation's location in rural Michigan provided opportunity to extend the work, particularly in the public health field, into a new frontier. Haven Emerson, a leading proponent of improvements in public health who collaborated with Josephine Goldmark on a national study of nursing, later began his prolonged involvement at the new Kellogg Foundation as adviser and board member. The Rockefeller Foundation had initiated a program to support the development of rural health departments just at the time the Kellogg Foundation had expanded its school health program. It was a natural development that the Kellogg Foundation would link with the Rockefeller Foundation in rural south-central Michigan, initiating a health department in Barry County, followed by other health departments in rural Michigan. The public health movement was fueled by the Great Depression, and the seven county areas in Michigan where the Michigan Community Health Project developed became a national model for the delivery of cost-effective preventive health services in rural areas, and a training ground for public health professionals. Nursing involvement in the early school programs and in the newly developing structures of county health departments was highly visible. As part of the Michigan Community Health Project, 26 nurses were employed by the Kellogg Foundation that met the following qualifications: completion of a baccalaureate program of Arts or Sciences, completion of a nursing educational program, and specialization training of at least one year in community-health nursing. Because of the interest in preserving private-practice medicine and dentistry, few doctors or dentists were employed by the Kellogg Foundation other than in administrative roles. Thus, the Michigan

Community Health Project became a training ground for community health nurses and for public health professionals. It was also a "field university," drawing students from throughout the country to gain experiences in this model setting.

The Michigan Community Health Project became a training ground for community health nurses and for public health professionals. It was also a "field university," drawing students from throughout the country to gain experiences in this model setting.

The War Years

Shift in Focus. In 1940, the board of trustees decided to expand the focus beyond local programming and operating to become a national and international grant-making foundation. The declaration of war in 1941 catapulted the Foundation on a new pathway. The decision was made to direct all possible resources to the cause of winning the war. Working closely with the secretary of state, the Department of Defense, and the professional community, it was determined that the most logical contribution the Kellogg Foundation could make to support the war effort would be to provide scholarships and loans to help produce adequate numbers of doctors, dentists, and nurses to provide medical care. In consultation with medical school deans, acceleration of education was to be accomplished by running educational programs year-round. Shortly after this policy decision, it was discovered that this approach might actually reduce the numbers of graduates of medical schools, because students relied upon summer employment to fund their education. While the federal government recognized the problem and identified the need for a scholarship and loan program to support students, it was impossible to get such a program approved rapidly, given the legislative process it would entail. The program was a natural recipient for Kellogg Foundation support until federal funds came through. A study was done to determine the amount of money that students earned from summer employment as the basis for calculating the amount of scholarship money that they would need to attend school year-round. A loan fund was also initiated, providing a permanent fund within the participating universities. This model was extended to other areas of the Foundation, such as nursing and education.

Developing the scholarship and loan program in medicine and dentistry was relatively clear-cut: provide funding to all of the existing university programs in the United States and Canada. While the intent was to treat nursing education equitably, the scholarship and loan approach did not "fit" nursing. First, much of nursing education was still being conducted in hospitals, and there were relatively few college-based programs. The challenge: Which programs should receive scholarship and loan funds? Secondly, since much of nursing "education" was conducted in hospitals, student nurses often received their education relatively free of charge in exchange for the work they provided to the hospitals. While nursing was treated "equitably" within the context of the Kellogg Foundation, these funds were not as easily targeted in nursing as they were in the medical and dental professions. Perhaps more importantly, the funding for nursing scholarships and loans did not provide as clear a focus on nursing education as it did in medicine and dentistry. In medicine and dentistry the efforts were also coordinated through professional organizations that focused solely upon education, such as the American Association of Medical Colleges and the American Dental Association.

While nursing was treated "equitably" within the context of the Kellogg Foundation, funds were not as easily targeted in nursing as they were in the medical and dental professions.

Divided Concerns. Unlike medicine and dentistry, where resources and attention were focused solely upon helping the war effort, nursing focused also on the health of citizens on the home front. The Michigan Community Health Project was decimated by the loss of nurses to the war effort. MCHP nurses were among the best-prepared in the nation, particularly in the area of community health, and were quickly recruited into military service. Those employed by the Kellogg Foundation had played a primary role in providing comprehensive health care within the communities surrounding Battle Creek, and now they too were called away for the war effort. The nursing leadership within the Kellogg Foundation felt responsible for doing something to fill the void and embarked on programs like teaching family members first aid so they could care for basic problems. Additionally, the development of rural hospitals, particularly in Michigan, emerged as a significant concern. The director of the hospital division focused his attention on this issue. Most of the small rural hospitals of this era were large, converted houses or mansions, many owned and operated by physicians.

The concern was that the quality of care for rural people in these facilities be comparable to that available in more urban settings. A considerable amount of support during the war and shortly after was provided to improve diagnostic services, to build rural community health centers, and to improve the quality of care in these settings. Within this context, nursing was caught between the interest of contributing to the war effort by producing adequate numbers of nurses, maintaining the health of people on the home front, and also staffing the emerging rural hospitals. It is important to note the significance of advisory committees in helping develop priorities and providing professional input into the programs to be funded. Advisory committees were appointed by the board of trustees in each of the divisions of the Kellogg Foundation: medicine and public health, dentistry, nursing, and education. Each of these committees was composed of leaders in their respective fields. Advisory committees to nursing were composed of the prominent nursing leaders of the time. And they were instrumental in determining the direction of funding within the Foundation. The advisory committees, working in concert with the division director in the Foundation, each had an annual budget, and within that framework recommended what should be funded. In medicine and dentistry the focus was to speed up the educational process to produce as many doctors and dentists as possible in the shortest period of time. In hospital administration, the emphasis was on improving rural hospitals and building a system of hospitals throughout the country. The nursing focus was not as clear-cut as it was in other areas. Nursing was pulled in a number of directions simultaneously. With training still occurring mostly in hospital settings, the challenge of producing more nurses for the war effort was fraught with complexity. Add to this the sense of responsibility for helping to maintain the health of people on the home front, and the burgeoning hospital field. It is curious that during the war years, the hospital division of the Foundation

Nursing was caught between the interest of contributing to the war effort by producing adequate numbers of nurses, maintaining the health of people on the home front, and also staffing the emerging rural hospitals.

did not concern itself with the war effort but rather focused upon the development of community hospitals. And unlike medicine, which worked solely with one professional association—the American Association of Medical Colleges—funding to nursing causes was compromised by the multiple nursing associations and lack of one voice for the field. While the Kellogg Foundation enthusiastically supported the development of a unifying structure for nursing through the National Nursing Council for War Service (NNCWS), these efforts were in vain. As Lynaugh points out, the organizations would come together with considerable financial sponsorship from the Kellogg Foundation to address production of sufficient numbers of nurses for the war effort, but once this external force was no longer present, the organizations reverted to pursuing their separate agendas. Although the Council did not succeed in bringing the nursing organizations together, one of the most important outcomes of this effort was the merger of the National League for Nurses Education, the National Association of Public Health Nurses, and the National Association of Colored Graduate Nurses. This merger was significant in facilitating racial integration in the profession.

The lack of a unified sense of direction for nursing is even more evident as the Foundation moved into the post-World War II era. The federal government quickly moved to correct the problem of the lack of scholarships to support medical and dental students through the legislative process, so that by 1943, the Kellogg divisions of medicine and dentistry shifted their focus to planning for the postwar period. Nursing remained enmeshed in responding to multiple demands of the here and now. With no clear-cut focus, there was no unified agenda around which to organize support for the post-World War II era.

The Post-World War II Years. With the war's end, a major challenge was to reintegrate the health personnel from the war effort into the civilian health care delivery system. The hospital field was rapidly emerging, and many physicians who had been in the Armed Forces wanted to build new careers. Not surprisingly, the field of hospital administration emerged as a major component of health care. Surveys of military health personnel indicated significant interests in pursuing health administration as a field once they returned to civilian life. Based on the responses, the Kellogg Foundation invested heavily in the development of graduate programs in hospital administration in the United States and Canada. Additionally, physicians and dentists expressed interest in pursuing graduate preparation in specialty fields. After World War II, the Foundation's medicine, dentistry and hospital administration divisions all focused on developing graduate education. With the G.I. Bill as a source of financial support, it was a logical decision to use philanthropic resources to build the educational programs in which health professionals returning from the war would enroll. Medicine and dentistry, required baccalaureate-level preparation for entrance into these professions. What happened to nursing in this era? Remember, nursing had not yet made a significant move toward collegiate-based education. And development of baccalaureate programs in nursing entailed moving into the mainstream of the general university. This issue was complex because universities were not accustomed to dealing with programs that included an intensive clinical component; nursing programs were considered very expensive to finance. And nursing programs had to meet the academic standards for baccalaureate education while also preparing competent, clinically prepared nurses. With the baccalaureate base so tenuous, following in the footsteps of the other fields was difficult. Ten U.S. graduate-nurse education programs did receive funding. Additionally, four colleges in upstate New York were provided funding for baccalaureate nursing education programs to serve rural areas. However, with the increased demands for nurses to staff the rapidly emerging hospital field, the pressure was on to produce nurses to staff hospitals and nursing administrators to provide leadership. And once more, the Foundation's nursing resources were diverted to addressing the problem of adequate numbers of personnel. The first such effort was to produce practical nurses to increase hospital staffing. One proposed model for nursing, as Lynaugh noted, was to develop two categories of nursing: the professional nurse and practical nurse as assisting personnel. Focusing first on Michigan, a model practical-nurse educational program was developed and tested. This model program was then expanded to four southern states—Alabama, Arkansas, Florida, and Mississippi—and later expanded to

Surveys of military health personnel indicated significant interests in pursuing health administration as a field once they returned to civilian life. Based on the responses, the Kellogg Foundation invested heavily in the development of graduate programs in hospital administration in the United States and Canada.

Illinois. As part of this orchestrated effort, the focus was on sustainability. In all privately funded efforts, a major concern is how programs can be maintained once private funding ends. Philanthropic funds are typically used to develop and test new approaches, and it is assumed that if the models succeed, a source of funding will be developed to sustain such programs over time. In the case of practical nursing, lodging these programs under vocational education assured such a long-term funding base. Unlike the post-World War II baccalaureate and graduate nursing education programs that had to compete within the context of the university for funding, practical nursing programs tapped into an ongoing funding stream that continues to this day.

1960–1980

Until 1964, the Foundation was organized around professional fields: education, dentistry, hospital administration, nursing, and public health. Each of these fields was led by a single program director working with nationally recognized leaders as advisory committees. As leadership transitioned in the early sixties, a concern was that the program areas were isolated and operating independently of one anther. In 1964, the newly appointed CEO, Dr. Russell Mawby, recommended—and the board approved—a change in structure disbanding these divisions, dismissing the advisory committees, and moving toward three integrated fields of emphasis: agriculture, education, and health, with each individual program director reporting directly to the CEO. The program director in nursing, for example, was part of the health group and as such played a role in setting priorities and directions, but when putting forth individual projects for funding, negotiated directly with the CEO to place proposals before the board. Before this, the nursing program director turned to a Nurse Advisory group to help determine programming directions and operated within the framework of a yearly budget established for the division of nursing. Now she had to negotiate with her colleagues in the other health professions for approval of programming and funding. Nursing was now in direct competition with hospital administration, medicine, public health, and dentistry for the Foundation's health-related funds. Not only did the playing field change, but so did the game itself. The Foundation's internal changes coincided with a change of nursing program directors when Mildred Tuttle, who was a nationally recognized leader in community health nursing, retired. Tuttle was replaced by Barbara Lee, who had a non-traditional nursing background. Lee had received a master's degree in hospital administration from Yale University and served as a faculty member in that program before her appointment to the Foundation. Not surprisingly, the focus of funding for nursing shifted toward developing leadership in Nursing Service Administration and increasing the supply of nurses to staff the burgeoning hospital field.

Nursing was now in direct competition with hospital administration, medicine, public health, and dentistry for the Foundation's health-related funds.

Major funding for graduate programs in Nursing Service Administration was provided throughout the country. A second generation of projects in which schools of nursing joined with business schools or schools of public health to give nurse administrators a strong background in finance, business, and marketing was supported. This coincided with broader health interests in the Foundation in health care financing, hospital efficiency, and staff productivity.

In the sixties, the Kellogg Foundation played a key role in the burgeoning community college movement across the country. Interest in community colleges and their role in health careers training focused on allied health workers such as technicians, dental hygienists and assistants and other areas of program support. Before the Nursing Advisory Committee disbanded, Mildred Montag presented the results of her study comparing associate degree nursing education with hospital-based diploma schools. The advisory group, interested in moving nursing education out of hospital settings and into educational institutions, supported the further testing of this concept and a very extensive strategic plan was developed. Four states were targeted for these models: California, Texas, Florida, and New York. A four-pronged strategy was devised to meet these goals:

1. Develop model programs in a community college in these states;
2. Support graduate education to prepare faculty to teach in these community-college programs;

3. Devise a public-policy strategy to work with the community college structure in these states to secure the ongoing funding base to sustain these programs;
4. Initiate consulting to encourage development of additional programs.

Through the years, Kellogg funded a number of other initiatives involving community college programs in nursing. Within the Foundation, support for associate degree nursing education was popular because it furthered the concept of community colleges and responded to needs to produce adequate numbers of nurses to staff hospitals. However, while nursing leaders on the advisory committee supported the associate degree as a way to move nursing education into the mainstream, baccalaureate and graduate education lacked similar support. None of this was helped by the lack of a unified voice for nursing from the professional organizations. Despite the amount of money, time, and effort invested in trying to bring the groups together, without a strong unified voice for nursing from outside of the Foundation, nursing program directors had little leverage inside the internal political arena of the Foundation. In contrast, the hospital division, for example, had a long tradition of working with the American Association of University Programs in Health Administration to develop graduate education in health and hospital administration. The American Association of Medical Colleges played a comparable role in the field of medicine. In nursing, requests for support of projects came from individual institutions or from one of many professional associations. With the exception of associate degree nursing, there was no developed strategic plan for moving nursing education and practice forward. In lieu of a unified voice for nursing, one strategy that was effective was to work with regional educational compacts—the Southern Regional Education Board, the Western Interstate Commission on Higher Education, and later the Midwest Alliance in Nursing. These compacts engaged in a number of projects aimed to improve nursing education and practice. Throughout the sixties and seventies, the health group focused increasingly on issues surrounding the improvement of practice, health promotion, and disease prevention. Where hospitals were concerned, the emphasis was often on cost and productivity, availability and access to care, and developing systems of care extending beyond the hospital. Medicine was increasingly focused upon primary and family health care. Support for dentistry was waning, with no strong voice either internally or externally. In this context, nursing allied itself with the hospital interests by focusing on long-term care and the elderly, and with medicine by focusing on developing joint-practice models and promoting nurse practitioners. During this era, nursing within the Foundation melded into the coordinated work of the health group. Outside the Foundation, alliances focused narrowly on trying to unify nursing and on specific areas such as the geriatric nurse practitioner movement. The longstanding relationships with community colleges continued. One characteristic of the community college funding initiatives through the years has been the careful planning of a national strategy that produces sweeping changes nationwide. No other level of nursing education has ever brought forward such strategic approaches, and from the perspective of investment of private funds to facilitate change, these programs have been hugely successful. The extent to which philanthropy can foster positive social change depends on a number of forces, most importantly, how private funds can launch and support innovations, but not be required to maintain them. The process is facilitated if there is a unified sense of direction that comes from the field seeking out support to test new ideas and new approaches. Without unified direction, the agenda is set by the internal politics of the philanthropic organizations instead of by those outside in the field. The Foundation clearly tried to bring some sense of unified direction to nursing that would in turn lead to philanthropic support. And, while some within the nursing profession may fault the Foundation for a lack of support for particular professional interests—in particular baccalaureate and higher degree educational programs—the lack of consensus among the numerous professional nursing organizations made it very difficult to sort out "who speaks for nursing." Despite these difficulties, the Foundation maintained a commitment to nursing as integral to health care—a commitment that would be sustained into the future.

In nursing, requests for support of projects came from individual institutions or from one of many professional associations. With the exception of associate degree nursing, there was no developed strategic plan for moving nursing education and practice forward.

9 Years of Change: 1980–1994

Helen K. Grace

The Kellogg Foundation moved into the 1980s with increased assets for more grant-making and a mission to refocus program activities and integrate efforts in the health field. The Foundation's shift expanded its focus from hospitals to the community, and emphasized prevention as intervention. The nursing profession as a whole did not follow this shift and remained aligned with hospitals as the center of health care, clinical specialization that paralleled high-technology specialized care in hospitals, and nursing administration as part of the big-business orientation of the health care industry. This "parting of the ways" changed the relationship between nursing and the Foundation in significant ways.

In the formative years, the Foundation's community health nurses had been the backbone of the Michigan Community Health Project, linking private-practice medicine and dentistry, the schools, and the community. As hospitals developed, demand increased for nurses and nursing leadership, and the Foundation and the nursing profession worked hand-in-hand to meet the need. But as the Kellogg Foundation changed direction back to the community and toward integrated efforts within health care, nursing faced increased competition for programming support.

1980–1987

As the Kellogg Foundation changed direction back to the community and toward integrated efforts within health care, nursing faced increased competition for programming support.

Nursing at the Crossroads. In the transition years of the late seventies into the eighties, some of the funding for nurse-related projects remained even as new program goals were put in place. Five major funding streams continued: (1) nursing service administration education, (2) long-term care, (3) joint practice, (4) non-traditional nursing education, and (5) differentiated practice. Each is described below:

- *Nursing Service Administration.* In the last phase of Nursing Service Administration grants, several graduate educational programs were built collaboratively between graduate nursing educational programs and business schools. These programs were based on the premise that nursing administrators needed to have a business background to do their work effectively. The University of Pennsylvania and the Wharton School of Business, the University of Washington, Virginia Commonwealth University, and the University of Michigan were among the grantees.

- *Long-Term Care.* An emphasis on support for the preparation of geriatric nurse practitioners extended into the eighties. In a collaborative program linking geriatric nurse practitioners into long-term care settings, the Mountain States Health Corporation worked with four university-based programs in the western states: the University of Washington, the University of California at San Francisco, the University of Arizona, and the University of Colorado. In these programs, nursing faculty worked with long-term care settings to prepare geriatric nurse practitioners. An extensive evaluation of this program found that placement of geriatric nurse practitioners in nursing-home settings improved the quality and outcomes of care while reducing costs.

- *Joint Practice.* Building upon recommendations made by the Lysaught Commission, a national commission for the study of nursing and nursing education, the Foundation provided extensive support to establish the Joint Practice Commission, also sponsored by the American Nurses' Association and the American Medical Association. The intent of the Commission was to bring medicine and nursing together in hospital-based committees to improve patient care and reduce costs. The concept was tested at a number of hospitals across the country. These test sites showed improved quality of health care, and the Commission was extremely active as long as it received external funding. But once this funding ran out, there were limited efforts to continue the committees, despite the valuable lessons learned. In addition to joint practice in hospital settings, Kellogg also provided support to prepare nurse practitioners at several sites. An evaluation of the collaborative practice conducted by the University of Missouri-Columbia clearly documented the value of physician-nurse efforts in improving the quality of care in primary-care settings.

- *Non-Traditional Education.* Late in the seventies, the Foundation had provided substantial support to the New York Regents external degree program in nursing. This program model helped diploma and associate degree nurses attain bachelor's degrees through a flexible distance-learning approach. Building upon the New York State model, the California State university system refined the clinical performance testing component and evaluated this non-traditional approach as compared with more traditional nursing education. These approaches were not readily received by the nursing profession because they challenged some of the traditional values that had guided nursing education (or training, depending upon one's point of view). Nurses within these non-traditional programs demonstrated that they learn in a variety of ways, that sitting confined in a classroom was not preferable, and that they were equally competent in providing quality clinical care.
- *Differentiated Practice.* To address the conflict between associate- and baccalaureate-degree nursing education, a nationwide effort was made to differentiate the practice skills of nurses prepared at these levels. The effort was fostered through regional nursing compacts of the Southern Regional Educational Board (SREB), the Western Interstate Compact on Higher Education (WICHE), and the Midwest Alliance in Nursing (MAIN).

The Eighties

In preparation for the 50th anniversary, for at least three years, the board and staff worked to refocus the programming goals of the Foundation.

These efforts were led by a newly appointed president, Dr. Robert Sparks. Dr. Sparks was a medical doctor who specialized in internal medicine, and had most recently been the chancellor at the University of Nebraska. He brought to the Kellogg Foundation an academic approach to strategic planning and embarked upon an extensive review of all of the policies and procedures within the Foundation, and an extensive formulation of programmatic goals.

In 1982, the board approved the document "Programming for the Eighties," which provided the guidelines for this next step into the future. "Programming for the Eighties" included seven goals to guide the work of the Foundation—with two specific to health care: (1) Improving human well-being through community-wide, coordinated, cost-effective health care services, and (2) Improving human well-being through health promotion/disease prevention. Nursing could gain support through other goals such as "expanding opportunities for adult continuing education" and "fostering leadership capacity," but in this new context, because integrated programming was particularly valued, nursing had to show evidence of collaboration with other health professions to be competitive.

A report to the board on "Programming for the Eighties" states, "The Foundation's work is carried out in the fields of agriculture, education and health, which broadly conceived, provide the three knowledge bases in which its activities are rooted. These three are not compartmentalized so far as programming is concerned, but . . . are integrated in order to find solutions to problems" (W. K. Kellogg Foundation, *Programming for the Eighties,* p. 2).

Following are some key components of the plan:

- Demonstrate health promotion services in sites where there can be continuing contact with people who can be motivated to improve their health habits (health institutions, work sites, and schools and colleges).
- Demonstrate health promotion programs emphasizing both changes in lifestyle and self-care practices that are targeted to adolescents and the elderly and that focus on specific behaviors (alcohol abuse, smoking, drug abuse, accidents, and suicide).

- Prepare practitioners and aides to provide health promotion services.
- Support efforts that define needed health promotion public policy and evaluation and networking of health promotion activities.
- In the area of disease prevention and public health, four strategies were emphasized:
    (1) Support the development and use of compatible computerized data collection and analysis systems by state and local health departments and other governmental agencies so they can be used by all appropriate health professionals and public health officials.
    (2) Prepare all health professionals to participate in and support public health by including public health principles in profes sional education.
    (3) Increase the number of professionals prepared in environmental and occupational health through cooperative programs of schools of public health and state departments of health.
    (4) Implement continuing education programs to teach health practitioners about environmental health hazards.

The other health goal—to improve the well-being of people through coordinated, community-wide, cost-effective health care systems—emphasized the following strategies:

- Demonstrate models in which community hospitals provide leadership to develop and coordinate community-wide, cost-effective, comprehensive medical and dental care and health services by improved inpatient management and discharge planning, services for the elderly, health promotion services, and client tracking.
- Demonstrate horizontal or vertical integration of health organizations to form and extend multi-health institutional systems.
- Demonstrate alternatives to institutionalization and new models of ambulatory primary medical care and health services to take the place of acute or long-term inpatient care where appropriate.
- Develop joint practices between doctors and nurses in hospitals and other medical-care settings to provide improved, comprehensive, and continuous medical care and health services in the most efficient and cost-effective manner.

- Develop new methods and measurements to improve decision making for cost-effective and more appropriate use of health services.
- Support health professional education programs that aim for cost-effective and appropriate use of resources.
- Encourage development and application of policies that support provision of appropriate medical care in the least costly setting.

In the area of adult continuing education, one strategy was particularly relevant to health professionals:

- Encourage the introduction of systems of collaborative continuing education (a) for health professionals and closely allied workers in universities, hospitals, and professional associations, or (b) for comprehensive professional programs in universities.

The Foundation made a conscious decision to continue to value the contribution that nurses made to improved health care.

Two strategies of leadership were particularly relevant to health professionals, including nurses:

- Develop leadership traits among a wide variety of U.S. citizens to include young leaders from business, the professions, political life, and academic fields who have high technical skills and/or humanistic interests (through the Kellogg National Fellowship program).
- Develop well-informed leaders in key agricultural, educational, youth-serving and health organizations through programs conducted by universities and other appropriate organizations.

Kellogg's change in direction coincided with my recruitment to the Foundation to succeed Barbara Lee, who retired in 1982. I joined the staff of the Foundation as a program director in October 1982 as part of a "health team" composed of Dr. William Grove, former vice chancellor for the medical center at the University of Illinois, and Robert DeVries, a hospital administrator who had been a program director at the Kellogg Foundation for 10 years before I arrived. In this programming transition, the Foundation made a conscious decision to continue to value the contribution that nurses made to improved health care. Indeed, my recruitment was based on a board decision that nursing was an essential component of the health care team. However, the board's commitment fell within Kellogg's new integrated approach to health care funding, and I was charged with representing nursing's perspective on that health care team. And while the Foundation was encouraging multidisciplinary practices, the nursing profession was on a course of "go it alone." Some of the following programming areas show how nursing fared during this time period:

Integrated Health Programming. Kellogg's programming shift occurred at the same time as interest increased in the development of Nursing Practice Centers that combined opportunity for clinical work with education and research. Within nursing, there was much talk of "independent practice," and the profession focused on removing itself from the control of medicine. As a newcomer to the foundation world, and most recently a part of the nursing establishment, I was sought out by many of my former colleagues to support the development of nursing centers. I would explain the priorities of the Foundation, and encourage my nursing colleagues to frame their requests in a way that enabled them to compete for funding in the framework in which I had to operate. Many discussions centered on my view that nursing centers were a "means" to a goal of improved health services for people rather than an "end" goal. In many of these discussions, I was encouraging nursing to take the lead, and engage other health professionals to work with them in their proposed projects. Few rose to the challenge, and I am sure that many saw me as a traitor to the nursing cause. Despite the challenge, some stepped forward and developed proposals that were ultimately funded.

Health Promotion Services. Although this is an area that nursing has long espoused as one of its strengths, the number of projects funded in which nursing played a primary role was very limited. Although the Kellogg Foundation sought to assure multidisciplinary practice and look for the presence of nursing in the projects that it funded, it rarely funded nursing projects directly. However, there were some notable exceptions: In 1985, Kellogg granted $1,330,071 to Pace University to conduct a university-wide, nurse-coordinated, health promotion program for students and staff. With funding from this grant, student health services for the five campuses were organized, directed, and provided by nurses with physician backup. In 1985, Kellogg granted $743,692 to Hampton

Although the Kellogg Foundation sought to assure multidisciplinary practice and look for the presence of nursing in the projects that it funded, it rarely funded nursing projects directly.

University to help establish health promotion services for adolescents and elderly in medically underserved areas of Virginia by establishing a nurse-managed clinic and faculty/student-operated mobile unit. In 1984, the foundation granted $695,700 for Michigan State University to develop a comprehensive health promotion program (Healthy U) that engaged a wide range of faculty and students across the entire university in an array of health promotion activities and research. This project was led by a faculty member of the College of Nursing. Another university-wide project that involved significant nursing participation was funded at the University of Tennessee at Memphis. In 1984, Kellogg granted $1,844,500 to "develop a national model for incorporating health promotion into the education of health professionals." In these settings, the superb leadership of nursing was highly visible and valued. These projects provided a wealth of opportunities for faculty research, and in both settings the academic programs in nursing were enriched.

Coordinated, Community-Wide, Cost-Effective Health Care Systems. While the Kellogg Foundation was interested in improving coordination and cooperation to control rising health care costs, federal and state approaches emphasized competition among providers instead. Setting limits on the amount of care that would be reimbursed created a highly competitive environment among health care providers. Within this context, nurses were caught in the middle of these competitive struggles, though they were largely invisible as the architects of systems of coordinated, community-wide health care systems. One small step in supporting these systems was a grant provided to the North Avenue Women's Center, which was based in the Foundation's home community of Battle Creek. When the locally based Battle Creek Area Medical Education Committee disbanded and medical residents were no longer placed in the community, the facility no longer provided maternity care for low-income mothers. Local physicians agreed to take responsibility for this care, provided support services such as nutrition counseling, and social services could be arranged for through a newly created women's center. Kellogg provided support to establish the North Avenue Women's Center and, not surprisingly, the center soon became the major provider of maternal services rather than just a support network. Directed by a nurse, with services provided by nurse midwives and a physician, the center at one point provided nearly 50% of the deliveries in the Battle Creek community, which saw a dramatic reduction in infant mortality. Later, the center expanded into a comprehensive, federally funded Family Health Center.

In 1986, a grant to the Maternity Center Association of New York City helped integrate a birthing center into a comprehensive health center in a low-income, largely Hispanic community in Morris Heights. Both of these projects supported an expanded role for nurse midwives in the delivery of health care services. While nurse midwifery has taken the lead in carving out expanded practice for nurses, the nursing profession has not embraced this as a part of the mainstream of nursing. Consequently, organized nursing did not view Kellogg's support for projects led by nurse midwives as support for nursing.

Introducing nurse practitioners into nursing-home settings helped reduce costs through such things as a reduction in the medications taken by the elderly, and prompt discharges from nursing homes back into the community. Other evidence of improved quality of care included a reduction in ailments such as bedsores, as well as improved nutrition among nursing home residents.

Alternatives to Institutionalization. Under prior funding priorities, Kellogg provided substantial support to projects directed toward care of the elderly and made a series of grants to develop geriatric nurse practitioner certification programs with the educational programs based in four western universities: University of Arizona, University of Colorado, University of San Francisco, and University of Washington. These programs were linked to nursing homes, and coordinated by a project funded at Mountain States Health Corporation. Introducing nurse practitioners into nursing-home settings helped reduce costs through such things as a reduction in the medications taken by the elderly, and prompt discharges from nursing homes back into the community. Other evidence of improved quality of care included a reduction in ailments such as bedsores, as well as improved nutrition among nursing home residents. This push to improve care for the elderly included a series of grants in 1986-1987 and in 1990 that led to a major change in the education of associate degree nurses to include content specific to long-term care. Throughout the history of funding for associate degree nursing projects within the Kellogg Foundation, leaders in the field consistently worked to create changes that would permeate ADN education nationwide.

The first phase in the effort to incorporate long-term care into the curriculum was to build six demonstration projects in highly respected community college programs nationwide: Ohlone College in California, Community College of Philadelphia, Valencia Community College in Florida, Triton College in the Chicago area, Weber State College in Utah, and Shoreline Community College in Washington. Project leadership was provided by Ohlone College and the Community College of Philadelphia. In the dissemination phase, these colleges trained faculty from other settings who wanted to make similar changes. This project led to the incorporation of elderly care into the classroom and clinical education of community college nurses. Community colleges, then, linked with area nursing homes. The National League for Nursing, the accrediting agency for nursing education programs at that time, was engaged so that the accreditation criteria would keep pace as nursing education programming changed.

Cost-Conscious Professional Education. Most of the funding for this strategy went to schools of medicine, medical-practice settings, or health care administration programs. One exception to this was a project led by the College of Nursing at Texas Tech University. This project aimed to extend the ability of the health professionals at Texas Tech to improve care in outlying areas. A major problem in West Texas was providing quality health care in small community hospitals that saw a range of problems, from emergency care for accidents to long-range care for the elderly. Computer technology enabled resources from the medical center to be available to caregivers in these remote settings. A second part of this project involved working with the largely Hispanic community in El Paso. This project was initiated by the College of Nursing and led by nursing faculty.

Health Care Policy Making. After four previous efforts to bring the major nursing organizations together, Kellogg made one more attempt in the early eighties. Despite reluctance on the part of other members of the health team and of the board of trustees, I remained concerned over the divisions within nursing, which were even more visible as I attempted to master my new role in the philanthropic world. In 1983, based on a request from the American Nurses Association, Kellogg made an initial grant of $1,174,840 (later supplemented to total $2,351,490) to the American Nurses Foundation on behalf of four national nursing groups (the American Nurses Association, the American Association of Colleges of Nursing, the National League for Nursing, and the American Organization of Nurse Executives) to organize a coalition to implement recommendations from nursing education and practice studies. To accomplish this work, the National Commission on Nursing Implementation Project (NCNIP) was formed. Recognizing that nursing must be viewed within the broader context of the health field, an advisory committee to NCNIP was formed with representatives from the American Medical Association, the American Hospital Association, the Business Roundtable, the National Consumers League, the Congress of Hospital Trustees, and the Health Insurance Association of America. This commission made considerable progress in identifying basic education and practice settings that appropriately use the skills and competencies of nurses from different educational backgrounds. It advocated helping nurses move up through the educational system. The grant intended that nursing organizations would agree to be responsible for maintaining the work and that NCNIP would no longer be necessary. Unfortunately, once the NCNIP structure dissolved, the agreements that had been reached were largely forgotten, and the divisions between the nursing organizations remained, and in some instances grew worse. The work of NCNIP was complemented by another grant of $1,868,954 made to the National Council of State Boards of Nursing. This grant was similar to one made to the National Board of Medical Examiners to incorporate computer technology into the testing system in order to validate clinical competence. With the two grants in this area, the Foundation encouraged collaboration between nursing and medicine. Thus in the early eighties, with the Foundation's shift to multidisciplinary collaboration, very few

The Foundation encouraged collaboration between nursing and medicine. Thus in the early eighties, with the Foundation's shift to multidisciplinary collaboration, very few projects headed by nurses were funded.

projects headed by nurses were funded. While these programming goals were approved for the decade of the eighties and were not implemented until late 1982, less than five years later, additional sweeping changes occurred.

1987–1994

Back to the Community. Kellogg's changing leadership led to dramatic changes in programming. Dr. Sparks resigned as president of the Foundation, and Dr. Norman Brown, an agricultural educator, assumed the position. Dr. Brown had joined the Foundation in 1985, coming from the University of Minnesota. He had previously been a faculty member at Michigan State University and had extensive experience in agricultural extension and 4-H. Dr. Brown brought to the Foundation considerable experience in working with communities and with youth. This background was important in providing leadership in a major swing back to the philosophical roots of the Kellogg Foundation: youth and community. Many of the documents generated during this time reflected on the origins of the Kellogg Foundation. In formulating the new direction, documents referenced the 1947 publication *The First Eleven Years of the Foundation,* which noted that problems in the Battle Creek area were similar to those of communities across the nation:

> These communities are typically rural and possess a little better than average economic resources. They have active and interested community leaders. They have the usual quota of physicians, dentists, school teachers, and all other human resources. Notwithstanding all of this, health programs were practically non-existent. . . . The distressing thing was that no better use was being made of the opportunities afforded by the American system to provide really adequate answers to these challenging community needs.

In 1987, some 56 years later, not much had changed. Were we to revisit these communities, the same observations might be made: that we have the usual, or perhaps a surplus, of human resources, health programs, and hospital beds, but in the midst of this plenty, a significant portion of the population has difficulty in obtaining adequate and appropriate health services. To address these problems in the eighties, as in the thirties, the foundation reaffirmed its commitment to look for solutions at the community level. This philosophy guided the Foundation's direction: (1) emphasize people, not systems, buildings, or technology; (2) start with the problems that people recognized; (3) exchange experiences; (4) focus on leadership; and (5) concentrate on the community. Shortly after Dr. Brown was appointed president, he asked me to serve as coordinator (a new position) for health programs, and I was charged with responsibility for reshaping the programming goals and strategies for the health area. In this role, I was to coordinate health programming, but not the program directors. The program directors in the health area still reported directly to the president. However, the composition of the health program team changed. Dr. Grove retired and only Robert DeVries and I remained as program directors. At this point of transition, the concerns were that the health programming of the Foundation had become driven by the interests of the health professions and of hospitals, and that there was a need to look at issues from a different perspective. Secondly, trustees were concerned about being able to evaluate the impact of funding on specific problems. They viewed much of the work in the health area as diffuse and unfocused, and were seeking greater clarity. In my early years at the Kellogg Foundation, I had the opportunity to work closely with Dr. Mario Chavez, who provided leadership for health programming in Latin America and the Caribbean. Through the years, the health priorities had largely been formulated from the perspective of the United States, and Dr. Chavez had to fit his programming into this framework. He retired in the mid-eighties, and leadership for health programming in Latin America was also changing. It was a good time to work collaboratively with the Latin America staff to formulate health goals that would guide programming across the Foundation. Having come to the Foundation at a time when programming priorities were shifting in the early eighties, I also had realized that it was very easy to change the rhetoric and maintain the status quo. As a sociologist, I was pessimistic about the likelihood of

We have the usual, or perhaps a surplus, of human resources, health programs, and hospital beds, but in the midst of this plenty, a significant portion of the population has difficulty in obtaining adequate and appropriate health services.

achieving any substantial social change when there is so much inertia to overcome in social systems. It was comfortable to continue to work with professional associations such as the American Hospital Association or the American Academy of Medical Colleges—the grantees were known, and relationships had been built over time. I realized that if any real change was to occur, it was necessary to change direction so it would be impossible to continue doing business as usual.

Working collaboratively with the Latin American program staff, it was proposed that the Foundation focus on problems that people face in obtaining adequate, affordable health services and in reaching community-based solutions to these problems.

Working collaboratively with the Latin American program staff, it was proposed that the Foundation focus on problems that people face in obtaining adequate, affordable health services and in reaching community-based solutions to these problems. The focus became Community-Based Primary Health Care, which meant different things to different members of the health care program team. For medical educators, it meant community-oriented primary care. For those coming from the perspectives of hospitals, it meant the hospital reaching out into the community and expanding its scope. Many discussions were to clarify these differences in perspective. Over time the health program directors came to understand that community-based programming entailed a significant power shift, with the community as the focal point. The shift established priorities for particular age groups (infants and children, adolescents, and the elderly) and for particular segments of the population (families in rural America, urban poor, and minorities). By focusing on communities and their health care problems, funding to health care institutions and health professionals would be based on the relationships that were established rather than upon institutional or professional interests. With this general framework, projects were funded under five strategic directives:

1. Implement community-based comprehensive health services addressing high-priority health problems.
2. Foster use of information technology to enhance the services.
3. Foster educational programs to prepare health professionals to administer services.
4. Support leadership development to oversee delivery of services.
5. Encourage new approaches to inform local, state, regional, and national policymakers of interventions that support the services.

Although Lynaugh's historical analysis extends only to 1982, she was conducting her work as these changes were occurring. She noted, "I can't think of any organization in the private sector which has had such a far-reaching, influential effect on nursing as the Kellogg Foundation" (Nursing and the Kellogg Foundation, p. 174). She also signaled that the changes proposed for the 1980s were reminiscent of the initiatives of the 1930s:

> . . . The public health nurses who worked in the Michigan Community Health Project would feel right at home reading about links between problem-focused community demonstrations and educational programs to reorient professional health workers. They might be astonished to see the demographic changes in our aging population, the change in its distribution from rural to urban, and the incredible rise in resources committed to health care. They would be saddened and humiliated to see the difference in infant mortality rates between white and black babies and the high incidence of births to girls under 15 (Nursing and the Kellogg Foundation, p. 174).

In focusing on the health needs of people rather than the self-interests of any one professional group or set of institutions, Kellogg recognized that nurses working at the community level played a key role in any long-range solutions to the problems that plagued health care delivery in this country. With an emphasis on health maintenance and early identification of problems rather than disease treatment, nurses in the eighties, as they did in the thirties, became an important link between people, health care institutions, and providers. This work was facilitated by the addition of two program associates to the health care team: Dr. Henri Treadwell, the first African-American program staff member in the Kellogg Foundation, who had a background in public health, and Dr. Robert Hodge, an M.D. with a primary care background, were employed to assist me in my work as program director and coordinator. Both Drs. Treadwell and Hodge had been Kellogg National Fellows, so they had

With an emphasis on health maintenance and early identification of problems rather than disease treatment, nurses in the eighties, as they did in the thirties, became an important link between people, health care institutions, and providers.

some prior exposure to the Foundation. Soon after the shift in direction, Dr. Thomas Bruce was recruited as a program director. He was dean of the medical school at the University of Arkansas with a background in internal medicine, and an interest in public health. Though working loosely as a team, program directors within the Foundation had traditionally operated as independent entrepreneurs. This was reinforced through the maintenance of a structure in which program directors negotiated directly with the president for project funding. The appointments of Drs. Treadwell and Hodge helped facilitate the organization's philosophical shift, because they were able to devote themselves to seeking out new grantees in a wide array of community-based organizations.

Community-Based Health Services. Although there was no predesigned plan to give nurses preferential treatment, nurses stepped forward with a number of innovative proposals developed in concert with communities. By 1990, 15 of 40 community-based health service projects were headed by nurses. In the remainder of the projects, nurses played key roles in a variety of projects and in very diverse settings. A 1989 report to the board on projects funded during this time included the following:

- *Wayne County, MI.* Reduce infant mortality by improving access to prenatal and pediatric care for black urban low-income women and their children by removing transportation barriers: $889,152.
- *Medical College of Georgia.* Reduce infant mortality and improve maternal and infant health in rural east-central Georgia through community outreach and individual case management: $795,697.
- *Alcorn State University, MS.* Provide access to health services for adolescents in rural and urban Mississippi communities with mobile health screening: $1,123,547.
- *Block Nurse Program, Inc., St. Paul, MN.* Demonstrate a nurse-managed model for effectively using community-based resources and the existing health care system to provide in-home care for the elderly: $496,878.
- *City of Atwater, MN.* Companion grant to the Block Nurse grant: $297,0029.

- *Highland Area Community Council, MN.* Companion grant to the Block Nurse grant: $315,192.
- *Lake Superior State College, MI.* Improve health and prevent disease by providing ambulatory and home-nursing services to elderly residents in Michigan's Chippewa, Luce, and Mackinac counties: $412,743.
- *University of Arkansas for Medical Sciences, AR.* Deliver community-based comprehensive health care to older adults with a multidisciplinary team of gerontologic specialists: $997,254.
- *Evanston Hospital Corporation, Evanston, IL.* Demonstrate effective ways to help the frail elderly by providing training for caregivers and offering respite care in community hospitals: $679,531.
- *Auburn University at Montgomery, AL.* Establish an integrated model to improve the well-being of elderly in rural Alabama through holistic health assessment, health risk appraisal, and human-resource development: $1,376,623.
- *Columbia University, NY.* Improve inner-city residents' access to primary health care by staffing an academic medical center's outreach program with students in the health professions: $933,069.
- *Howard University, D.C.* Improve health care for the homeless through a nurse-run clinic in a Washington, D.C., homeless shelter: $1,040,520.

The 1989 report also described the difficulties in doing community-based programming. This report distinguished between community-based ("of and with") and community-oriented ("to and for") approaches and noted:

> While the concept of community-based services is a simple one, development of partnerships is difficult to achieve. Health professions and health-professions educational programs frequently take a prescriptive approach to community health problems. That is, they define high-priority health problems from epidemiological data and from their own sense of priorities. The concept of meeting with community groups and seeking from them

> their sense of problems and priorities is a foreign one. On the other hand, community groups have little experience in community decision-making processes involving health problems. Community residents tend to defer to the health professionals and accept a paternalistic relationship to the providers.

All of these projects demonstrated innovative approaches to address health care problems of underserved populations. The solutions, such as providing prenatal care at an early stage of pregnancy, or supportive services to the elderly, significantly improved care and reduced reliance on expensive technology to treat problems, because they were prevented in the first pace. The public-policy challenge was to find ways to maintain support for these programs once Kellogg funding ended.

Information Technology. In addition to their community-based health services work, nurses led several information technology projects. The College of Nursing at Texas Tech University extended the resources of the health science center to rural areas of West Texas. Nurses led a Joint Commission on Accreditation of Health Organizations project, which involved two practice sites for designing standards of community-based health services ($330,658). They also led a Kaiser Foundation Hospital ($328,398) in Portland, Oregon, designing standards for adolescent care; and a Battle Creek Health System ($328,398) program, designing standards of care for the elderly.

Education of Human Resources for Community-Based Health Services. Continuing its support of projects preparing human resources for the care of the elderly, Kellogg helped Johns Hopkins University School of Nursing ($622,366) develop a graduate program that bridged institutional and community-based care. It also helped the San Francisco Institute on Aging ($982,830) offer fellowships to health professionals in geriatrics, in a project under nursing leadership. Funding to the University of Michigan ($980,973) helped improve care for the elderly in six communities by providing interdisciplinary geriatric education for teams of health professionals. This project was also led by nurses. Preparation of human resources for underserved populations received particular attention. The University of Illinois at Chicago ($575,929), headed by nursing faculty, developed a training program for community health advocates to provide health services in inner-city communities. These advocates then went to work for a wide range of community-based health service agencies throughout the Chicago area. Kellogg also helped develop nursing education for Native Americans in tribal colleges. Two initial programs were supported: Salish Kootenai ($964,044) in Pablo, Montana, and Sisseton-Wahpeton ($66,950) in Sisseton, South Dakota. With a leadership development grant ($464,200), Indiana University developed governmental and health leaders in six Indiana cities to provide community-oriented primary care for area residents. The American Nurses Foundation was awarded several consecutive awards to foster leadership development for nurses of color. Nurses were major players in the health programming of the Kellogg Foundation. Even with the change in direction, nurses still played a major role in fostering innovative community-based health care projects.

Nurses were major players in the health programming of the Kellogg Foundation. Even with the change in direction, nurses still played a major role in fostering innovative community-based health care projects.

While the nature of projects that Kellogg funded had changed, the next step was to leverage these community-based demonstrations to create change in the health professions and the health care institutions.

One of the problems identified in networking meetings was that the community-based projects and health-professions educational programs seemed far apart. As in the Michigan Community Health Project of the thirties, those working at the community level recognized the importance of health professionals' being prepared to work with communities, and sought ways to connect the two through experiential learning. The 1989 board report addressed other issues as well:

> Concern was expressed for the lack of minority representation of students and faculty in relation to the population being served in community-based projects. . . . Issues that arise in terms of ownership and control [were discussed]. . . . In entering the communities, the importance of students' having appropriate preparation in the culture and context of communities, and coming to communities to learn from them, rather than to impose upon them, was

> stressed. Fear of communities being "used" for research and for teaching purposes was discussed (W. K. Kellogg Foundation, 1989).

Program advantages were judged to outweigh the concerns, and staff worked to develop a strategy to address the concerns.

The Early Nineties

As the board moved into the nineties, it evaluated the work of the prior decade and set a direction for the future. In February 1991, a board report identified the following objective: "Building from the lessons that have been learned from the current set of community-based health services projects, some special initiatives that are being considered are directed toward broader systems change in health services delivery and in the education of health professionals." The need for more systemic approaches was based on the following observations of problems at the community level:

> Fragmentation of health and human services; the need to better allocate health care resources; the inability of primary health care services to prevent problems; underserved populations' limited access to basic health services; exorbitant costs and inefficiency; the focus of health professions education on high technology tertiary care; and the isolation of public health education from health-professions education, and of public health practice from comprehensive health services. All these cry for innovative and bold approaches to broad systems change.

Kellogg's programming staff was taking a much more hands-on approach to generating proposals than it had in the past.

From this, three initiatives were proposed: (1) modeling and testing of an alternative comprehensive model for community-based health services; (2) community partnerships with health-professions education; and (3) community-based public health. In proposing these initiatives, Kellogg's programming staff was taking a much more hands-on approach to generating proposals than it had in the past. Here are highlights of those initiatives:

- *Alternative Comprehensive Model for Community-Based Health Services.* The directive was to design a prototype "to provide a clear vision of what might be the end product of major system change," the report stated. Once the prototype was developed, the plan was to select a demonstration site that might be interested in redesigning health and human services within their communities. The plan was also to recruit a national advisory panel composed of "community representatives (drawn from our projects), employers, insurers, health care providers, human service experts, legal experts to address issues related to litigation, ethicists, health economists, and representatives of state and local government." Ultimately, three sites in Michigan became testing grounds for the Comprehensive Community Health Models (CCHMs) of Michigan. This project engaged a broad cross-section of three counties in Michigan to identify and address health care priorities and problems within their communities. But it did not achieve the overall goal. It is difficult, if not impossible, to get people to think clearly about total redesign of health care, because of the pressing needs of the current system. A basic underlying premise of the CCHMs was to demystify health care for lay community leaders so they could be more effective decision makers. This assumed that "experts" would be available to help demystify the process. This premise was also not valid because the gulf between the public policy experts and the leaders of communities could not be bridged easily. And finally, Kellogg program staff had become so committed to solutions coming from communities that the intent of the CCHM project to provide community members with a better knowledge base was viewed as a top-down approach. While nurses were active throughout this project and provided good leadership, their effectiveness was lost in the politics of dealing with communities.
- *Community Partnerships with Health Professions Education.* The intent of this initiative was "to direct change in health professions education toward community-based, problem-focused health care," as the report noted. Kellogg made a nationwide announcement of this initiative, informing the health-professions education community. From this initial step, it received 111 pre-proposals. From them, Kellogg chose 15 sites that it saw as having the most potential to develop projects. These sites then underwent a 14-month leadership and model development seminar series.

Thirteen of the 15 institutions involved in the seminar series developed proposals for multi-year, multimillion dollar initiatives to reshape health-professions education in their sites. Of those, six sites were chosen that were judged to have the potential to be successful. Each of these sites required collaboration among health professional schools including "medicine and nursing" as well as "significant community participation and involvement." This initiative was intended to assure that nursing would have a place at the table alongside medicine. Nursing was especially visible in the projects involving Boston and Northeastern Universities, where the project leader was a nurse faculty member, and in East Tennessee University, where nurses worked with a number of rural communities to provide primary health care. This initiative had a well-orchestrated approach to garner high visibility and a public-policy strategy to alter the financing of medical education.

- *Community-Based Public Health.* In the original design of the Community Partnerships Health Professions Education initiative, it was anticipated that schools of public health would participate in this effort, but that did not occur. In the observations drawn from the Community-Based Health Services Projects, the lack of participation with local health departments was obvious. Based on these concerns, a new grant-making initiative was developed "to assist communities, academic public health programs, and local health departments who wish to come together for the purpose of developing a public health model demonstration program for community-based education, research and service," according to the report. Criteria for this initiative were as follows:

    (1) Identify one or more communities, urban or rural, for which an academic public health program will become an active partner in providing community-based, problem-focused, comprehensive public health services.

    (2) Structure applicants as a consortium or team, with team members drawn from academic programs in public health, local public health agencies, and community-based organizations that represent the recipients of public health services.

    (3) Propose a model of public health practice that recognizes the social, economic, and cultural differences of local communities; emphasizes community empowerment and

> capacity-building; makes use of community networks, and shares responsibility of improving community health and well-being.

By 1992, more than $64 million had been invested in community-based demonstration projects in the United States. A July 1992 report to the board concludes that

> seeing people in communities around the world who are actively engaged in solving problems at the local level creates an exciting and important new dynamic in the health care field ... we continue to be struck by the two different worlds of community-based projects and our health-related institutions. ... The sad part is that frequently there is no awareness or desire that these worlds intersect and understand one another. The people within the community have profound distrust for the "intellectuals" who come out to their community for their own purposes. Many professionals are motivated by a real concern for people, but they do not realize that doing for, and unto people, without seeing them as human beings with value, needs, and aspirations that go beyond their need for treatment, become part of the problem rather than part of the solution.

The early nineties was a period of intense activity aimed at addressing community-based health care. This approach ran counter to the predominant focus upon health care as a business and the intense competition for "market share." In nursing and nursing education, the preoccupation with gaining respect as an academic discipline continued and ran counter to the strategy of focusing on communities and the health care needs of people. Nurses played a substantial role in a wide range of projects funded by the Kellogg Foundation during the 1980s and 1990s, but they received little recognition by the nursing profession for those roles, and the Kellogg Foundation received little recognition for the substantial support it has provided to nurse-run projects. The Foundation continued to be the most substantial donor to nurses during this time, and its programming goals in this era were a natural fit for nurses to connect communities and institutions, as they did in the thirties. A few stepped forward and demonstrated their ability to do this, but professional nursing, maintaining its preoccupation with hospital-based nursing and with being part of the big-business establishment, did not join this movement. The development of the Association of Nursing Centers, an independent organization, is perhaps closest to being supportive of new roles for nursing in community-based health care.

Through the mid-eighties into the nineties, the Kellogg Foundation continued to support nursing and nursing leadership in providing affordable, appropriate health care for people in diverse communities. Foundation support has enabled the profession to grow and develop. Indeed, the profession—at least initially—welcomed the Foundation's investment in helping it move its educational enterprise from hospitals to colleges and universities. But the profession seemed unable to develop a system to differentiate practice by levels of education, despite a substantial investment from the Foundation to help it do so. The Foundation supported a second major movement that helped large numbers of practicing nurses gain baccalaureate degrees in nursing, thus enabling the profession to have a more adequately prepared workforce. The Foundation also supported advanced training programs, curricula model development, continuing education models, and leadership development, as well as the development of various organizations to serve the interests of the nursing profession and unity among organizations. As nursing remained aligned with the agendas of the health care industry and worked to gain legitimacy as a scholarly profession and compete with other health professionals in the academic and hospital environment, the Foundation began its journey back to its roots in the community. And to some, the Foundation's emphasis on community-based approaches to addressing health care problems led to the perception that it had abandoned its support for nursing. That could not be further from the truth.

As nursing remained aligned with the agendas of the health care industry and worked to gain legitimacy as a scholarly profession and compete with other health professionals in the academic and hospital environment, the Foundation began its journey back to its roots in the community.

Since starting the Foundation, Mr. Kellogg believed that "the application of knowledge" should address "the problems of people." Within this framework, funding for basic research or research programs fell outside of Kellogg's programming priorities. While the Foundation did not support the nursing profession's agenda per se, it gave nursing a golden opportunity to capitalize on its strengths and excel on a different playing field—in the context of the community. As our technology-driven, highly specialized medical care system becomes more fragmented every day, the only long-term solutions will be found within the community. The work of nurses in community-based projects supported by the Kellogg Foundation shows how the nursing profession can lead the way in solving the problems that plague us. As it headed into the nineties, the Foundation recognized that. But as long as the nursing profession continued to devalue the work that nurses did, it ran the risk of missing a critical opportunity.

Notes and References

Nursing and the Kellogg Foundation. (No date.) *Nursing and Health Care, 16, no. 4.*

W. K. Kellogg Foundation. (No date.) *Programming for the Eighties.* Battle Creek, Michigan: W. K. Kellogg Foundation.

W. K. Kellogg Foundation. (May 1989). Board Report. Battle Creek, Michigan: W. K. Kellogg Foundation.

Appendix : Nurse-Related Projects Funded 1980-1990

Community-Wide, Coordinated, Comprehensive Health Care Systems

Community-Wide Multi-Institutional Arrangements

7/90	Community College of Philadelphia	$ 683,560
7/86	Montgomery General Hospital	$1,040,948
7/90	Morris Heights Health Center	$ 75,000

Betterment of Health

Health Promotion Services

6/85	Hampton University	$ 743,692
9/85	Pace University	$1,330,071
9/90	Pace University	$ 306,034
7/86	Aleutian Pribilof Islanders	$ 292,763

Education of Health Promotion Professionals

11/84	Michigan State University	$ 695,700
2/87	Michigan State University	$2,227,370
8/84	University of Tennessee	$2,532,275
7/86	University of New Mexico	$ 900,910
8/90	University of New Mexico	$ 307,556

Community-Wide, Coordinated, Cost-Effective Health Services

Community-Wide Multi-Institutional Arrangements

1/85	North Avenue Women's Center	$ 566,962

Alternatives to Institutionalization

3/86	Ohlone College	$ 822,700
7/90	Dissemination/Ohlone College	$ 233,500
4/86	Community College of Philadelphia	$ 537,700
7/90	Dissemination/Community College of Philadelphia	$ 233,500
6/94	Dissemination/Community College of Philadelphia	$ 40,800
5/87	Valencia Community College	$ 288,900
7/90	Dissemination/Valencia Community Coll.	$ 233,500
7/87	Triton College	$ 288,600
7/90	Dissemination/Triton College	$ 233,500
7/87	Weber State College	$ 289,000
7/90	Dissemination/Weber State College	$ 233,500
6/87	Shoreline Community College	$ 288,900
7/90	Dissemination/Shoreline Community Coll.	$ 233,500

7/86	Montgomery General Hospital	$ 1,040,948
10/86	Maternity Center Association	$ 1,891,370
1/89	University of Missouri-Columbia	$ 254,527
9/90	University of Missouri-Columbia	$ 1,880,165

Collaborative Practice/Role Change

3/84	Mountain States Health Corporation University of Washington	$ 279,042
6/94	University of California-San Francisco University of Arizona University of Colorado	$ 89,000

Cost-Conscious Professional Education

9/85	Texas Tech Health Science Center	$ 1,578,490

Health Care Policymaking

83	American Nurses' Foundation	$ 1,174,840
5/85	American Nurses' Foundation	$ 2,351,245
5/89	American Nurses' Foundation	$ 28,640
4/90	American Nurses' Foundation	$ 243,000

Community-Based, Problem-Focused Health Services

Comprehensive Health Services

Adolescent

10/87	Alcorn State University	$ 1,199,232
1/92	Alcorn State University (Evaluation)	$ 44.265
11/93	Alcorn State University (Dissemination	$ 114,379
10/88	Texas Tech University Health Sciences	$ 103,962

Elderly

6/88	Auburn University at Montgomery	$ 1,474,158
1/88	City of Atwater	$ 297,029
7/88	Evanston Hospital Corporation	$ 690,531
1/89	Highland Area Community Council	$ 315,192
1/88	Living At Home/Block Nurse Program	$ 481,903
11/91	Living At Home/Block Nurse Program	$ 85,000
9/92	Northwest Aging Association	$ 599,060
2/88	University of Arkansas for Medical Sciences	$ 1,059,999
5/92	University of Arkansas for Medical Sciences	$ 18,024
8/90	University of Virginia, Charlottesville	$ 1,117,791

Family

1/88	Auburn University at Montgomery	$ 1,414,158
6/91	Auburn University at Montgomery	$ 1,295,630

6/94	Auburn University at Montgomery	$ 1,026,329
5/89	Columbia University	$ 992,132
8/90	CHIP of Virginia	$ 2,328,500
5/95	Elmhurst College	$ 1,180,000
7/89	Howard University	$1,108,816
2/91	Lake Superior State University	$ 971,225
4/91	Medical College of Georgia	$1,736,762
12/94	Respite Care Incorporated	$ 848,602
6/89	Texas College of Osteopathic Medicine	$1,087,156
2/94	University of Texas at Arlington	$ 24,266
10/94	Vanderbilt University	$ 814,084
2/94	Vanderbilt University	$ 440,896

Infant/Child

6/90	Family Health Center of Battle Creek	$ 1,841,790
1/91	Family Health Center of Battle Creek	$ 1,075,400
7/86	Medical College of Georgia	$ 1,032,390
1/91	Morris Heights Health Center	$ 456,412
3/90	Region II Community Action Agency	$ 567,396
9/90	Texas Tech University Health Sciences Center	$ 2,145,513
2/91	Texas Women's University	$ 598,838
12/91	Texas Women's University	$ 79,172
6/90	University of Texas Medical Branch at Galveston	$ 1,331,083
3/88	Wayne County Health Department	$ 1,184,310
7/91	Wayne County Health Department	$ 1,108,640

Health Professions Education

Elderly

1/88	Johns Hopkins University	$ 622,366
1/88	University of Arkansas for Medical Sciences	$ 1.059,999
1/88	University of Michigan	$ 980,973
10/89	Northwest Aging Association	$ 599,060

General

89	University of Illinois at Chicago	$ 575,929
9/91	California State University-Long Beach	$ 720,929
1/94	Grand Valley State University	$ 368,429
1/93	University of Michigan	$ 142,716
2/91	University of Michigan	$ 17,000
11/91	Western Institute of Nursing	$ 50,000
12/93	Western Michigan University	$ 2,000,000

Underserved Communities		
9/89	Salish Kootenai College	$ 1,034,064
89	Sisseton-Wahpeton Tribal College	$ 66,950
9/90	University of Texas-El Paso	$ 940,172
5/89	Columbia University	$ 992,132
Informing Policymakers		
4/93	American Psychiatric Nurses' Assoc.	$ 48,589
Information Technology		
8/90	Advocate Health Care	$ 240,000
10/93	American Nurses' Foundation	$ 50,000
2/93	National Council of State Boards of Nursing	$ 100,000
8/89	National League for Nursing	$ 1,124,968
10/88	Texas Tech University Health Science Center	$ 872,078
10/88	Joint Commission on Accreditation of Health Care Organizations	$ 330,658
10/88	Battle Creek Health System	$ 422,248
	Kaiser Permanente Foundation-Portland, OR	$ 328,398
Leadership Development		
9/81	American Nurses' Foundation	$1,556,080
8/88	Indiana University	$ 534,291
11/91	Indiana University	$ 174,728

From its founding in 1930 to the beginning of World War II, the W. K. Kellogg Foundation concentrated on local health and education issues in Michigan. The Michigan Community Health Project (MCHP, so named in 1935) became an excellent vehicle to provide leadership in community health nursing. Prominent nursing leaders were on the staff of the Foundation, and many nurses came to Michigan as part of their educational programs to learn from these leaders.

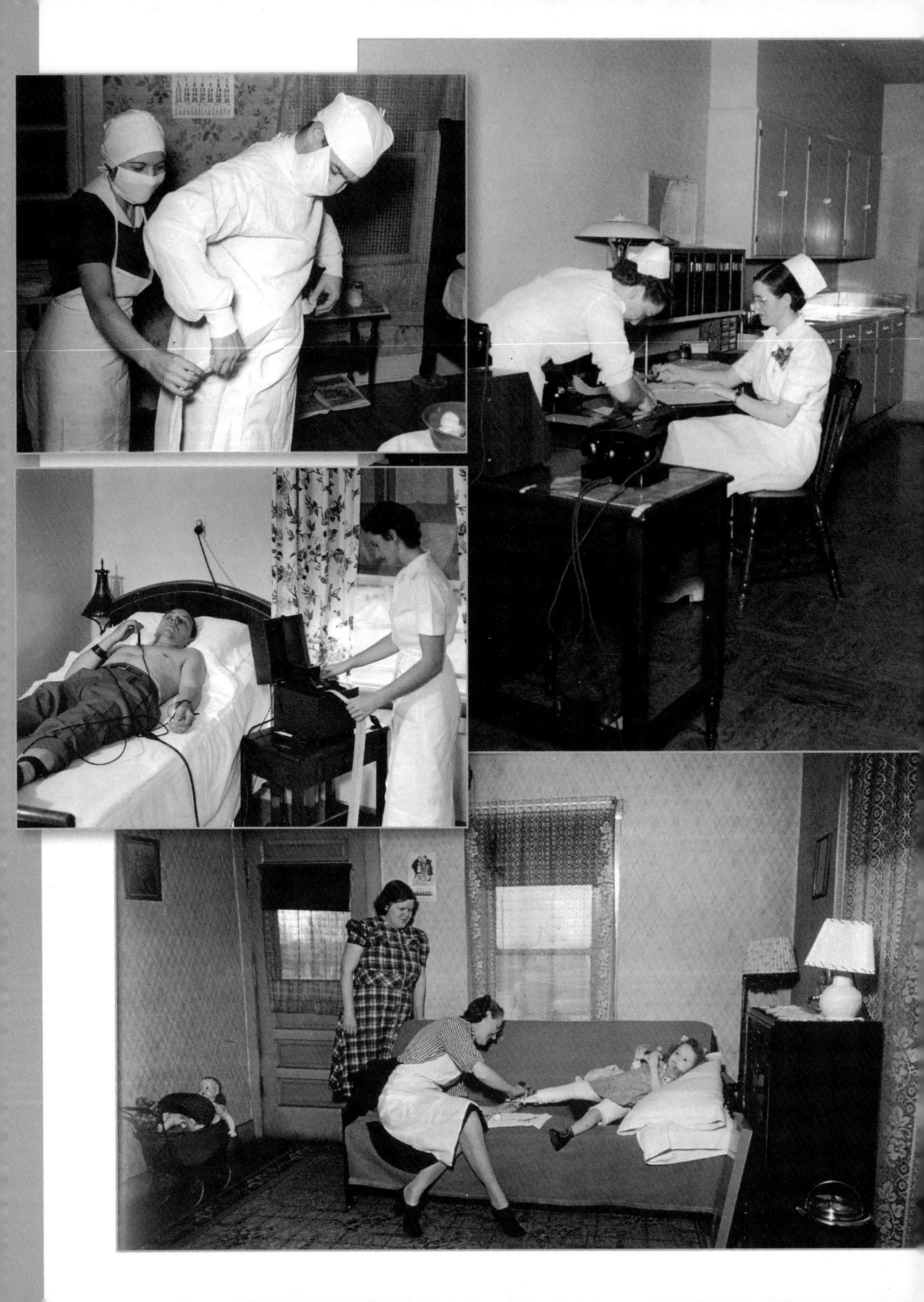

The W. K. Kellogg Foundation's work with nursing in Latin America and the Caribbean began in 1940 and continues to this day. During the early years, the major emphasis was on the health professions, and through the years nursing has received ongoing support for its integral role in providing quality health care for people.

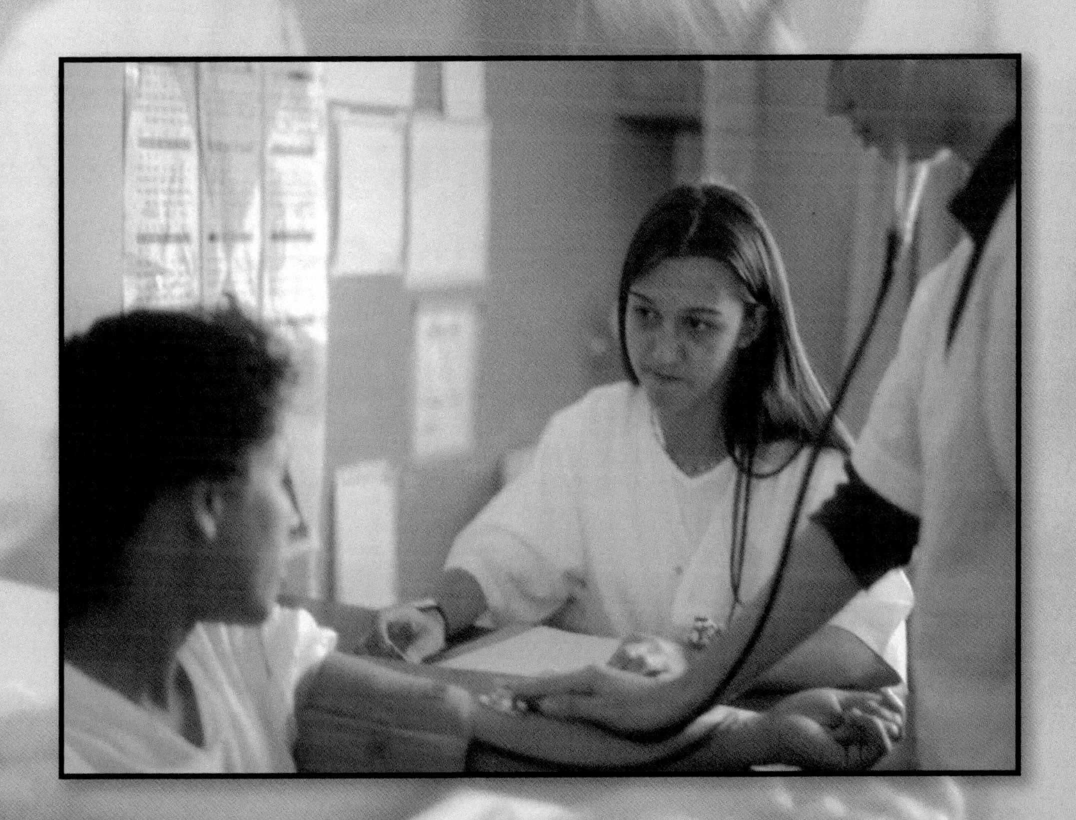

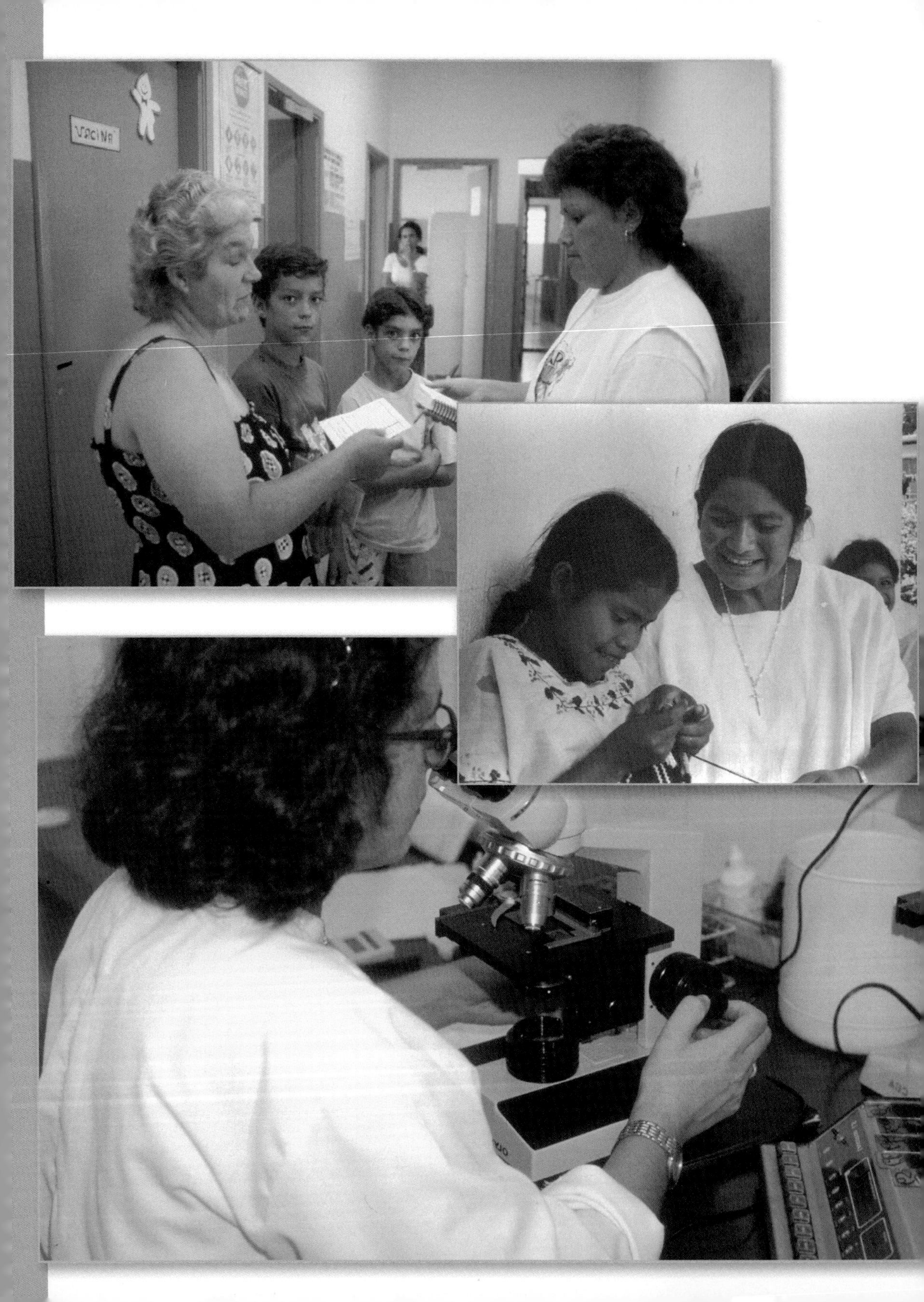
VACINA

The first grant made in southern Africa was in support of a nursing project in Botswana in 1986. The decision was based on the results of an earlier investigation into the feasibility of extending grant making to this part of the world. The Foundation's funding to nursing in the southern African region extended over a 10-year period of time in which the seeds were planted for development of the nursing profession.

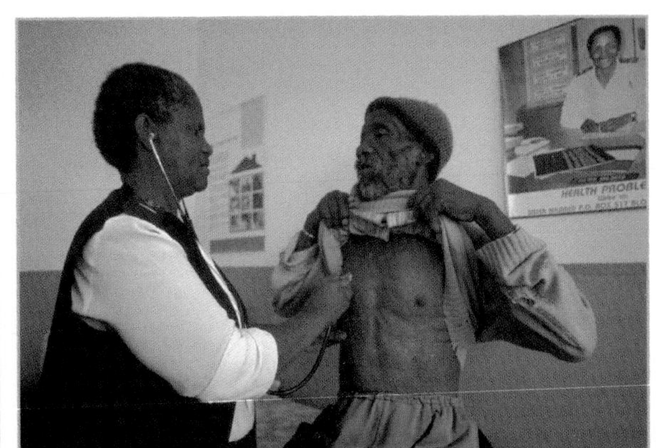

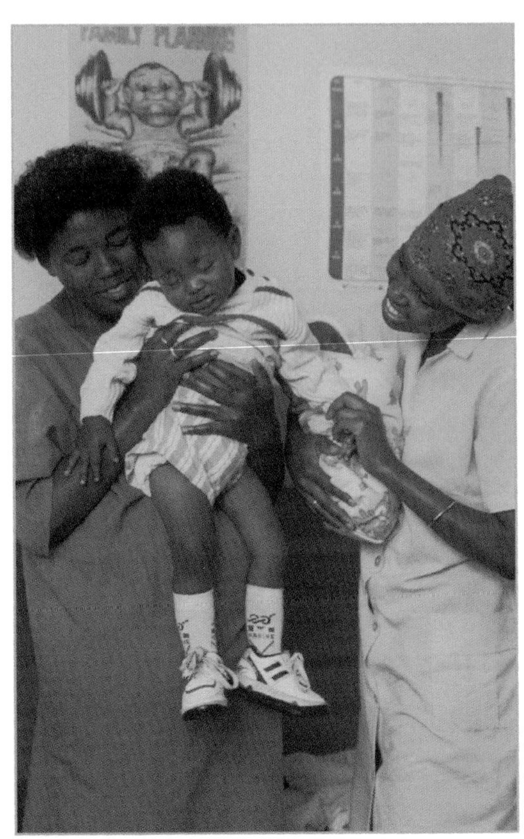

10 From Opportunistic to Strategic Grant-Making: 1991–2001

Gloria R. Smith

When I joined the health team at the Kellogg Foundation in November 1991, the team had grown to 10 members. Foundation assets had increased rapidly, and the health team was leading the Foundation in its return to a focus upon community. It was a beehive of activity with a large portfolio of community-based projects and a number of major initiatives that had just been started. In 1992, I was appointed coordinator of the health team. Helen Grace had risen from coordinator to vice president for health programs in 1991 as part of the Foundation's new structure for senior leadership. Her assumption of the leadership role for health programming had ushered in a new era and new ways for approaching work. By formally recognizing health program directors as being part of a "team," the Foundation's senior management had begun the shift from an entrepreneurial mode of operating to that of a "team" of members working together toward commonly shared goals.

Helen's selection as vice president signaled that leadership would no longer be granted by virtue of standing in the medical hierarchy. All staff members were expected to contribute as equals to overall program goals. Program staff were encouraged to think about problems and problem solving holistically, unbounded by disciplinary lines. The health program staff represented the basic disciplines: medicine, nursing, dentistry, public health, mental health, health care administration, health and human services policy, and social work. Later, our staff would also include an expert in health law. Further, the staff experiences included health-professions education, hospital and community-based practice, and cabinet-level agency administration. Among the staff, three of us were nurses, with Helen and I in the lead roles of vice president and coordinator respectively.

The original premise behind the problem-focused approach was that addressing high-priority health problems would lead to further community development. Many of the projects were poised to take a next step toward addressing more comprehensive health issues within their communities.

In the process of reviewing the health programming goal, which had been initiated in 1986, adjustments were being made. "First generation" projects had been highly successful and were growing. The original premise behind the problem-focused approach was that addressing high-priority health problems would lead to further community development. Many of the projects were poised to take a next step toward addressing more comprehensive health issues within their communities. Second-stage funding was provided to a number of these projects. Additionally, the health team recognized the need for leveraging the power of the community-based projects to lead to health system reform, change in health-professions education, and linking of public health to community-based health care. Three major initiatives were started: Comprehensive Community Health Models of Michigan, Community Partnerships in Health Professions Education, and Community-Based Public Health.

This chapter tells the story of how developments in the field combined with developments in the Foundation to transform our work and enable health programming to program for systems change. Making progress toward systems that meet needs involved us in the practical application of lessons from the field and much more. Here are some specifics:

- Conceptualizing theories of change to address the systemic prob-lems that were leaving a gaping hole between the health status of the vulnerable and disadvantaged and the health resources of the nation.
- Discerning readiness for a higher order of change in potential grantees.
- Maintaining, amid the transformation of our work, our bedrock values of working closely with people in projects, flexing to meet their needs and circumstances, and honoring their ability to help themselves.

- Undertaking bold experiments that capitalized on the investment niche of foundations as risk-takers, the perspectives of multiple disciplines, and rich experience grounded in work with grantees.

This account tells the story from the perspective of the intersection of health programming and nursing. It is organized into four sections. The first section is a brief history of the evolution of practice and thinking in health programming and the Foundation. The second section examines grants related to nursing in the health portfolio in the 1990s. The third section reports findings from the unique, large-scale evaluation by a national firm of the portfolio of grants made under health programming's 1994-1999 strategic plan. The final section gives my concluding remarks.

Section I.

From Individual Grants to Strategic Programs

In its first 60 years, the Foundation's organizational structure had encouraged program directors to be entrepreneurial. Until the late 1980s, each program director negotiated directly with the president for the grants he or she wished to put forward to the board for funding. With changes in 1987, Helen Grace became coordinator for health programs, but it was clear that she was to coordinate the health *program* and not *program directors* (how one does this is hard to fathom, but those were the ground rules). And the model of grant-making up until the late eighties was to have a broad set of over-arching goals and then find potential grantees who were proposing work that would fit within this general framework.

This section of the chapter reports on the evolution of the concepts and theory of change that guided health grant-making in the 1990s, and the evolution of the team's capacity and practice. Both the thinking and the teamwork elevated the importance of strategic grant-making and made it more intentional. This section concludes with a discussion of the integration into our work of services designed to increase the impact of grants and developments in the Foundation that affected all programming.

We had found wisdom in the beliefs of Mr. Kellogg that if we helped people to help themselves by learning to solve one set of problems, they would develop the skills that would enable them to solve other problems.

Evolution of concepts. In January 1994 the five-year strategic plan, *Future Directions in Health Programming*, was presented to the board. We argued in this document that the new plan was based on lessons learned from previous programming. Under the 1986 plan, we had targeted underserved and underrepresented communities and vulnerable populations. We had worked to move away from institutional dominance. We had found wisdom in the beliefs of Mr. Kellogg that if we helped people to help themselves by learning to solve one set of problems, they would develop the skills that would enable them to solve other problems. We affirmed our belief that health programming could be an avenue for community development.

We had learned that the creative solutions crafted by communities could be sustained if they were linked to larger systems that had ongoing sources of funding. We had observed that the gap between institutions and communities was immense. Community representation in decisions related to health care was minimal, whatever the issue: access to care, range and type of services, or health careers training. Low-income and minority consumers had almost no voice. Out of our experiences, we had evolved the concept of partnership as a way of expanding resources and maximizing effect. We had learned that the community had resources and skills to contribute; it was a matter of helping both people and their institutions to identify how to accept and use these resources. We saw partnerships as the mechanism for building bridges between institutions and communities.

A new goal emerged from the old. The new plan to achieve this goal was based on the vision of a reformed health care system that would result in improved health status and access to appropriate, affordable care for all—initially in the regions where the Foundation worked, and, ultimately, more globally. The strategic plan reflected the staff's multidisciplinary and global perspective. The proposed new health goal was stated as follows:

> To improve the health of people in communities by increasing their access to integrated, comprehensive health care systems that are organized around primary care, prevention, and public health and that are guided, managed, and staffed by a broad range of appropriately prepared personnel.

Movement to a comprehensive approach. Throughout the early 1990s, many of the initial community-based health projects became more comprehensive, demonstrating important lessons that could apply to broader issues. Instead of remaining focused on single issues, they emphasized community-based comprehensive health care. While the institutions for which the Foundation had supported capacity-building in the early 1980s had not performed well at system-building, some community-based demonstrations had moved strongly in that direction. They moved, for example, from centering on one age group to centering on the whole family or community. Or they moved to a broader array of services. One moved from being a small bootstrap project to gaining status as a broader-based clinic eligible for federal reimbursement.

The developments in the field informed our thinking. As we saw projects become more inclusive in their clientele and the organizations to whom they related, we began to incorporate "inclusiveness" into our concepts. As we discerned that new working models of "comprehensiveness" of service had appeared in the field, we realized that our programming, too, had to move to the next stage—the stage in which comprehensiveness, inclusiveness, and system-building are the focus *in communities with capacity*, whether they came to us with the capacity or we helped them develop it. In observing developments in the field, we learned that the growth of community capacity and nascent systems is organic. One stage grows out of another. As we became more aware of the kind of natural experiment we had become a party to, we saw that the process of development must be understood, respected, nurtured, and not hurried.

In observing developments in the field, we learned that the growth of community capacity and nascent systems is organic. One stage grows out of another.

In contrast to community-based organizations, health-professions education institutions did not move toward system-building. Our work with them informed reformulation of the 1986 health goal and the design of new ways of working. Health-professions schools—including schools of nursing—that created access points in communities with Kellogg Foundation dollars tended not to sustain them after the grant periods were over.

Theory of leveraging systems change. As we moved into the early 1990s, we wanted to increase the power of the community-based demonstrations. The nation had elected a new president in 1992. Early in his first term of office, Bill Clinton gave health care reform a high priority. He made an unprecedented decision to assign leadership for this undertaking to his wife, Hillary Rodham Clinton. An elaborate structure was established for gathering input, data, and information on a very broad basis. The Kellogg Foundation was not among the consultants, analysts, and organizations invited to provide information or lessons learned. Foundation staff were not able to gain recognition for the work we had funded. Foundation policy had always been to keep a low profile in the public-policy arena and to not cross the line between "informing public policy" and attempting to "influence public policy." That line had been established as a result of federal investigations related to other foundations' funding policies. Both the attempt to gain greater visibility and the recognition that institutional changes were needed led us to develop strategies to try to harness the power of community-based models. We saw partnerships between institutions and communities as a way for them to share power. Foundation staff remained convinced that any true reform of health care would draw on some of the approaches in primary health care and public health that had been piloted and tested by our grantees. And nurses were at the center of community-based primary care.

Foundation staff remained convinced that any true reform of health care would draw on some of the approaches in primary health care and public health that had been piloted and tested by our grantees. And nurses were at the center of community-based primary care.

Our approach to system change was anchored by seven underlying assumptions:

(1) The current system was not working cost-effectively for the majority of people.

(2) Fundamental changes to refocus the system on health and human services were needed.

(3) A multifaceted approach toward an integrated, comprehensive system of services and health-professions training was required.

(4) Strong community involvement was an essential element.

(5) Institutions were already realizing the imperative to change and were reaching out to communities.

(6) Educational institutions played a significant role in effecting system changes.

(7) A cadre of new visionary leaders was required.

In its review, the board considered what unique contribution or insights might result from undertaking the new plan. The staff stressed the fact that the Kellogg Foundation had developed a niche in using systems change to improve health by fostering two types of development: first, reorienting health services toward public health, primary prevention, and primary care; second, working with communities to develop active participation in partnerships with institutions in which they play the role of articulating broad values, needs, and priorities that transcend institutional interests.

Seeing the whole, acting within the context of the whole. No grantmaker, public or private, has the resources to do, by itself, everything necessary to improve the health of people in communities. The interest of the Kellogg board and staff in our grant-making niche reflects this truism. But the term *niche* should not be misconstrued as thinking small. Small thinking makes a niche into a hiding place instead of a spring from which transforming forces flow. Work can be confined to a niche, but thinking should not be; it should range out into the world to grasp reality and shape the niche within reality. Working effectively within a niche means seeing the whole and choosing and guiding every action with an awareness of the whole.

We changed our focus from underserved communities to people in all communities. We envisioned a reformed, integrated system of delivery that would have a more fully developed range of appropriate services at every level of care; that meant strengthening public health and primary care.

Our learning from the field about the capacities of communities and institutions enlarged our thinking and vision. We changed our focus from underserved communities to people in all communities. We envisioned a reformed, integrated system of delivery that would have a more fully developed range of appropriate services at every level of care; that meant strengthening public health and primary care. From our work we reached three key conclusions:

(1) You cannot get to health for all unless you go via comprehensive systems.

(2) You cannot have systems with the right attributes unless you have the right people to staff them.

(3) You cannot build the right system unless you do so collaboratively, through communities and institutions in partnerships.

We saw improving health care delivery by improving the preparation of providers as a means to the end of improving health status. We believed that the huge investments by the American people in health-professions education were not being leveraged well, particularly the investments in bachelor's and master's degree nursing. Nurses were not being used to full capacity, and the need for their use in primary care delivery was evident.

We saw that reforming a system required work in four sectors: communities and institutions as a pair, and services and education as a pair. Communities and institutions were the actors, services and education were the activities.

Making grants through initiatives. Under this approach, Kellogg had to choose grantees strategically. One option for doing this was initiative grant-making, in which the Foundation would define the problem it wanted to work on and invite proposals from organizations that wanted to work on solving it. In the late 1980s an internal debate within the Foundation focused on the possibility of adding this other mode of grant-making to its practices. The Foundation leaders had an unwavering conviction about the ability of people to help themselves and their innate authority to choose the solutions to problems that are right for them. Two towns might have the same problem but two different solutions. The Foundation would support and respect each solution. Telling people what to do could be a sacrilege. Yet, there was another part of the history too. The Michigan Community Health Project, described in another chapter of this book, nudged people in rural communities into coming to grips with problems and finding solutions. Expert advisory committees had also been a part of the culture.

There was a great deal of tension about the relative merits of supporting field-generated proposals only versus diverting a significant portion of the allocation to fund responses to Foundation-directed requests for proposals. Those who opposed the emergent strategic grant-making argued that the opportunity to support good or creative ideas would be diminished. Those who supported strategic grant-making argued that, while it was true some good ideas might be screened out, the proposals making it through the screen could potentially aggregate to a more focused attack on a problem area.

At its team meetings and elsewhere, the health team periodically struggled with the issues around what we called "prescriptive versus non-prescriptive" grant-making.

Health programming was the first programming group to develop an initiative, which was Community Partnerships in Health Professions Education. Starting one initiative did not mean ending the debate. At its team meetings and elsewhere, the health team periodically struggled with the issues around what we called "prescriptive versus non-prescriptive" grant-making.

During the course of the 1990s, initiatives became a standard part of all programming groups. Experience with initiatives has confirmed that they can—sometimes—generate momentum, visibility, evaluative power, and critical mass for informing policy and practice that constitute a good return on investment. The Foundation board has been careful, however, to require that the groups' grant portfolios maintain a balance between initiatives and field-generated projects.

The irony is, of course, that it was field-generated projects that taught the lessons that inspired the thinking that led to the perception of the need for another mode of grant-making.

Reformulated goal, evolved strategic plan. The field experience and evolution of concepts I have discussed gave rise to a goal statement that carried its theory of change within it. The goal, again, was "to improve the health of people in communities by increasing their access to integrated, comprehensive health care systems that are organized around primary care, prevention, and public health and that are guided, managed, and staffed by a broad range of appropriately prepared personnel." The goal assumes that increasing access is the way to improve the health of people, and that increased access will improve health. But access is not enough. Everything depends on what people have access *to*.

The language describing the "what"—the envisioned systems—was actually sharpened during the course of the board's review and ultimate endorsement of the reformulated goal and associated plan. Initially, they were described as "systems of health and human services that are inclusive, appropriate, affordable, and effective." The language was sharpened to articulate the structural features of the systems that would make them inclusive, appropriate, affordable, and effective. Saying that systems should be inclusive, appropriate, affordable, and effective does not say how to make them so. The description became "integrated, comprehensive health care systems that are organized around primary care, prevention, and public health." At the same time, the description of who would guide, manage, and staff the systems remained the same; it did articulate that the appropriateness of preparing health personnel was the powerful determinant of the attributes of the systems in which they came to practice. So, our effort to envision and articulate what kind of systems would have the power to improve the health of the public continued even into the period of board review. As it happened, that effort to envision and articulate enabled our work and our thinking to continue to develop as we stayed focused on our subject but grew in our understanding of it and our skills in addressing it.

Integrated, comprehensive health systems organized around primary care and public health are clearly more than medical care. Health grows out of many forces in people's lives.

Integrated, comprehensive health systems organized around primary care and public health are clearly more than medical care. Health grows out of many forces in people's lives. Medical treatment alone is not the force with the biggest impact. The systems envisioned by the goal reach beyond medical care to promoting health, preventing disease, improving community conditions, and addressing the social determinants of health; beyond individual action to community and public health action; beyond reducing the risks people take to reducing the risks to which society exposes them.

The 1994-1999 Strategic Plan for Health had nine strategies at the time of its endorsement by the Kellogg Foundation board. They fell under services, education, and support. Our imagination was of dynamic development. Our understanding was that where communities and institutions, services and education came together, the envisioned health system would arise. Movement and development had to occur in all four sectors (community, institutions, services, education). Some staff referred to their point of coming together as the "center." This was a good term. This was where all the paths of development came together to give birth to a system that no single path of development could have produced by itself.

The configuration of the strategies was intended to display the dynamic of development of the systems we envisioned.

Services strategies. The three strategies under this area represented developmental stages in movement toward the integrated, comprehensive health systems at the "center":

Problem-focused models. In our vision of communities and institutions, services and education, this stood for the intersection of services with communities.

Institution-initiated change. This strategy was to support institutions that combined signs of more progressive thinking with prominence among their peers in their efforts to initiate system-building. It stood for the intersection of services with institutions.

Comprehensive health care systems. This strategy was to support partnerships around building comprehensive service-delivery systems between communities and institutions. Standing for the intersection of communities, institutions, and services, it represented a stage of development approaching the "center."

Education strategies. Three strategies in education were as follows:

Recruiting and supporting community members. This strategy for bringing people from underrepresented minority and other vulnerable communities into health-professions education institutions addressed building the pipeline to, and retaining students in, health-professions education. Changing the racial, ethnic, and socioeconomic composition of student bodies was expected to improve the delivery system, increase access to services through larger numbers of professionals practicing in underserved communities, produce change over time inside the educational institutions themselves, and—perhaps—help the economies of low-income neighborhoods.

Changing the racial, ethnic, and socioeconomic composition of student bodies was expected to improve the delivery system, increase access to services through larger numbers of professionals practicing in underserved communities, produce change over time inside the educational institutions themselves, and—perhaps—help the economies of low-income neighborhoods.

Educational reform. This strategy for redirecting health-professions education toward practice in communities stood for the intersection of education with institutions.

Public health partnerships. This strategy for partnerships among communities, health-professions education institutions, and public health practice institutions stood for the intersection of communities, institutions, and public health services and represented a stage of development approaching the "center."

Support strategies. Three strategies were intended to support the work in services and education: information technology, leadership development, and informing policy. While major grants were made under leadership development and informing policy, the information technology strategy was not implemented as a separate strategy. As our work developed through time, we found that integrating leadership development and informing policy into grants in services and education was essential as well. (Support for information technology in grants was occasionally provided.) The place of leaders in the envisioned system is explicit in the goal statement, that is, the systems must be "guided" and "managed" by appropriately prepared personnel. Attention to leadership, however, permeated our work beyond what those words connote. Development of community leaders was embedded in our work. They enabled vulnerable communities to inform public and institutional policy. Fellowship programs of various types were developed in support of the missions of various initiatives. The place of informing policy is not explicit in the goal statement. Policy change is, however, an enabler and a driver in systems change. Systems change cannot occur without it. The Foundation and its grantees invested substantial effort in informing policy.

Development of community leaders was embedded in our work. They enabled vulnerable communities to inform public and institutional policy.

Although the plan was not endorsed by the board until 1994, in many ways it represented the work throughout the 1990s. All initiatives that had begun in the early nineties were incorporated into the plan. More importantly, they were conceptually aligned with and had actually informed the plan. All open project grants were incorporated into the plan.

Conceptualizing nine strategies was a good idea. Having one goal and nine strategies to reach it broke the Foundation's programming mold and produced clarity of thinking and an understanding of the lines of development converging in systems change. But with experience, we no longer needed to distinguish every element with that particularity. We later requested, and the board gladly approved, a change in the strategic plan that collapsed the three services strategies into one and the three education strategies into one. Thus we went from nine strategies to five. Another change was to rename the education strategy "workforce development."

In the 1990s the national government's failure to enact any health reform deepened our worries about vulnerable people. This also highly contrasted the system we envisioned and the troubled system of the day. A 1999 report to the board discussed the significance of partnerships between community-based organizations and health services institutions in the following way:

> Health care costs are so high and the delivery system is so tilted toward treatment and technology, that *managed care is mostly managed cost* and often a matter of limiting access to more expensive treatments or longer hospital stays. To actually *manage* care—so that people live in conditions that promote health and have routine access to primary care, prevention and public health services—would be a boon for cost control and human health alike. But the leaders of managed care and most health care delivery systems do not know how to make this shift. Many do not have the vision, the experience and the skills necessary. Among the tasks essential to making the shift is that of linking the traditional system with community-based organizations that do not have roles in health care delivery, but can have tremendous impact on health improvement. Schools, churches and neighborhood organizations can play important roles.

To overcome barriers in health occupations and management for diverse people, we directed resources to address the problems of recruitment, retention, and advancement of members of underserved groups in the health field, and particularly in nursing.

In the areas of workforce and leadership, development of the plan acknowledged that health care delivery can be most responsive to the needs of the people it serves if its workforce and leadership have community roots and reflect the diversity and experiences of those being served. To overcome barriers in health occupations and management for diverse people, we directed resources to address the problems of recruitment, retention, and advancement of members of underserved groups in the health field, and particularly in nursing.

The 1999 report to the board summarizes the broad concerns regarding workforce development:

> The delivery system must be staffed, managed, and led by people whose education and training have been reoriented toward an emphasis on promoting and maintaining health, and on partnering with people and organizations in communities for the improvement of health. Visionary and skilled leaders to guide, manage, and inspire change in health systems are clearly essential. Also essential are health workers who can practice with skill in a changed system and partner with those who have not traditionally participated in health care delivery. In order for workers and leaders to be prepared for practice in the reoriented delivery system, change is also needed in health-professions education institutions. Unfortunately, neither flexibility nor responsiveness to community needs and changing social conditions are hallmarks of higher education in the United States. Like managed-care plans and health care delivery systems, schools of the health professions largely do not know how to reorient health care.

Intersection with nursing. The health system we envisioned and our theory of strategically leveraging it into being have a very special relationship to nursing. The vision and the hope of this system are similar to the vision and the hope of the Alma Ata Declaration of the World Health Organization of "Health for All." "Health for All" is to be achieved through primary health care, and primary health care for all is absolutely dependent on nurses, the most numerous caregivers in the United States and the world.

It is hardly only their numbers that make nurses count in the vision of better health and better health systems for all. Cost-effective, affordable systems are mandatory if coverage for all is to be gained through social and public policy. The nurse's place in such systems is as a cost-effective practitioner. Several reasons explain the cost-effectiveness of nurses' care Training nurses is less expensive than training physicians. Yet the quality of primary care by nurses equals or exceeds the quality of comparable physicians' services. A significant expansion of primary care would be impossible to achieve without nurses—even though the current coverage of the nurse shortage focuses on hospital care. By remedying the imbalance between primary care and specialist or tertiary care, the nation's savings can become enormous over time.

Training nurses is less expensive than training physicians. Yet the quality of primary care by nurses equals or exceeds the quality of comparable physicians' services. A significant expansion of primary care would be impossible to achieve without nurses.

This point is significant because the nurse works with the whole person: nursing practice is aligned with the vision of comprehensive services that encompass public health, community conditions that promote health and well-being, and the social determinants of health. Viewed in this light, the nurse is the central caregiver in the integrated, comprehensive health system that is organized around primary care and public health and truly meets the needs of people.

It will not be possible to reach and serve vulnerable and disadvantaged people in communities without the nurse as primary caregiver. Without nurses, the health system we envisioned cannot come into being.

Evolution of team capacity. The Foundation's senior management team in 1991 created the context for how we would do our work. This senior management team, composed of three vice presidents (agriculture, education/youth, and health) and directed by the president, Dr. Norman Brown, was driven by a belief in the power of multidisciplinary approaches to solve complex societal problems. The crux of the debate about putting this belief into practice revolved around how the staff was to organize to accomplish its work within the Foundation. Were they to continue being independent entrepreneurs, or did the new form of strategic grant-making that was emerging mandate a different working relationship? In response to the mandate of the senior management team, Helen Grace and I worked to transform the health staff from a group of entrepreneurs who were engaged in parallel play to a multidisciplinary team of creative professionals who worked together effectively. The lessons learned from the community-based projects and our emphasis upon multidisciplinary work pointed to the importance of teamwork. How could we be advocates of the merits of multidisciplinary work while, in our programming group, we were engaged in such parallel play? How could we require prospective initiative grantees to apply as partners if we did not expect working in partnership inside the Foundation as well?

This shift in thinking and working had its challenges. Nowhere was the tendency to revert to past practices more evident than in our three large initiatives, which were more often referenced by the lead program director's name than by the initiative's name. Our monthly staff meetings were lengthened to allow in-depth progress reports and discussion. Decisions on budget expenditures and corrective actions evolved from a process of group input. The staff of individual entrepreneurs came to recognize the value of collective analysis and problem-solving. There was less certainty about sharing power, but power-sharing did increase. Over time, the staff group process moved from being purely political to being an actual partnership. Increasingly, our thinking and discussions were focused on the problem of increasing access to health care for all people. We became less protective of our turfs, disciplinary or otherwise. We were applying what we were learning externally to our internal functioning as a group within the Foundation. By 1994, when the five-year strategic plan was approved by the board, the health team was positioned to work collaboratively toward its implementation.

We became less protective of our turfs, disciplinary or otherwise. We were applying what we were learning externally to our internal functioning as a group within the Foundation. By 1994, when the five-year strategic plan was approved by the board, the health team was positioned to work collaboratively toward its implementation.

The greatest test of the team's abilities came with the process of development of our Community Voices Initiative and, then, its management. A great deal was at stake in Community Voices. We believed we had to respond in a very significant way that would inform national policy that led to the increase in the number of uninsured, the closure of safety-net hospitals, and the diminution of resources for uncompensated care. We went through a 20-month developmental process that engaged the whole team. Every team member was in the field when we did site visits (26 in three weeks). Every team member participated in managing grants, with one program director still designated as lead to assure accountability. The model of how we did our work had truly changed.

The ability to work as a group took on a different kind of significance as the size and composition of the group changed. After $55 million had been appropriated for Community Voices, the work multiplied while the number of people to do it declined.

Some change stemmed from natural turnover. However, in the latter half of the 1990s the Foundation went through a major adjustment in the size of its staff because of fluctuations in the stock market affecting its assets. It had done the same in the 1980s, but at that time the change had been upward. The mix of disciplines changed along with team size. Two nurses, including myself, were on the staff. Dentistry and medicine were no longer represented. Another public health program administrator and another former cabinet-level administrator joined the group.

Integration of communications, evaluation, and policy. During the 1990s the Foundation gradually evolved into a matrix organization. When the evolution was completed, communications, evaluation, policy, and information technology were allied in a unit called Impact Services in the office of the senior vice president for programs. Directors for each of the four areas reported to her.

Communications, evaluation, and informing policy have the power to increase the impact of funded work. Hence the name Impact Services. (Information technology is not dealt with in this discussion since, during the 1990s, information technology as an impact service did not figure significantly in our work with grantees beyond email and submission of electronic documents.)

Staffing of Impact Services evolved as the organizational matrix developed. In the completed model, an evaluation manager and a communications manager who both worked in Impact Services were incorporated into the health team. The Foundation's policy director did not supervise managers, however. Instead, each programming group had one program director designated as its policy coordinator. The policy coordinators and policy director worked together as a group to keep a focus on policy in all work.

The human resources for Impact Services are important. But health team members' thinking about impact services as activities is equally important to the successful deployment of these services. The roles of communications, evaluation, and policy were integrated into our own thinking about strategic grant-making.

In addition to calling for team approaches to programming and spurring on grant-making initiatives, the Foundation fostered a fertile climate for creative thinking in other ways during the first half of the nineties.

From grant-making to change-making seminars. In addition to calling for team approaches to programming and spurring on grant-making initiatives, the Foundation fostered a fertile climate for creative thinking in other ways during the first half of the nineties. "From Grant-Making to Change-Making" was the subject of seminars that Helen Grace led for all Foundation program staff. While these supported development of skills in designing and managing initiatives, they pushed Foundation staff to think in broader terms in project grant-making as well. The seminars also complemented the Foundation's enhancement of policy work.

Cross-goal programming report. A great deal of discussion among Foundation program staff focused on a classic organizational problem that we experienced—working in silos. If we entrepreneurs were being transformed into members of multidisciplinary teams and if grantees were being funded to partner, what were we doing with walls around our respective programming groups' turf? Wasn't it time to partner? Wasn't it time for multidisciplinary approaches across program lines? A committee and group of subcommittees were formed and produced a cross-goal programming report. This certainly did not lead programming teams to abandon protection of their respective spending allocations in the Foundation's budget, but it did produce a few results. The most significant one for health programming was initiating the cross-goal devolution initiative, in which health staff were very active. The initiative developed tools for communities to be informed advocates as major changes were made in the place of control over federally funded human services programs.

Evolution of cross-cutting themes. Another development was the evolution within the Foundation of formally recognized cross-cutting themes that were intended to inform grant making in all programming groups. These themes were in various ways identified through work that was already being done. The theme of Social and Economic Community Development grew out of Families, Neighborhoods, and Communities, which had received a time-limited grant-making allocation as an "emerging programming" area. Leadership evolved from a programming area—indeed, a signature programming area for the Foundation—into a cross-cutting theme. Capitalizing on Diversity made some of the interests and practices in programming groups explicit and intentional. Information and Communications Technology grew out of work that had originated in an earlier unit of the Foundation that supported all programming groups.

The notion of cross-cutting themes was laudable, but putting it into practice challenged our way of doing business. Building the infrastructure to support their implementation presented some difficulties. The themes, however, resonated with the health team and our work. We had invested in all the theme areas and had demonstrated special commitment to diversity and leadership. Our understanding of what creates health had also moved us in the direction of social and economic community development from time to time. For example, Auburn University School of Nursing created a stellar project in two of the poorest counties of rural Alabama in which the power of people to help themselves blossomed. One of local people's many accomplishments was to bring a potable water system to residents who lacked this basic necessity.

The notion of cross-cutting themes was laudable, but putting it into practice challenged our way of doing business. Building the infrastructure to support their implementation presented some difficulties.

Section II.

Increasing Access to Improved Health through Strategic Grant-Making

Increasing access required strengthening the capacity in both the services sector and the education sector. It required strengthening the capacity of communities to exert influence in those sectors. And it required supporting and developing leaders with the vision of a better system from underserved communities and traditional institutions.

Access is not only a matter of coverage but of the strength and capacity of the system to deliver primary care and public health services with the goal of improving health. Coverage alone is not sufficient.

A crucial premise was that access is not only a matter of coverage but of the strength and capacity of the system to deliver primary care and public health services with the goal of improving health. Coverage alone is not sufficient. It does not overcome non-financial barriers to care; for instance, language, transportation, service hours, and child care. Coverage does not overcome the limited availability of primary health and oral health prevention and treatment services for underserved people. And it does not overcome the inappropriateness of care in a system heavily overinvested in tertiary care.

We saw the number and quality of primary care and public health providers as crucial to system capacity. Health-professions education programs needed to be supported to prepare leaders and caregivers for integrated, comprehensive, community-responsive health systems.

Communities also should have a say in setting the direction for health system improvement and the preparation of health professionals and leaders for practice. Communities, further, should have strengthened capacity to inform and engage in shaping public policy.

The intersection of nursing and initiative grant-making. Six health initiatives were funded by the Kellogg Foundation in the 1990s. To put into context the two initiatives in which nursing was central, I will give a brief description of all six. The initiatives can be viewed almost like members of a team. Each player has a different role, but a common purpose.

Community Partnerships in Health Professions Education. The first initiative was designed to reform educational programs for medical students and undergraduate nursing students. The objective was a sustained increase in the number of students who chose to enter primary care practice. At the time, 70% of physicians were specialists, and 30% were generalists. Training in community-based sites was intended to influence students' choices. The initiative's strategic roles were to demonstrate how (1) to increase the supply of primary care practitioners and prepare them for practice in community-centered systems, and (2) to partner communities and institutions in ways that enabled communities to push for changes within institutions. Grants were made to educational institutions in seven sites. At each, a community-university board was responsible for project oversight. Benefiting from increased access and stirred by the hope that newly trained professionals would choose to serve them, communities developed expectations for the institutions. Among the results were measurable changes in institutional policies and several ongoing partnerships with communities.

The first initiative was designed to reform educational programs for medical students and undergraduate nursing students...at the time, 70% of physicians were specialists, and 30% were generalists.

Comprehensive Community Health Models of Michigan. The second initiative explored how a community could determine what kind of health system it wanted. In three mid-sized Michigan communities, a neutral convener (a community foundation) brought together consumers, purchasers, and providers to envision their desired health system and to plan how to develop it. The Foundation was a partner. In addition to funding, it provided hands-on support from an operations office and technical assistance from consultants. The communities implemented plans that both encompassed and ranged beyond the health care delivery system to financing, community information technology, and health improvement. This initiative helped to carry out the Comprehensive Health Systems strategy in the services sector of the plan in which system development is the work of communities and institutions together. The team saw it as close to the "center"—the place where the goal was achieved—because its approach to a community's health system was so comprehensive (although it did not explicitly incorporate the element of appropriately prepared personnel). One result is a coverage option for uninsured workers that is being replicated.

Community-Based Public Health. The third initiative was designed to strengthen public health by reorienting education and practice toward communities and giving communities a voice in the education and practice of professionals. Academic partners (the school of public health with another health-professions school at each site), practice partners (local health departments), and community partners (community-based organizations and groups) came together in consortia in seven sites. While classified in the education sector of the strategic plan, this initiative also encompassed practice in the services sector. It was very intentional about community-institutional partnerships. The theory was that communities would leverage change in the practice and educational institutions from the outside. Measurable changes occurred, including changes in tenure policy that accorded more value to community-based participatory research. Program staff did not anticipate how active and forceful the communities turned out to be. The initiative has legacies, including the community-based public health movement it prompted participants and like-minded colleagues to launch.

Graduate training has a strong influence on trainees' practice choices. It had to be part of demonstrating how to produce a sufficient number of appropriately prepared personnel for the envisioned health systems.

Graduate Nursing and Medical Education. The fourth initiative carried the design of Community Partnerships in Health Professions Education into graduate training for nurses preparing for advanced practice and for physicians in residencies. Three of the six grantees had participated in Community Partnerships. Graduate training has a strong influence on trainees' practice choices. It had to be part of demonstrating how to produce a sufficient number of appropriately prepared personnel for the envisioned health systems. Training at this level also helped to build such systems. An illustration is that one university combined faculty practice with graduate training to create a system of care serving thousands of patients in rural communities. The initiative took community-institutional partnerships to a new level: the partnership boards had oversight of Graduate Medical Education (GME) funds and not just Foundation funds; one grantee was a community-based organization and not a university. When anticipated change in the federal policy that directs GME funds to hospitals did not occur, this hurt the initiative's chances of informing policy and practice nationwide. However, one grantee was able to have policy changed in its state to redirect GME funds to follow the resident so that non-hospital training sites could receive them.

Turning Point: Collaborating for a New Century in Public Health. The fifth initiative responded to the lesson from Community-Based Public Health that local public health agencies especially needed strengthening. Its premise was that public health was genuinely everyone's responsibility, not just the agencies'. Multisector engagement around a community's public health concerns could, it was thought, infuse strength into public health. The Foundation funded 41 local partnerships that included community-based organizations, churches, hospitals, businesses, police and fire departments, schools, parent organizations, youth groups, tribal members and governments, county commissions, and public health agencies. Using relationships and capacity they built by working on unique projects, the partnerships developed plans to improve systems; 17 were funded for the first year of implementation. The initiative demonstrated community engagement in integrating public health into community life and incorporating it into comprehensive-system development. The Robert Wood Johnson Foundation funded state-level partnerships in the seven states where the local grantees were located.

The Foundation funded 41 local partnerships that included community-based organizations, churches, hospitals, businesses, police and fire departments, schools, parent organizations, youth groups, tribal members and governments, county commissions, and public health agencies.

Community Voices: Health Care for the Underserved. The sixth initiative grew organically from established projects and initiatives. Thirteen communities tested solutions to the intertwined problems of high costs and high numbers of uninsured. The goal was to achieve a sustained increase in access, preserved safety net, stronger service-delivery system, and models to inform practice in other communities and policy at all levels. Participants included health systems and clinics, local public health, universities, grassroots groups, and community-based health and social services organizations. The intent was to demonstrate the access envisioned by the goal, and development of comprehensive, community-centered primary care. Projects resulted in new coverage options, enrollment in public programs, outreach for coverage and care, new primary care services typically unavailable to the underserved, and community engagement. In 2003, eight projects received second-stage funding.

There was another initiative sometimes associated with the Kellogg Foundation's name. The Community Care Network National Demonstration was an initiative of the Health Research and Educational Trust

Thirteen communities tested solutions to the intertwined problems of high costs and high numbers of uninsured. The goal was to achieve a sustained increase in access, preserved safety net, stronger service-delivery system, and models to inform practice in other communities and policy at all levels.

(HRET), the research arm of the American Hospital Association. The Foundation funded this effort as its premier work under the Institution-Initiated Change strategy in the services sector of the plan. The grantee funded 25 partnership projects (two of them with additional funding it received from the Duke Endowment) to build local demonstrations that had four elements: (1) community health focus, (2) community accountability, (3) seamless continuum of care, and (4) management within fixed resources. This produced a number of benefits for the communities involved. More significantly, it showed traditional health care providers the importance of making new linkages.

Initiative Work with Schools of Nursing

The two initiatives in which schools of nursing participated were Community Partnerships in Health Professions Education (seven consortia) and Graduate Medical and Nursing Education (six consortia). Three schools of nursing were involved in both initiatives. The two initiatives, then, showed us educational reform within schools of nursing that would contribute to development of the health systems envisioned by the goal. By taking a look at this experience in some depth, this section will explore our initiative-related grant-making, as seen through our work with nursing.

Community Partnerships in Health Professions Education. A decade of spiraling costs fueled a clamor for change in the health care delivery system. While most Americans agreed that the United States had developed one of the finest health systems in the world—indeed the flagship system—there was growing fear that uncontrolled costs would reduce the availability of health care. After World War II, most Americans had come to believe that basic health care was a right and that the best should be available to them upon demonstrated need. Health care expenditures from public and private sources tripled between 1983 and 1993 to over $1 trillion. There was much public commentary about the potential impact of too little money to sustain the system. The discussions about a two-tier system of care, one for the haves and one for the have-nots, seemed un-American or certainly clashed with the prevailing notion that the United States was a society dedicated to assuring equity in basic services for all citizens.

After World War II, most Americans had come to believe that basic health care was a right and that the best should be available to them upon demonstrated need.

Publicly and privately supported institutions tackled the question of how to reduce health care costs. At the Kellogg Foundation, health program staff took the position that the health system needed reform. They observed that the system was so expensive because it was inefficient. People received care at inappropriate points of entry. For instance, high-cost specialist entry points were too often points of access for primary care services. The practitioners best prepared to provide primary health care worked in systems that prevented them from being used appropriately. The staff pondered how the system might be strengthened, made more efficient and effective. Could the nation benefit from a coordinated system of care that effectively integrated public health and primary care? If people accessed care more appropriately, could costs be spread across the care continuum? The data on health staffing showed that there was an imbalance in the number of primary care practitioners, and the forecast showed that imbalance would continue. The distribution for physicians was especially out of balance: 70% were specialists while 30% were generalists. Nurses and physicians entering the field increasingly selected specialty training. So in order to support a reformed system that appropriately integrated primary care and public health, the system needed more and better prepared primary health care and public health practitioners.

One approach to stimulate reform would be to influence the way health care professionals are trained, and thus harness the resources of training institutions to inform policy change. Exploring the question of *how* to stimulate reform raised the question of *how* to do grant-making. As noted before, grant-making at the Kellogg Foundation had primarily proceeded based upon reactions to proposals as they came in. Programming priorities were established and publicized, and potential grantees contacted the foundation in response. Program staff reviewed proposals as they came in. Throughout its history, the degree to which program staff should be involved in shaping its programming was a matter of much

Exploring the question of how to stimulate reform raised the question of how to do grant-making.

debate. The debate was rarely around what approach was the more strategic but one of facilitating the broadest access to the most creative ideas and approaches. But if Kellogg Foundation-sponsored work was going to have an impact and inform the public debate about spiraling costs and declining access, it was clear that a more strategic approach was needed. To address these pressing public issues through an *initiative* was a clear break with past practice. With Community Partnerships in Health Professions Education, then, the health group both hoped to stimulate reform *outside* the Foundation and did stimulate reform *inside* the Foundation.

Launched in 1991, the bold initiative of Community Partnerships in Health Professions Education sought to influence change in the clinical training of both physicians and nurses, the two primary providers of primary health care. It was bold also because it aimed to reduce the distance between academic institutions and the communities in which they were located. The $47.5 million initiative was a five-year program to change primary health care curricula in schools of medicine and nursing and increase the number of new graduates who selected primary health care as their field of practice. This program funding followed a two-year process beginning with a major national announcement made to representatives from more than 600 training programs, which was followed by review of 100 applications and selection of 15 applicants for an initial development phase. A one-year leadership development phase required participation by a team from each applicant institution. The team members were representative of the partnership: deans or appropriate department heads of medicine and nursing, a senior academic administrator, and a community representative. At the end of the leadership development phase, comprehensive proposals were submitted for review, site visits were conducted, and seven partnerships were chosen to implement the vision.

The partnerships were in seven states: Georgia, Hawaii, Massachusetts, Michigan, Tennessee, Texas, and West Virginia. Each project was awarded $6 million to redirect health-professions education at the entry level—wundergraduate for nursing, pre-doctoral for medicine—to increase focus on primary care. In recognition of the essential contribution of impact services to carrying off this bold initiative, the $47.5 million included

$5.5 million earmarked for initiative-level evaluation, leadership development, the activities of health-professions organizations, communications, policy education, and networking.

Each project committed to increase course content and clinical training in out-of-hospital sites. In order to gain increased understanding and a better perspective on the knowledge and skill of other health professionals, each partnership committed to developing opportunities for multidisciplinary practice. Thus, each of the proposals included plans for developing community-based practice sites. One of the requirements at all stages of this endeavor was that at a minimum, medicine and nursing had to sit at the same table in planning and implementing their projects. The Foundation had been informed by community grantees that academic institutions had been fickle partners in that, once grant funding ceased, community-focused programs and alliances vanished. This knowledge, coupled with a deep belief that any real change toward judicious use of health care resources required an informed community, spurred the Foundation to also require evidence of real and sustained community participation in the partnership. In all projects medical and nursing faculty and students worked side by side in true partnerships with communities.

In all projects medical and nursing faculty and students worked side by side in true partnerships with communities.

As the five-year initiative progressed, it became clearer that in order to penetrate the cultures and structures of medical and nursing education, we had to required extended project goals to graduate-level education. The project grantees were able to increase the amount of undergraduate and pre-doctoral curriculum content in primary care-related subjects and time spent in related clinical practice. The differences in educational methodologies between medicine and nursing were apparent. Medical schools approached change by developing blocks or components that could be inserted into existing courses with ease (and just as easily removed). The nursing school approach was more of an integration requiring an overall curriculum adjustment. The effect was an increased number of units and courses in primary care added to curricula. New clinics were created extending the reach of academic health centers into underserved communities. Through these sites faculty and students served people closer to their problems. This allowed them to see close up the many factors affecting health recovery. Both students and faculty developed a greater awareness of the value of collaborative practice. Students did learn more about the practice of professionals in other disciplines. There were opportunities for students to function on multidisciplinary teams. But the structural differences in educational approaches and the reticence to question one's ways made it impossible to have real multidisciplinary education. The hard work of reaching consensus on a definition, philosophy, and framework for "multidisciplinary" was for the most part underdeveloped at the site level. Without this process experience, the sites were left with a distant flame to guide their work.

In both medical and nursing schools, sharing power with laypeople seemed anathema to professional status; a diminution of authority to direct the healing process.

Connecting to communities appeared more daunting for medical schools than for nursing schools. In both medical and nursing schools, sharing power with laypeople seemed anathema to professional status; a diminution of authority to direct the healing process. This response seemed contradictory to the reams of medical literature on the role of the self in health and healing and the patient as part of the curative team. As a condition of the funding, grantees were required to have advisory boards that gave the majority to community representatives. All boards achieved a community majority by the end of five years; 74 of 120 board members were community representatives. The boards advised on budget, administrative structure, program, personnel, evaluation, contracts, and more. Thus the advisory boards were avenues for engaging and enabling community members to participate in governance and decision making. The schools learned to see and appreciate the community as a resource. They developed deeper appreciation of the issues and problems facing communities. Schools and individuals became engaged in efforts to improve community health status that often went beyond direct clinical care to the single patient. They had evidence that people, regardless of their social class and cultural differences, were interested in contributing to solutions for their own health improvement. The schools vested, so to speak, in the communities where previously they had been disinterested harvesters.

To support the Community Partnerships innovations in health-professions education, the academic institutions had to make changes in curricula and institutional policies, manage complex logistics for placing students in community-based sites, and seek reallocation of funding. A challenge was to coordinate schedules for students, 600 faculty from universities and 1,200 faculty from communities, and 44 community training centers. Curriculum change (194 new or revised courses) affected 3,300 students—60% of the enrollment in the 28 schools in the initiative. A total of 63 policy changes were connected with admissions, curriculum, or faculty roles and responsibilities. Strengthening the new approach and carrying it forward into the future depended, in part, on reallocating resources. The Foundation's funds leveraged $62 million from schools, communities, and states to Community Partnerships in Health Professions Education projects.

Lessons learned. Over the years the Foundation observed, as had other funders, that grant proposals were often the brainchild of individuals who hoped to demonstrate ideas and approaches in such a compelling way that others would follow their lead. It often appeared that followers emerged, based as much on the standing of the individuals or their institutions as on persuasive evidence of innovation and efficacy. In that regard, the selection of the cohort for Community Partnerships in Health Professions Education ran counter to conventional wisdom. The cohort was selected from among institutions whose university and school missions were more aligned with primary care and community partnerships. The staff reasoned, and the experience affirmed, that it would take clear vision and strong leadership to stay truer to institutional mission than to professional aspiration. No matter what goal statements faculty set as a group, each longs to be number one, the best. The tendency is for health-professions schools, no matter the setting, to compare themselves with the top research universities. It takes strong, effective leadership to keep such unrealistic desires in check. East Tennessee and West Virginia University were both experienced in developing approaches to improve life in their rural communities; they were held accountable by their communities, through legislative bodies, to be responsive. The people in these communities expected the universities to solve their growing problems of lack of access to health care. West Virginia's response led to a legislative solution to ensure a continued supply of primary care physicians. East Tennessee's response created changes in health care delivery

The tendency is for health-professions schools, no matter the setting, to compare themselves with the top research universities. It takes strong, effective leadership to keep such unrealistic desires in check.

as well as in medical, nursing, public health, and allied health education. The nursing faculty developed a primary care clinic that was open during evening and weekend hours. The need for such a clinic had been determined through gathering information. The clinic was in a plaza near the hospital and co-located near a physician primary care practice. In both of these instances the initiative goals were consistent with the grantee's goals. We learned a lesson here that would be repeated again and again in our work—the closer the fit between the vision and goals of an initiative and the vision and goals of grantees, the more likely the implementation will stay on the agreed-upon course. Both universities have been responsive to their community and in doing so have raised expectations by their community. It will be difficult in the future for these institutions to retreat from their engagement in helping to solve community problems even if they want to. East Tennessee University was so impressed with its results that a decision was made to expand community partnerships to include the non-health-professions schools.

Throughout the world, nurses were making significant contributions to the delivery of primary health care. But structure and hierarchy made it difficult for the systems to use nurses effectively or efficiently. No place was this more evident than in unserved and underserved communities in the United States.

Intersection with nursing. Throughout the world, nurses were making significant contributions to the delivery of primary health care. But structure and hierarchy made it difficult for the systems to use nurses effectively or efficiently. No place was this more evident than in unserved and underserved communities in the United States. Academic health centers often had statutory or historical obligations to serve indigent and needy populations. Increasingly, care for the poor was felt to be an albatross that would eventually sink the centers or hamper their ability to pursue a research agenda. Academic health centers needed, for their own survival, to find ways to serve the indigent. They had the standing, capacity, and need to develop models for cost-effective health care delivery. In fact, no institution is better poised to develop and demonstrate models for efficacy and effectiveness than the academic health center that has co-located training programs.

The tensions between medicine and nursing often surfaced during the demonstration; they could be pushed down but were never pushed away. Developing trust and appreciation for the work of colleagues from a discipline other than one's own required a lot of energy and good will.

The projects demonstrated the range of roles for nurses from clinical nurse to primary care clinician to senior executive. The faculty at East Tennessee developed and staffed a primary care clinic. The faculty at the University of Texas-El Paso developed a network of family health clinics in schools that linked to the primary care system. The School of Nursing at the University of Texas El Paso changed its policies in order to institutionalize faculty service in the primary care training centers that the consortium established in schools. Such faculty service can substitute for research and publications toward meeting tenure requirements. The nursing school at Northeastern University, Boston University College of Medicine, and four federally qualified health centers in Boston created a new formal structure to guide their effort. A nurse faculty member was selected as their consortium's CEO to lead the entire project. In their model, nurses and physicians delivered clinical care.

Graduate Medical and Nursing Education. While the Community Partnerships in Health Professions Education projects had been successful in preparing graduates for practice in primary care settings and in engaging communities in partnerships with educational programs, the long-range influence was limited to undergraduate education for nurses and to pre-doctoral education for physicians. Reforming the system also required developing models for graduate education in order to sustain the out-of-hospital process. But public funding of graduate education for physicians flows into hospitals and maintains the culture of hospitals, providing role models and financial incentives that deeply influence ultimate career choices.

Graduate medical education funding, accordingly, became an increasingly important policy target. *The cost of buying care has the hidden costs of training people to provide it. Medicare and Medicaid cover the cost of residency training in their payments to teaching hospitals.* In 1990-1991, teaching hospitals received approximately $6.5 billion—or $97,000 per year per physician in training—without regard to the practice field for which the person was being trained. In contrast, nursing education programs received no such special supports, but were faced with the challenges of preparing graduate nurses nonetheless. Based on these concerns the Graduate Medical and Nursing Education initiative was launched in 1996.

In 1990-1991, teaching hospitals received approximately $6.5 billion—or $97,000 per year per physician in training—without regard to the practice field for which the person was being trained. In contrast, nursing education programs received no such special supports, but were faced with the challenges of preparing graduate nurses nonetheless.

The Graduate Medical and Nursing Education Initiative sought to shift a significant portion of the total graduate experience to multidisciplinary work in out-of-hospital settings and to redirect funding streams in order to sustain community-based graduate education. The initiative-level evaluators described the purpose this way: "The long-term goal is to help create healthier communities through a transformation in the current system of health-professions education, which relies heavily on specialist training in hospital settings." Transformation was to be brought about through:

- Enhanced training of health care providers for practice in multidisciplinary primary care in community settings where they would experience the diversity and range of primary health care needs.
- Policy dissemination to inform decision makers about out-of-hospital community education as a way to meet the health professions needs of their communities.
- Integration of graduate health-professions education with community needs.
- Development of long-term partnerships between health-professions education and communities to influence decisions about planning and implementing graduate training.

Six projects were chosen for funding under the Graduate Medical and Nursing Education Initiative—three of the original Community Partnership grantees (East Tennessee, Boston, and El Paso), along with three other sites: the University of Minnesota, the University of New Mexico, and a consortium of George Washington and George Mason Universities led by Zaccheus House, a community-based organization. This initiative was designed to build upon Community Partnerships. It is significant to note that the three Community Partnerships projects that had the most involvement of nursing in all phases of the project had made the greatest advances and were prepared to take this significant second step. A report to the board in 1997 noted that

> [t]he GMNE initiative represents a powerful idea because it implements a new way to practice and is doing so for a stage of professional training that is heavily subsidized by Medicare and Medicaid. Directing the flow of those subsidies away from the hospital setting and into community-based settings and multidisciplinary training will, unquestionably, influence how health care is delivered in America. *The GMNE effort focuses, therefore, on changing public and institutional policy rather than solely on changing curricula.*

The changes in nursing education and practice in the three settings of Boston, El Paso, and East Tennessee were profound, but the nursing profession seemed to ignore the potential for using these demonstrations and the tremendous investment in trying to inform public policy to make dramatic changes in nursing itself. Instead the nursing profession remained focused on hospitals and acute-care systems and faulted the Kellogg Foundation for its perceived lack of support for nursing. The Kellogg Foundation insisted that nursing have a "seat at the table" and the opportunity to be full participants. Some wonderful nurse leaders rose to the occasion, but they received little recognition for their work in nursing's professional forums.

Changing Policy and Building Systems

Changing policies for financing undergraduate and graduate medical and nursing education to support communities' needs for primary care was a monumental task. Throughout the course of the two initiatives, extensive work was done to inform policymakers and the media about issues and options.

When the Clinton administration's health reform effort came to nothing, the chance for fundamentally rethinking financing and direction for medical and nursing education as an integral part of creating affordable, universal coverage was lost.

When the Clinton administration's health reform effort came to nothing, the chance for fundamentally rethinking financing and direction for medical and nursing education as an integral part of creating affordable, universal coverage was lost. But there was debate about incremental changes in use of graduate medical education dollars, and some changes did occur. Market forces were not paralyzed by policy inertia, and produced effects with greater speed. The decisions made by purchasers, including government, to turn to managed care for control of costs increased demand for primary care practitioners. Academic health centers also found patient care revenues squeezed off for faculty in subspecialties. In this environment researchers found that, while community partnership models did not cost more than other models for out-of-hospital training that had increasingly come into use (indeed, they cost a little less), they provided a much richer experience. This experience included some multidisciplinary training.

Upon the failure of Clinton's health reform proposal, efforts were redirected toward state policy. Results were good for a number of projects, among them were those at Michigan State University and East Tennessee University. Following is a look at those.

Interestingly, although the Community Partnership project in Michigan was not selected as a Graduate Medical and Nursing Education site, it prompted the development of a strong model led by nurses. Somewhat apart from the project's policy work, the broad Community Partnership concept was recognized by key state executives. The result was a reallocation of Michigan's Medicaid funding for graduate medical education, including establishment of a set-aside of $10 million to fund especially creative ideas for out-of-hospital training. In 1997, Helen Grace and the lead program director for Community Partnerships in Health Professions Education were asked to serve on the review panel for awards from the $10 million fund. The effort was part of the state's process to begin moving graduate medical education from hospital-based care to communities. The process had the potential to transfer some funding to disciplines other than medicine. One proposal was from a consortium of nursing education programs in Michigan to develop a network of nurse-managed centers that would work collaboratively with medicine. The nursing proposal was not judged to be competitive within the framework of the state initiative, but the prospective grantee was encouraged to submit a proposal to the Kellogg Foundation.

The Foundation funded the Michigan Academic Consortium for Nurse-Managed Primary Care for work that is described in more detail below in the section on projects. What will be noted here is that the consortium's model has become part of the tip of the wedge for systematically incorporating nurse-managed primary care into delivery systems and recognizing it for purposes of reimbursement. A graduate training site developed by one of the four participating schools of nursing was even recognized by the Veterans' Administration as a clinic provider for its clients. During the course of the grant the consortium documented its work and developed a plan to become a resource for diffusion of its curriculum and strategies for financing patient care and preparing nurses as managers and community-access entrepreneurs.

The Michigan Academic Consortium for Nurse-Managed Primary Care model has become part of the tip of the wedge for systematically incorporating nurse-managed primary care into delivery systems and recognizing it for purposes of reimbursement.

East Tennessee State University was a Community Partnerships grantee with a competitive and successful application for Graduate Medical and Nursing Education. Although adding Graduate Medical and Nursing Education bolsters a Community Partnerships project, the three grantees that moved to the second stage were actually selected *because* of the sustainability of their Community Partnerships projects. Each was prepared to reallocate dollars to sustain Community Partnerships in Health Professions Education *before* Graduate Medical and Nursing Education came along.

Participation in both initiatives enabled the consortium to sustain system-level change. The consortium's system-building resulted in educational institutional change, service delivery change, institutional accountability to the community, increased access, health status improvement, and state policy change. The starting point was the first grant directed toward health-professions education. Through the cooperation and mutual trust that grew between community and institution, the development of training sites evolved into the development of a service delivery system. Improvements in health status were documented, including a 19% reduction in deaths from heart disease in one county from 1991 to 1994. When the TennCare program (the state's alternative to Medicaid) terminated reimbursement to hospitals for training recent medical school graduates as hospital residents—which created a crisis in the state's teaching hospitals—East Tennessee State University was able to contribute to a unique policy change adopted in the resolution of the crisis. A plan was adopted to phase out TennCare payments to hospitals for graduate medical education and phase in payments to medical schools. The medical schools planned to use the money to "follow" the resident to wherever he or she is training. In this way, training in community-based, out-of-hospital primary care could come to receive the same kind of support once available only to hospitals, with their emphasis on tertiary and specialist care.

The intersection of nursing and field-generated grant-making. The Kellogg Foundation funded many projects involving nurses and nursing during the period covered by this chapter. This section highlights many of them, showing their impact on nursing's development. I have grouped the projects by theme or topic in order to demonstrate their relationship to working strategically and attaining goals.

Recruiting and supporting community members. As the health programming of the Kellogg Foundation became increasingly focused on communities, one glaring problem stood out: the racial and cultural imbalance between the students, faculty, and practitioners in the health professions and the people in the communities being served. In nursing, the lack of diversity compared with the general population is persistent. About 12% of people in the United States are of African descent, about 12% are ethnically Hispanic or Latino/a members of any race, and about 1% belongs to indigenous groups, according to Census data. The composition of the nursing workforce has remained consistent through the years, with only 10% of practicing nurses coming from any underrepresented group as of 1996.

The composition of the nursing workforce has remained consistent through the years, with only 10% of practicing nurses coming from any underrepresented group as of 1996.

The health team turned its attention in the 1990s to this racial and cultural imbalance with an education strategy for Recruiting and Supporting Community Members. In our vision, all workers, no matter what their race or ethnicity, should have among their colleagues people whose races, cultures, and backgrounds are different from their own. The contributions of diverse workers and leaders would, together, increase the strength of the systems and their capacity to meet the needs of people. The same would hold true for students in training and the educators who were preparing them for practice.

For example, we recognized (and research showed) that physicians of color were more likely than other physicians to devote their practice to serving people on Medicaid and other vulnerable populations. But ours was not just a strategy to increase the supply of practitioners of color as a source of care for the underserved. Our thinking, as always, was in terms of the whole. In the health system we envisioned, any practitioner serving any patient should be appropriately prepared to serve people from many backgrounds. Our interest was not in a system to serve just one group, but to systems to which *leaders* in practice, education, research, and management brought *the diversity of perspectives and experiences and the vision necessary* to move the system toward high quality health care for all. But bringing the possibility of health careers into the awareness and the dreams of disadvantaged youth is a daunting task. Supporting members of underserved communities in their educational enterprise and helping them stick it out against the odds demanded skill, conscientious effort, and commitment. The field had tried out programs, but their results were disappointing: the racial and ethnic composition of student bodies in health-professions education stayed the same. Below is one example of a program that worked.

We recognized (and research showed) that physicians of color were more likely than other physicians to devote their practice to serving people on Medicaid and other vulnerable populations.

A case study in training Native American nurses. Native Americans are severely underserved. They have extraordinarily high rates of diabetes, obesity, alcohol, and substance abuse, and associated conditions. Care provided by Indian Health Services facilities to reservation residents often requires long-distance travel and long waiting times and is viewed as culturally insensitive. Even when tribes assume responsibility from the federal government for managing care, they must overcome fragmentation and limitation in services. Urban Indians, who live off reservations, face serious access problems as well.

Beginning in the late 1980s the W. K. Kellogg Foundation funded several projects for preparing Native American nurses. The Association of American Indian and Alaska Native Nurses documented fewer than 80 Native American nurses in the United States in 1972. No effort had been able to overcome this longstanding shortage when the Kellogg Foundation began its work on the problem in 1989 with the establishment of associate degree nursing programs in the tribal college network, starting with a program at Salish Kootenai College in Montana and the Sisseton-Wahpeton Community College in South Dakota, which serves tribal communities along the South Dakota-North Dakota border. By 1994, the Salish Kootenai program had graduated 76 nurses with associate's degrees, 33 being Native Americans or of Indian descent.

The Foundation then funded "Strengthening the Pathways to Indian Nurse Leadership" so that Salish Kootenai could work with other institutions to develop a rural education model that offers several pathways to make obtaining associate's or bachelor's degrees in nursing more feasible for Native Americans to earn. To create this model, the tribal college partnered with other tribal colleges, hospital training programs in nearby states, the Indian Health Service, the National Science Laboratories, and community-based services on reservations and other surrounding underserved areas. To further enhance educational opportunity, bachelor's level graduates could also be mentored to enter master's degree programs. Since bachelor's- and master's-prepared nurses could become faculty and health care administrators, they could, by their presence in these roles, in turn, foster more rapid development of the field. By June 2000, six students had earned baccalaureate degrees and 200 students had earned associate's degrees in nursing. Also during the year 2000, two teaching posts were assumed by Native American graduates with master's degrees.

The many barriers that slowed the developmental process included: (1) students' need for financial support, (2) off-campus students' lack of access to faculty and computer technology, and (3) failure to recruit Native American master's prepared faculty (due, in part, to the low tribal college salaries).

We funded another partnership that probed into the issue of historic cultural barriers to the success of Native American students. Cultural and socioeconomic barriers in traditional nursing programs had led to dropouts. Were these programs suitable for people from a culture for which they had not been designed? The University of South Dakota and Oglala Lakota College partnered on a program designed for people of the Native American culture. The program served Native American students from the Pine Ridge and Rosebud reservations and surrounding areas. The project focused on attracting Native American students to nursing education and adapting existing nursing programs to better serve those students. To increase the rate at which their students succeeded, the university and the tribal college implemented cultural awareness training sessions for faculty and staff, made curricular changes to facilitate learning by Native Americans, provided students with stipends and Native American nurse-mentors, and developed approaches to meeting Native American student needs, including culturally appropriate ceremonies to honor these students. Called the Piya Wiconi New Life/New Beginnings Nursing Education Program, this initiative increased enrollment by 500% to achieve the maximum training capacity of 20 students per year at the university and 17 per year at the tribal college.

The Piya Wiconi New Life/New Beginnings Nursing Education Program increased enrollment by 500% to achieve the maximum training capacity of 20 students per year at the university and 17 per year at the tribal college.

The innovations designed and implemented by this partnership project throw into high relief a tenet of our strategic plan—that students from disadvantaged and minority groups cannot just be dropped at the door of institutions of higher education and left to fend for themselves. They must be actively supported through academic-community partnerships. Institutions of higher education, of course, are staffed and organized to support students on their journeys. But the statistics on the composition of health-professions student bodies show that traditional institutional methods of support help to attract and retain traditional students in these student bodies but do not go far enough in attracting and retaining racial, ethnic, cultural, and socioeconomic minorities. Our perspective was that the principle of community-institutional partnership should be applied at this point in the health system also, that doing so might make a difference. The grants for Native American nursing

The statistics on the composition of health-professions student bodies show that traditional institutional methods of support help to attract and retain traditional students in these student bodies but do not go far enough in attracting and retaining racial, ethnic, cultural, and socioeconomic minorities.

were made under a cluster for academic-community partnerships. The fact that we received proposals that were in alignment with our understanding of how the health system could be strategically improved was evidence that we were on the right track and that there was readiness in the field. What support that connects to students' communities and their cultures can mean is exemplified by this project. The health goal evaluators reported a series of innovations from which I have made a few excerpts:

- Native values of bravery, courage, wisdom, and respect are infused into the . . . [tribal college] curriculum. . . . A circular classroom seating arrangement incorporates the Native American values of inclusion, respect, and the primacy of the group rather than the individual. Native faculty and guest speakers at . . . [the university] integrate Native culture into lectures.
- Non-traditional testing options (oral testing, group testing) are used to address the learning style of Native Americans.
- Special student support is provided for the female Native nursing student who, on the average, is 30-35 years old with four to six children. Flexible schedules and an educational support system . . . enable students to successfully complete nursing coursework. Students are also allowed flexibility to miss classes in order to care for sick relatives. Recognition of the primacy of the family reportedly created a more comfortable environment for Native students.
- The . . . mentoring program provides Native American role models in nursing and a student support system. This mentor program pairs traditional and acculturated Native American role models with students according to background, personality, and level of acculturation. Alumni employed on the reservation and in urban settings provide counseling, tutoring, and moral support for students. Travel to national professional conferences exposes students to successful Native Americans in nursing. Native guest speakers and faculty also serve as role models for students.
- Social activities and joint dinners with . . . [university] and . . . [tribal college] faculty, . . . [university] and . . . [tribal college] students and Native American nursing alumni provide a sense of connectedness and address the cultural shock of leaving the reservation, especially for students located in the urban areas.

- The establishment of distance learning through newly obtained technology has increased student access to education and training.

Projects in support of educating Native American nurses contributed toward the goal of recruiting and retaining members of underserved communities. This is critical to assuring that health systems are staffed, managed, and led by appropriately prepared personnel. The projects increased the woefully low number of Native American nurses, seeded ongoing production of new nurses trained at different levels, stimulated development of institutional capacity, and demonstrated how to lower cultural, socioeconomic, and geographic barriers to recruitment and retention of members of isolated, minority groups.

The projects increased the woefully low number of Native American nurses, seeded ongoing production of new nurses trained at different levels, stimulated development of institutional capacity, and demonstrated how to lower cultural, socioeconomic, and geographic barriers to recruitment and retention of members of isolated, minority groups.

Reorienting Practice to Communities

In the two initiatives discussed earlier, we saw outstanding work to reorient practice to communities. The Northeastern University College of Nursing, for example, created "a model of clinical education in which nursing students receive 50% of their clinical experiences in the community—often in settings where no other kinds of health care services are available." This type of work arose in grantee-initiated projects as well and not only in initiatives.

Grantees were varied, reflecting our conceptual framework of four sectors that came together in the envisioned system—services and education, communities and institutions. As a matter of course we funded educational institutions to increase the links between nursing education and community-based practice by reshaping the nursing curriculum at both the undergraduate and graduate levels. Most basic nursing education has

Most basic nursing education has its roots in hospital care of patients. But we also funded services organizations for some very original ideas that have taken hold in health care delivery.

its roots in hospital care of patients. But we also funded services organizations for some very original ideas that have taken hold in health care delivery. This subsection shows some of the grantees' work and accomplishments.

Rural health outreach program. The Medical College of Georgia established its first rural health program through two successive Kellogg grants (1988 to 1996) to its school of nursing. Together the grants resulted in academic institutional policy and practice changes, including use by the Medical College of Georgia of indigent care funding it receives from the state for post-grant support of an outreach team for maternal and infant-care in the county where the Rural Health Outreach Program operated—a small but important incremental increase in interdisciplinary cooperation between medicine and nursing; an attempt to replicate parts of the project's model in other counties; nursing faculty practice in a high school-based health clinic in the rural county; training of nursing students in the county; and use of the Rural Health Outreach Program as a training site by the Area Health Education Center (part of a long-established, federally initiated network).

The school of nursing learned a major lesson about how to enter communities, since community members had not trusted outside institutions to begin with and their mistrust of the nursing school had grown because of an earlier program that was unworkable and unacceptable to community members. The school redirected its work to become part of a collaborative endeavor that resulted in a comprehensive, multifaceted system of maternal and infant care that reportedly contributed to reduction in infant mortality in the county.

Other university projects. The Foundation had the chance to support other institutions of higher education that were willing to try things they had not done before.

Valparaiso University in Indiana converted an older building on its campus into Hilltop Neighborhood House in the mid-1990s. In doing so, this private, church-affiliated, traditional institution in a historically non-diverse town of Valparaiso reached out to an underserved, demographically changing community. Hilltop House, set up as a separate organization with its own board, made a serious effort to engage the neighborhood community in setting the direction of the organization. The Kellogg Foundation provided a grant to help the new organization launch programs, which include a neighborhood health center, child care, and services for youth. Students and faculty in nursing and a number of other disciplines are involved. Students serve as interns and as volunteers. University employees pledge financial support. The local hospital and health department also provide support.

The University of Virginia, a public institution, collaborated with public agencies in the first half of the 1990s to try to make health services more accessible and responsive to the living situations of underserved people. The collaborators focused on promoting mental health among the rural elderly—one of the most important and intractable geriatric health issues. The approach, which included nursing outreach, gained positive publicity and worked so well that it was institutionalized by the collaborating organizations before the end of its three-year demonstration period. (Reportedly, however, public budget challenges later affected the future of the program.) A December 1995 Associated Press story described the Jefferson Area Rural Elder Health Outreach Program and made several important points: A university medical center and two community-based organizations collaborated on, sustained, and institutionalized an experimental model because *they* saw it met needs, reached an underserved and higher risk group, cost much less than emergency room visits and nursing homes, and improved emotional and physical conditions. Elders and families were described as collaborators, too, in this Kellogg-funded project.

Western Michigan University in Kalamazoo, with Foundation support, established a new baccalaureate program in nursing that was *designed* for the purpose of reorienting nursing practice toward communities. This program bases the clinical experiences of students primarily in community settings rather than in acute-care settings. Through strong links to the community and to community-based organizations such as the public school system, the program is recruiting students and building of pipelines into the nursing profession.

Western Michigan University in Kalamazoo, with Foundation support, established a new baccalaureate program in nursing that was designed for the purpose of reorienting nursing practice toward communities.

In 1997, when Hispanics and African Americans together comprised 38% of Texas's population, only 221 African Americans and 249 Hispanics were among the state's 6,000 advanced-practice nurses... now the number of underrepresented minority advanced-practice nurses in Texas will continually increase.

Prairie View A&M University, a historically black institution in Texas, inaugurated a master's degree program to prepare advanced-practice nurses for primary care and developed four primary care training sites. The college of nursing triumphed over a number of obstacles in founding this program, which admitted its first 15 students in June 1999. The goal of establishing a master's degree program in nursing at Prairie View A&M University had been announced more than 20 years before by the State of Texas, which saw such a program as one means to achieve greater equity in higher education. Their triumph addresses the inequity the state's recommendation spoke to. In 1997, when Hispanics and African Americans together comprised 38% of Texas's population, only 221 African Americans and 249 Hispanics were among the state's 6,000 advanced-practice nurses. The new Master's of Science in Nursing degree program in primary health care focuses on urban and rural minority populations. Now the number of underrepresented minority advanced-practice nurses in Texas will continually increase.

With numbers so low, the university had trouble finding appropriately prepared faculty to recruit to train more nurses. Noting our mission to help people help themselves, the college said it "grew its own" faculty for the program. Two of its faculty, for example, became Kellogg Family Nurse Practitioner Fellows in the grant-funded program discussed later in this chapter. The bureaucratic approval process of state boards for higher education and nurse licensure delayed the planned starting date, but the college's active program development effort produced an exciting result in the meantime: The Houston Endowment Foundation funded a minority health research chair that allowed three distinguished professors to work with faculty in developing research activities and implementation plans for the master's program.

Living at home/block nurse program. The Block Nurse Program was founded in the early 1980s in a St. Paul, Minnesota, neighborhood by six women. The founders designed a program to enable elders to "age in place" with the help of a coordinated support network of health professionals, community volunteers, and service organizations. This group included nurses who had direct experience in caring for people within the context of their families and communities. The founders had the insight and drive to work for change in practice and policy for services to seniors, including funding and reimbursement policies for services. Their vision was very much in alignment with Kellogg's.

The program's name communicates the concept of the neighbor who is a nurse. Block nursing is based on the concept that nurses live in communities across the country and can play an important role in developing and operating networks of care for their communities. The idea of neighborhood- or community-connectedness is key. The program began in the St. Anthony Park section of St. Paul and was next replicated in other St. Paul-Minneapolis area *neighborhoods*. The rural community of Atwater, Minnesota, with a population of a little over 1,000, also replicated the program. In the later 1980s the program merged with the Living at Home Program that had been created in 1986 when St. Paul became one of 20 sites for the National Living at Home Demonstration of ways to reduce premature institutionalization of the elderly.

Block nursing is based on the concept that nurses live in communities across the country and can play an important role in developing and operating networks of care for their communities.

Around 40 communities in the state now have the Living at Home/Block Nurse Program, which has also been replicated in states in the Upper Midwest and in Texas. Replication is now supported by an institute created for the purpose. Beginning in 1988 and continuing well into the 1990s, the Kellogg Foundation supported the expansion, dissemination, and replication of the program with two grants totaling more than half a million dollars.

Other grants to health care services organizations. The Visiting Nurse Association of Central New Jersey is a long-established services organization that embarked in the 1990s on an innovative model, Neighborhood Nursing. Innovators can encounter unexpected challenges. Trained in hospital-centered care, nurses in practice have not normally been appropriately prepared for a neighborhood-based approach and need time to learn how to work differently. The grantee turned to enriched orientation sessions and mentorships to address the training needs that emerged. Despite the "growing pains," the nursing association saw an increase in volunteers, people using their services, and community awareness. Our experience with this grantee demonstrated how a funder's resources can enable an organization to work through challenges on the path to change.

The parish nurse movement has become widely known and has spread across the country, facilitated by linkages with community hospitals.

The Foundation supported the concept of embedding health care in the social structure of churches from the early 1980s, first through the support of wholistic health care centers in churches and later through support of parish nursing. The parish nurse movement has become widely known and has spread across the country, facilitated by linkages with community hospitals. We supported Lutheran General Hospital in greater Chicago, where the model originated and from which it was disseminated through a dedicated resource center, and its successor organization, Advocate Health Care. Advocate Health Care was formed in the late 1980s when Lutheran General merged with Evangelical Health Systems, an early adopter of the parish nurse model. We continued support for parish nursing by providing Advocate with funds to take the model to another stage of development. To link congregational health services into its vertically integrated delivery system required developing a computerized documentation system to manage patient and other information through multiple sites. From the perspective of a grant-maker this work showed us that networking non-traditional community sites with traditional medical centers presents challenges that grant-makers can help innovators overcome. From the perspective of the development of nursing, another important aspect of this later grant to Advocate Health Care was that development of the documentation system dealt with the content of the nurse's entries into the patient record, not just with computerizing and electronically linking the nurse's entries to the hospital's records system. A "health values clarification" tool was also developed under the last Foundation grant. By using this tool, Advocate parish nurses helped members of their congregations increase awareness of their own definitions of health and values around health, including spiritual health.

Partnering with Communities, Fostering Community Development

The special gift of the discipline of nursing is to see the whole person. It is sometimes said that doctors treat body parts and nurses care for people. The person's reciprocal relationships with family and community are part of the whole that nursing is concerned about. The practice of nursing moved from home care—caring for people in their homes in their communities—into hospitals. Later nursing sought and gained status as a profession within academe, which it continues to guard jealously. Tertiary care institutions and academic institutions are characterized by distance from communities. The passage from community to hospital to university has put nursing at a distance from communities, thus insinuating a contradiction into the discipline, removing it from the living contexts of the people nurses care for, and diminishing its ability to see and serve whole persons.

We might wonder whether a source of some of the suffering of people who are leaving nursing today is the feeling that the profession has lost connection with its origins and inner resources, its mission, itself. To counter this possibility, should there be another stage of development of nursing—a stage both innovative and restorative? Our experience in health programming suggests that another stage is struggling to emerge in a hostile environment. I love the title nurses from Northeastern University chose for their book about their experiences in Community Partnerships in Health Professions Education—*Teaching Nursing in the Neighborhoods*—and the title of the chapter "Out of the Tower and onto the Streets." There is no question that some nurses have conceived of the next stage of nursing's development and poured their talents, labor, and intelligence into helping it be born. That visionary nurses are present in academe and ready to make the most of funding opportunities was demonstrated in the Community Partnerships in Health Professions Education and Graduate Medical and Nursing Education Initiatives. Their presence, persistence, creativity, and skill were demonstrated, too, in grantee-initiated projects. A sampler of projects shows how nurses can cut pathways of many kinds to bring their healing powers into communities.

We might wonder whether a source of some of the suffering of people who are leaving nursing today is the feeling that the profession has lost connection with its origins and inner resources, its mission, itself. To counter this possibility, should there be another stage of development of nursing—a stage both innovative and restorative?

Rural Elderly Enhancement Project. Started in 1988 with a grant to Auburn University in Montgomery, this project was designed to form and coordinate coalitions of volunteers helping elders to stay in their homes. Many of the volunteers were youth, and the project expanded to intergenerational services by using elders to tutor youth. The project, headed by nurses, then expanded to establish rural school-based family clinics. Based on experiences in this project, the School of Nursing at Auburn University implemented a new curriculum that emphasizes interdisciplinary practice. An Alabama foundation now supports a scholarship program for residents of the two impoverished rural counties in which this project was based.

The originator of the Rural Elderly Enhancement Project (REEP) has published her reflections on the processes of entering communities so remote from academe without being intrusive, of building trust, of awakening residents to greater awareness of their own assets and abilities, and of transferring ownership to the community.

The university's nine years' worth of grants did more than inaugurate services and improve physical infrastructure. The work in those years developed youth and adults. Stating that "[i]ndigenous leaders developed in their roles as parent volunteers or student participants in youth programming," the health goal evaluators told the following story to illustrate their point:

> Two African American women "reported that the home health aide/nursing assistant training they received under the first and second grants to Auburn brought them out of their jobs as a shirt factory worker and seamstress, respectively. Building on the training, one woman volunteered to aid the elderly as part of REEP, and then volunteered to help children, via BAMA Kids. Her volunteer role as parent coordinator of BAMA Kids turned into a paid position as administrator of a summer enrichment program and coordinator of the Super Sixes Enrichment Program. (Super Sixes was a research program of two other universities that had been attracted to one of the counties by the visibility of Auburn's project.)

Operating and sustaining family clinics in two schools faced obstacles. Studying the project near the end of the third grant period, evaluators found several reasons why the clinics had not achieved their full potential. In this phase of its work, Auburn did not manage the project in a way that transferred ownership to the community (although, for the time being, the clinics were not simply closing down). The evaluators noted, for example, that community participation in governance had not been achieved. We can wonder whether some of the obstacles might have been better addressed if the community had been more in charge.

Bilingual interdisciplinary health services for Korean immigrants. While the Auburn School of Nursing reached out to minority-majority rural counties, the College of Nursing at the University of Illinois at Chicago (UIC) reached out to a small minority group in a large city. The demographics were different, but both schools formed and relied on relationships to carry them out. The UIC College of Nursing formed relationships with the city health department and the Comprehensive Korean Self Help Community Center. Through this city-university-community partnership, public clinic services in Chicago became available in the Korean language for the first time.

While the Auburn School of Nursing reached out to minority-majority rural counties, the College of Nursing at the University of Illinois at Chicago (UIC) reached out to a small minority group in a large city.

The university established a clinic for Korean immigrants in an existing city clinic with the city providing space, utilities, supplies, equipment, lab tests, and appropriately translated public health brochures. A volunteer Korean-American physician provided medical services. The college of nursing employed a bilingual nurse practitioner and trained and provided a bilingual Community Health Advocate to assist the nurse practitioner in patient care and outreach. The Korean-born Community Health Advocate provided screening, translated materials, and educated patients in Korean. Hiring a community member as a Community Health Advocate helped gain community participation in the project directly and through the linkages the Community Health Advocate gave community members to the clinic.

The college's partnership strategy paid off. Studying the project at the time it was closing, the evaluators found that the Chicago Department of Public Health would institutionalize the clinic by employing the nurse practitioner, continuing its in-kind support, and maintaining brochures in the Korean language. The physician would continue to assist the nurse practitioner one day a week. It was unclear whether the position of Community Health Advocate would be sustained.

The Korean-born Community Health Advocate provided screening, translated materials, and educated patients in Korean. Hiring a community member as a Community Health Advocate helped gain community participation in the project directly and through the linkages the Community Health Advocate gave community members to the clinic.

The evaluators reported that the college's lack of planning for sustainability helped make the future of its school-based health clinics discussed later in this chapter shaky, but institutionalization of the clinic for Korean immigrants was explicitly sought from the beginning. Nursing leaders partnered with a division director at the city health department to locate the clinic in an existing city clinic for the purpose of adopting bilingual service.

The college built on an earlier effort in which professors received funding from the Kellogg Foundation in the late 1980s to train bilingual/bicultural Community Health Advocates. It expanded its effort from preparing workers for a more responsive delivery system to changing the delivery system to incorporate more appropriately prepared workers. This was very compatible with our concepts about building health systems. The college did not test a delivery-system change just for the sake of research and publication. That is what funders see too often: universities that walk away when they have gotten their data. Instead, the change that the project tested was driven by the intent to see it institutionalized.

Nursing careers for the homeless. One project, a program organized by the School of Nursing at Howard University to provide health care for the homeless living in a Washington, D.C., shelter, became a fertile training ground. The shelter became a site for community-based practice, the school became a site for training shelter residents to become health aides, and the project evolved to demonstrate that some residents can become registered nurses with baccalaureate degrees. The shelter now has an infirmary in which care can be provided to those who need periodic extended care.

The project grew out of the experience of a nurse with an extraordinary capacity to see good and potential in other people. She had been trying to influence her faculty colleagues to get more directly involved in the issues of access to care for poor and disadvantaged people. In doing so, she naturally encountered traditions that schools of nursing have worked hard to establish in order to ensure their viability in attracting faculty, students, and funding, and their status among their peers. Her colleagues felt the need to give priority to increasing sponsored research and scientific publications. One day she realized clearly that her personal goals were not entirely compatible with the school's. She decided that she had to be more effective in demonstrating that her vision was a direction that could both strengthen nursing education and forge new relationships between the school and the community. One day as she was driving around, she found herself in front of a homeless shelter. She decided to go in; the rest is, as they say, history. Through this healer-leader, the visionary partnership was established between the school of nursing and the homeless shelter. The project's founder went on to lead a new nursing school in another community during its beginning years. She maintained an ability to focus on serving communities and not worry about detractors.

One day as she was driving around, she found herself in front of a homeless shelter. She decided to go in; the rest is, as they say, history.

Creating New Service Options and Points of Access

Both nurse-managed primary care and nurse-midwifery are vital to increased health care access and affordability. Grantees pursued curricular innovations to help make these services available: Increasing the numbers of advanced-practice nurses for delivering these services required developing more urban and rural training sites in underserved communities, which, in turn, increased access. These innovations demonstrated the powerful link between the services and education sectors that contributed to the rationale of our strategic plan. For health professionals to be available to *practice* in communities tomorrow, community service sites have to be available for them to *train* in today. For the training to be suitable for the practice of tomorrow, the sites must be *community-based* not just *community-placed.*

For health professionals to be available to practice in communities tomorrow, community service sites have to be available for them to train in today. For the training to be suitable for the practice of tomorrow, the sites must be community-based not just community-placed.

The Kellogg Foundation also produced substantial impact in the area of nurse-midwifery and birthing centers by supporting health services organizations' innovations in service delivery. Another vital means we saw for embedding nurses in community settings was school-based health care, which we supported through educational institutions, health services organizations, and coalitions of organizations. Following is a look at a few of them.

The Michigan academic consortium for nurse-managed primary care. This was an interesting outgrowth of the Community Partnerships in Health Professions Education Initiative. State executives in Michigan sought proposals to redirect certain Medicaid graduate medical education dollars to test innovations in out-of-hospital training. Foundation personnel, who had been invited to serve on the state's review panel for these proposals, noted that a certain proposal that did not fit well with the state's priorities might fit very well with the Foundation's. They invited the applicant—a consortium of nursing education programs in Michigan—to send us its proposal.

The proposal was of definite interest. Community-based nurse-managed primary care centers were making services available where people live and held promise for improving care in terms of cost, quality, and access. Yet the model had not gained widespread acceptance or support. Insurers often did not recognize the centers as providers for purposes of reimbursement for services. Because of health insurers' policies, the relatively few nurse-managed primary care centers in the United

States had difficulty sustaining themselves outside some special sponsorship. The consortium proposed training that would encompass business and other skills needed to navigate the system of managed care. The proposal focused on preparing advanced-practice nurses to be savvy administrators who could make a case for a nurse-managed model of primary care delivery.

Community-based nurse-managed primary care centers were making services available where people live and held promise for improving care in terms of cost, quality, and access.

The proposal was funded, leading to the establishment of the Michigan Academic Consortium for Nurse-Managed Primary Care. Four public universities (Michigan, Wayne State, Michigan State, and Grand Valley State) work in partnership with the Michigan Public Health Institute. The consortium is preparing advanced-practice nurses for administration of nurse-managed primary care centers. The consortium is *both* increasing access through nine centers developed as training sites *and,* through curricula and practice, increasing the nurse practitioner's abilities to combine the humanistic, business, and scientific components of primary care to improve outcomes in a managed care environment.

The service delivery sites range from a school-based center to one serving the homeless to one reaching out to the residents of a low-income high rise to one recognized by the Veterans' Administration as a primary care provider. Two of the nursing schools partnered on establishing service through the Tried Stone Baptist Church, which is located in a racially and ethnically mixed Detroit neighborhood of 10,000 residents, of whom 40% are uninsured.

For the pivotal task of changing reimbursement policy it was very significant that the Veterans' Administration (VA) recognized the Michigan State University site in Lansing as a clinic provider for its clients. Through a unique capitated contract, this nurse-managed health center serves the local population of veterans. "Before this program, half of the 1,400 veterans we serve received no care whatsoever," reports the nurse who directs the Michigan State University program. "The others traveled to Battle Creek or Ann Arbor for primary care. The VA clients and their families are thrilled with our approach to care. Our client outcomes have been excellent. The recognition by the VA may establish a precedent that

The service delivery sites range from a school-based center to one serving the homeless to one reaching out to the residents of a low-income high rise to one recognized by the Veterans' Administration as a primary care provider.

will boost the chances for the growth of nurse-managed primary care in other locations."

Maternal and child health care. The Family Health Center in Battle Creek was established in 1986 as the North Avenue Women's Center. It was one of the more than 100 community-based, problem-focused models funded under the health team's 1986 goal and programming plan. The center began with a focus on the unmet needs of pregnant women and the problems of infant mortality and lack of health insurance. It saw that its scope of services needed to grow; older children, teens and, indeed, the entire family also should be served for both the viability of the health center and the enhancement of health within the whole family. The Kellogg Foundation helped the clinic become more comprehensive. Receiving both Stage 1 and Stage 2 grants, this grantee progressed from its original focus to serving families and people of all ages. The Family Health Center is headed by a nurse and the complement of staff includes nurse practitioners in pediatrics and family practice, along with physicians, dentists, social workers, and others.

I got to see the community connection first-hand when the Morris Heights Health Center in the Bronx, a long-established Federally Qualified Health Center that served lower-income people, aimed to establish a birthing center. I was asked by the Kellogg Foundation to provide consultation to the participants.

Family-centered birthing centers emerged in the 1980s as another option for middle-class families. In the Morris Heights neighborhood of the Bronx, the Kellogg Foundation was funding an early and significant test of whether birthing centers—with their capacity to deliver care of high quality at lower cost—could be incorporated into care for low-income families. I saw the process continue when I later joined the Foundation's staff. From 1987 to 1993 the Foundation supported the creation of a birthing center modeled on a free-standing service for middle-income families in Manhattan. The Maternity Center in Manhattan, a leader in the modern natural childbirth movement, was first funded to establish the Bronx birthing center. The next stage of development was to integrate the center and its midwifery services into the services at the Morris Heights

Health Center. The Morris Heights project represented the different stages of development along the continuum toward comprehensiveness in the strategic plan. The first stage was to establish and gain community acceptance of America's first birthing center for low-income mothers. The next step was to work with the middle-class group from Manhattan that set up the center and the Morris Heights group to integrate its established operations and staffing with those of the childbirthing center. The organizational relationships were not easy, but the outcome moved services toward more comprehensiveness.

In the Bronx, physicians and low-income families did come to accept this alternative service, which generated health care savings. The day came when any family coming to either the health center or the birthing center was able to choose either hospital or birthing-center delivery. The range of supportive services available also expanded through the merger. Midwifery is lowering the cost of pregnancy services while increasing use of early prenatal care. The center's location and welcoming environment helped engage an impoverished community with many barriers to prenatal care. A pregnant mother of two having trouble accessing prenatal care at a hospital said, "This is better than the hospital. It's right around the corner and I don't have to take the bus or the train with these kids." The Bronx model was replicated and began to be integrated into the New York City Maternal and Child Health system.

The challenge for us was to determine how the model would transfer to a variety of community settings. We devised a plan to support further development of nurse-midwifery centers for the underserved and apply the birthing center model across a spectrum of contexts and settings. Under this plan we funded a proposal for a statewide network of birthing centers and a proposal for a partnership of family-support organizations in a city.

We devised a plan to support further development of nurse-midwifery centers for the underserved and apply the birthing center model across a spectrum of contexts and settings. Under this plan we funded a proposal for a statewide network of birthing centers and a proposal for a partnership of family-support organizations in a city.

The statewide network was in Tennessee. We funded Vanderbilt University for five years to initiate a statewide maternal and child health program in collaboration with community hospitals and physicians, including four birthing centers in underserved communities. The centers are tracking infant mortality, birth weight, sudden infant death, and availability of and access to culturally acceptable perinatal and well-woman services. They are also working with community advisory groups to examine the biological, social, medical, and environmental factors that promote or detract from childbirth and infant health.

The partnership was in the District of Columbia, where an accessible birthing center was established and managed with other organizations. The partnership is providing a continuum of care to low-income families. The District of Columbia Developing Families Center is home to three community support organizations: Healthy Babies Project, National Child Day Care Association, and the D.C. Birthing Center, which Kellogg funded to develop a childbirth center in collaboration with Howard University and to strengthen partnership and service capacity. In this collaborative model, the three organizations created an administrative system to smoothly link patients and their families to holistic support. Located in a building donated by the head of a family-owned metropolitan business who saw the enterprising spirit of the partnership, the Developing Families Center helped to spark business ventures in the neighborhood.

Also of note is a small grant we made to the University of Puerto Rico to enable students to attend a new program in nurse-midwifery that had been developed with a federal grant. The new program had to grow its own faculty, and one graduate who we supported was mentored and trained for a faculty post.

School-based health care. Health programming developed a body of work around school-based health care. This subsection illustrates the range of the work and then explores in more depth a project funded to the College of Nursing at the University of Illinois at Chicago.

School-based health care figured in the Comprehensive Community Health Models of Michigan Initiative. As described earlier, the initiative developed a community-decision-making model to enable consumers, purchasers, and providers to collaborate on a common vision and comprehensive approach to improving the systems for public health in three mid-sized Michigan counties. In one, the local project brokered a community decision to institute the School Health Initiative to make school nursing available throughout the county. The agreement reached between the local project organization and the county intermediate school district depended on the commitment of the major regional utility company, the county health department, school officials, providers, and parents; and built on broad public support. At focus groups on the project, the idea for placing nurses in the schools consistently received enthusiastic approval. Piloted in an urban and a rural area and planned for all school districts in the county, the School Health Initiative made neighborhood schools an entry point to health system resources. The project focused on sustaining the program by maintaining the enhanced intermediate (county-level) school district funding for the services. The nurses provide health education to students, teachers, and parents; screen students for immunizations; provide triage in emergencies; and make referrals for ongoing care.

Piloted in an urban and a rural area and planned for all school districts in the county, the School Health Initiative made neighborhood schools an entry point to health system resources.

School-based health care figured in our work under a cluster (or substrategy) of field-generated projects called "Community-Based Organizational Networks" that was housed in the services sector in our strategic plan. It tapped networks of organizations to create new, *non-traditional entry points for access* to health care. In one project the Henry Ford Health System in Detroit partnered with the Detroit Public Schools to develop elementary-level school-based health clinics. The health system helped to finance the project, incorporated school-based care into its delivery system, and worked with two other hospitals and the local health department on adopting clinics and sustaining them after the grant period ended. Programmatic and health services—organized into four tiers ranging from health education, screenings, and immunizations to full clinic services—were delivered at 13 schools. This arrangement demonstrated that school-based health care can be networked and does not have to be place-dependent; children can have access, when needed, at another school if theirs does not have a full-scale center. The demonstration of school-based health care in Detroit models how urban elementary schools can be integrated into health care delivery systems. The special power in the demonstration was the public-private partnership between challenged inner-city schools and a strong, progressive health system that spans urban, suburban, and rural areas. The project secured an arrangement with the state Medicaid program for recognition of the clinics within its managed-care strategy.

In a field-generated project funded to California State University at Dominguez Hills, the university and other partners developed a model in which a clinic for young people would begin in a school in a developmental program started by United States corporate leaders. The clinic was intended to later begin serving families in the neighborhood.

We supported development of infrastructure for the field through funding to the National Assembly on School-Based Health Care, which was created in 1995. A priority was sustaining the delivery of health services in schools through funding streams and reimbursement for care. Accordingly, our support included retaining a consulting firm to provide technical assistance to Assembly members around this subject and helping to finance a pertinent survey. The *National SBHC Finance and Patient Revenue Survey* is the report of the Assembly's 2001 survey of every school-based health center in the United States through a cooperative agreement with a federal agency. The consulting firm's workshop provided technical assistance to school-based health center managers to identify and tap into funding streams they might otherwise overlook. By carefully describing what the work actually is and comparing it to a broader range of funding programs than are usually considered by school-based health centers, a program manager may be able to discern a potential fit between an aspect of the work and a funding source.

Primary health care for urban schoolchildren in Chicago. The College of Nursing at the University of Illinois-Chicago used a Kellogg Foundation grant to create and operate two comprehensive school-based health clinics in underserved communities in Chicago elementary schools—one predominantly Hispanic, one predominantly African

American. Evaluators included the project, which had recently closed, in their field study of different kinds of projects we had funded in the Chicago area. The evaluators found that, often in these projects, strong indigenous leaders were central to projects' success, nascent leaders emerged through community participation, and neighborhood members played key staff roles. In the case of this project, the schools were selected for their strong principals and residents were hired to link the children and community with the clinicians. The principals' long-time residence in the school neighborhoods, political savvy, and commitment to integrating clinic activities into their schools proved vital. For example, the skills of the principal at one school were pivotal in negotiating educational politics in the city to secure renovation of space.

These clinics were excellent demonstrations of what integrating health services into schools can achieve. The clinics provided services and access points to services that were of great value to the school children, their families, and the schools. Combined volume for both clinics was 70 to 95 people per day. Noting that one school was located at an intersection that divides the territories of two rival gangs, evaluators reported that people they interviewed described the school's clinic as a safe place for children to receive health care. The presence of the clinic at the other school reduced the need to seek care at the nearest city health clinic, which was three miles away and had waiting lists of two to three months. In some cases parents and siblings also received health services, parents received information about health and social services programs, and children from other schools received immunizations and health care.

According to the goal evaluation team, principals and clinic staff "attributed increased productivity among school children and increased attendance rates to the school based health clinics. Many students enrolled in school on time as a result of being able to get the physicals and immunizations required for school enrollment at the school health clinic."

By including Community Health Advocates on the health services teams the College of Nursing created a model for delivering services that reached well beyond clinic services. The Community Health Advocates were respected members of the communities hired because of their understanding of the backgrounds of the patients served. A Community Health Advocate at one school was called "Grandma" by the children because of her 30 years of connections to the neighborhood and the school. Trained in first aid and CPR by project staff, the Community Health Advocates greeted walk-ins, assessed problems and either referred the child to a clinician or themselves treated basic problems, such as a stomach ache resulting from skipping breakfast. Sometimes "[t]hey bathed children who were not washed properly at home and gave them clean clothes donated by community members and collected by parent volunteers." They communicated clinicians' treatment recommendations in ways patients could understand, and sometimes did follow-up home visits.

The Community Health Advocate model turned out to function as a pipeline to a nursing career for one worker who decided to pursue a degree in nursing at the university. In our extensive experience with the use of various types of community health workers it was not uncommon to see such interest in pursuing a career.

The Community Health Advocates greeted walk-ins, assessed problems and either referred the child to a clinician or themselves treated basic problems, such as a stomach ache resulting from skipping breakfast. Sometimes "[t]hey bathed children who were not washed properly at home and gave them clean clothes donated by community members and collected by parent volunteers."

Given these successes, we were concerned to read in the goal evaluators' report that the future of the clinics was unclear. At the time of the field study, the clinics had "neither an institutionalization strategy nor funding in place to continue very long." Because the experiences project leaders had are, unfortunately, all too common, they are worth noting as a source of useful lessons for nurse educators, funders, and others interested in strategic management of grant-funded models. The goal evaluators observed no strategy from the university for institutionalizing project activities into *mainstream* practice after Kellogg Foundation funding ended. When a reported effort to secure financial support to continue the clinics from within the university was unsuccessful, the plan was to retain control over the clinics in the College of Nursing and rely on "soft money," that is, grant-seeking, to keep them going rather than funding streams rooted in public health care programs or resources of the schools or the community. One of the lessons was the peril in anticipating or hoping for ongoing university support and falling back on further grant-seeking when university support does not come through.

Another lesson of equal importance about sharing information and power: The principals at the schools where clinics were housed in this project were known for getting things done. Evaluators observed that school personnel and clinic staff "wanted to take on more responsibility for . . . financial sustainability." One principal had been using certain state funds, fundraising, and in-kind contributions from companies to pay for an additional nurse and obtain supplies. Yet, despite the availability of able potential partners, the project design had not included the kind of partnering and transfer of ownership that had worked well for another College of Nursing project, the Korean Immigrant project described earlier.

Such lessons are important for grantors as well as grantees. We emphasized sustainability planning in our work with grantees and looked for it in grant proposals. Our increased attention to sustainability had been observed by evaluators and was reflected in their use of *institutionalization* and *financial sustainability* as specific criteria to measure progress toward attainment of the health goal.

Faculty Preparation and Development

To actually meet the needs of people, the integrated, comprehensive health systems we sought to program had to be staffed, managed, and led by diverse and appropriately prepared personnel. For the systems to achieve their purpose, "appropriate preparation" had to emphasize practice in communities in primary care, prevention, and public health. On the whole, health-professions education was not itself ready to provide this kind of preparation. Health professions education, instead, supported and reflected modern American health care's disproportionate and costly emphasis on tertiary and quaternary care in institutional settings. We used both initiative and project grant-making to redirect health-professions education toward preparing professionals for community-responsive practice. We tried to offer incentives and opportunities for change in institutional values and culture and in curriculum and educational programs. We also gave considerable attention to faculty—to their opportunities, incentives, and rewards for development, leadership, advanced preparation, and innovation that were in alignment with the characteristics of the health systems we envisioned.

The shortage of primary care practitioners and their misdistribution were paramount issues for us. For instance, specialists comprised 70% of the physician workforce at the beginning of the 1990s. In contrast, nursing was not seen then—and is largely not seen today—as a vital source of cost-effective primary care practitioners. The funding sources for graduate health-professions education provided scant resources to increase the supply of advanced-practice nurses prepared to become primary care practitioners, particularly in underserved rural and urban areas. One result of this policy marginalization was a shortage of faculty who were clinically and educationally qualified to train advanced-practice nurses for nurse-managed care. This shortage limited institutions' ability to start or expand training programs.

The shortage of primary care practitioners and their misdistribution were paramount issues for us. For instance, specialists comprised 70% of the physician workforce at the beginning of the 1990s. In contrast, nursing was not seen then—and is largely not seen today—as a vital source of cost-effective primary care practitioners.

We had very good experience with grantees who undertook faculty preparation and development on a multi-institutional, regional basis. If managed well, a regional approach can be more efficient, spread the benefits of investments more widely, and enhance quality because different perspectives and experiences inform the work. The Southern Regional Education Board, in affiliation with its Council on Collegiate Education for Nursing, participated in a project to develop faculty's skills in environmental health and the grantee for a project to prepare more faculty to train advanced-practice nurses. The Southern Regional Education Board serves 16 states and the District of Columbia.

The Faculty Development in Environmental Health project was supported by a grant to the University of Maryland-Baltimore. Besides the Southern Regional Education Board, the partner organizations were Howard University and the National League for Nursing. This collaborative serving nurse educators prepared nurses to be leaders for environmental health. Its project responded directly to the Institute of Medicine's 1995 report, *Nursing, Health and the Environment.* The Institute recommended that environmental health concepts be incorporated into all levels of nursing practice, education, and research.

The grant to the Southern Regional Educational Board increased the number of faculty prepared to teach advanced-practice nursing. More than 20% of the program participants were members of racial minorities. Twenty-two nurses who received postdoctoral training in the Kellogg Family Nurse Practitioner Fellowship program were certified to serve as faculty in Family Nurse Practitioner training programs.

This postdoctoral training of their faculty benefited 17 universities, three historically African American, across the South. At the time the grant was closing, two universities were starting new nurse-practitioner degree programs with a big boost from appropriately prepared faculty. The postdoctoral fellowship experience included seminars on health policy and regulation and the use of information technology. The fellows are combining teaching, curriculum development, and faculty clinical practice, which increases family nurse-practitioner services in the region. One, for example, became a volunteer at Camillus House in Miami, which was a Kellogg Foundation grantee for health care for the homeless.

Building the pipeline. Attracting young people from underserved communities into health careers is essential to creating health systems that meet the needs of people. The Kellogg Foundation saw the pipeline as extending from kindergarten through postdoctoral education, especially

Attracting young people from underserved communities into health careers is essential to creating health systems that meet the needs of people. The Kellogg Foundation saw the pipeline as extending from kindergarten through postdoctoral education, especially with regard to diversifying the workforce.

with regard to diversifying the workforce. Building the K-12 pipeline was one arena in which we worked. In this segment of the pipeline, one key to attracting young people from underserved communities into nursing specifically is the visible role models. In impoverished communities, young people may not have a favorable impression of nurses from their contacts with the health care system. They see nurses in practice only in hospital or office settings and they experience nurses as hurtful people—they give shots, they do procedures to you, and most of the time, it hurts.

In addition to exposure to favorable role models, exposure to the whole idea of health careers should also be integrated into young people's experiences. The idea is for students' interest to grow as the students grow and for them to see not only what the roles are but themselves in the roles.

We made a contribution to the state of the art through both individual grantees and a collaborative effort spearheaded by the Association of American Medical Colleges. The latter effort encompassed the range of professions in the health sciences, not just medicine. This subsection looks at work relevant to nursing and some of the lessons learned.

Health sciences and technology academy. Administered by West Virginia University, this project developed as an adjunct of the statewide university system's highly successful Community Partnerships with Health Professions Education project. Like that project, it focused on the state's needs for practitioners to locate in its many underserved areas and represented the policy thrust and investments of both the state legislature and the university system. As of the fall of 1999, more than 500 students and 57 Academy-trained teachers in 22 counties participated in the program.

The Health Sciences and Technology Academy demonstrated the effect of working with students and teachers over time. Over 80% of the students in the program its first year remained for the full four years of high school. Data collected by the project showed that 45% of the students earned three units of college credit in math before they entered the twelfth grade. It was further showed that 96% of Academy graduates went on to college and that 16 of the 57 teachers who participate in the program earned master's degrees in secondary education with a science focus through the Academy. The legislature authorized state universities to provide tuition waivers to Academy graduates.

S.M.I.L.E (Science and Math Investigative Learning Experiences). When the S.M.I.L.E. program began in 1988 it served 80 students in four middle schools. Its purpose is to increase the number of educationally disadvantaged students who graduate from high school qualified to enroll in college and pursue careers related to health, science, math, engineering, and teaching. Today, it brings year-round activities to more than 700 elementary, middle, and high school students in 35 schools. It provides workshops and support for 60 K-12 teachers who work with the students. The geographic areas served are poor, largely rural, and educationally underserved. They have significant numbers of Native American and Hispanic students. The Kellogg Foundation participated in the growth and development of the program by funding Oregon State University from 1990 to 1994 to provide S.M.I.L.E. programming in mathematics and science to elementary school students in eight rural counties through camps, field trips to see health professionals in training, clubs, and other means. The focus was on youngsters with little chance for exposure to these opportunities.

When the S.M.I.L.E. program began in 1988 it served 80 students in four middle schools...the S.M.I.L.E. program of today is a partnership among Oregon State University and 14 Oregon school districts—mostly rural—serving underrepresented and other educationally underserved students in grades four through 12.

The program was sustained with very high rates of retention, minority participation, high school graduation, and entry into college by participating students. The S.M.I.L.E. program of today is a partnership among Oregon State University and 14 Oregon school districts—mostly rural—serving underrepresented and other educationally underserved students in grades four through 12. Oregon Health and Science University and other state universities are also partners. The S.M.I.L.E. program today also provides transition support for S.M.I.L.E. graduates who are entering college, as well as scholarships to Oregon State University. It has garnered support from a wide range of public and private sources, and it has been replicated in Rhode Island.

Health Professions Partnership Initiative (HPPI). The Health Professions Partnership Initiative of the Association of American Medical Colleges brings together higher education, public schools, and community organizations. Its purpose is to build a pipeline from underserved communities to health-professions education to increase the number of underrepresented minority health professionals. The Association has made grants to sites, supports a national advisory group, and provides technical assistance, networking, evaluation, and dissemination. Kellogg Foundation funds are being used for seven of the 26 sites; the Robert Wood Johnson Foundation funds cover the rest.

The funded projects systematically develop students' interest in and academic preparation for entry into health careers. The Association of American Medical Colleges uses the grants program to implement its "3,000 by 2000" project. The project's name refers to the goal of annual matriculation of 3,000 underrepresented minorities in allopathic medical schools by 2000. Although the Association has embraced this goal since 1970, enrollment still falls far short of the desired level, however, signifying the difficulty of the challenge. The extent of the challenge is illustrated by the fact that 1,871 students from underrepresented minorities entered medical schools in 1998, while 1,732 such students entered in 1999. The Association responded to the persistence of the problem by broadening its Health Professions Partnership Initiative to encompass *the health professions in general.*

The project of Western Michigan University in Kalamazoo illustrates developmental approaches, reaching from seventh grade through medical school and other graduate training. The nursing program at the university that we had helped to establish in the 1990s had a strong orientation toward community and was an active participant in the project. Several other programs at the university were involved along with the Kalamazoo Center for Medical Studies, an off-campus health-professions education center of Michigan State University, which is located in East Lansing. In the project, two area hospitals serve as career-awareness sites for the public schools. The project is designed so that, at each stage of preparation—from middle school to the day of professional licensure—a young person benefits from special activities, studies, counseling, and mentoring. Parents are involved, and teachers can receive graduate credit for taking a summer course at the Kalamazoo Area Math and Science Center. The project's design exemplifies a principle our grant-making experience brings to the field of changing the composition of the workforce.

A wealth of lessons. Evaluators reviewed and evaluated several pipeline projects, including projects at the K-12 level. In a substudy on diversifying the pipeline, they suggested a framework grounded in experience that might be used for future work. They observed that the Foundation had "a vast reservoir of experience to draw upon from its history of investment. . . ."

As my very selective account of our work at the K-12 level suggests, the effort with school children and teens usually does not focus on attracting them into a particular health discipline. The broad idea is to interest them in math and science, make them aware of various careers in the health sciences, engage them through both didactic learning and practical experiences in real-world settings, foster their development in understanding, and support them in making decisions to advance into some form of higher education. Although K-12 work is more often broad-based rather than discipline-specific, it is important to remember that students' daily experience of K-12 education *can* include a health professional as a role model. This role model is the school nurse. One effective way of building a pipeline is through integrating a school nurse into the school setting and allowing that nurse to have an "open door" to be a counselor to young people.

From our "vast reservoir of experience" evaluators drew a wealth of lessons. I shall mention one on the theme of community development and one on the theme of institutionalization and sustainability. It is not just the products of community-institutional partnerships that improve health. In this case the product is greater workforce diversity. A more diverse workforce is a better workforce because the diversity itself is a factor contributing to the ability of the health system to improve the health of people. It is also the processes of community-institutional partnerships that improve health. They build social capital and develop the community. A key finding of the evaluators' pipeline substudy is that "[p]rojects intervening with children in public schools generate a broad range of benefits to children and families that can be viewed as contributions to economic development and public health, whether or not the children ultimately make decisions to pursue health careers."

A more diverse workforce is a better workforce because the diversity itself is a factor contributing to the ability of the health system to improve the health of people.

One of the evaluators' observations about institutionalization and sustainability arose through comparing two projects that were subjects of field studies. Brown University in Providence and Lake Superior State University in the Upper Peninsula of Michigan had different experiences with sustaining pipeline projects. Evaluators found that, while the Brown program, Access to Medical Careers, was thought good, the city schools had not institutionalized it and its survival was unlikely. The Opening Doors program of Lake Superior State University was designed to bridge to its area's Native American community and was directed primarily toward Native American young people. It was *not* institutionalized as a separate program *but it did survive* through incorporation of key elements into a tribal youth program.

Diffusion. Progress required that we demonstrate the strategic elements of the system, their lines of development, and the process of system building. Some demonstrations might concentrate on one or more elements. Some might focus on building systems, modeling not only elements of the system and lines of development but also how lines of development converge in integrated, comprehensive approaches. Our eye was always on the system to which the lines of development and elements would contribute. The ultimate responsibility was to model what was necessary and sufficient for the systems to come into being because it was the systems, not one component or another, that would make improving the health of people possible. It should be noted that systemic change was important within the different sectors involved and not just at their point of convergence. Systemic change within sectors was important to achieve change throughout the system at what, as I noted earlier, we sometimes called the "center," where services and education, communities and institutions came together.

Working in this way made diffusion doubly important for us. We needed diffusion at the level of components and we needed diffusion at the level of the integrated, comprehensive system that brought the components together. Optimal impact from a successful, grant-funded innovation often will depend on whether change lasts, whether change spreads, and whether change is built on to achieve a higher order of change. To attain the health goal, we needed this impact from the models and strategies our grantees were testing in the field. But we also needed diffusion of the system as a whole. What made this difficult was the fact that any given project—whether field-generated or belonging to an initiative—rarely encompassed all the necessary and sufficient elements and their convergence. People interested in learning from our work would have difficulty in finding and seeing in the field the "whole" that we said we envisioned. We needed the "whole" to be synthesized in order to communicate to policymakers, practitioners, and community members and leaders. More remains to be done in disseminating this synthesized "whole."

People interested in learning from our work would have difficulty in finding and seeing in the field the "whole" that we said we envisioned. We needed the "whole" to be synthesized in order to communicate to policymakers, practitioners, and community members and leaders. More remains to be done in disseminating this synthesized "whole."

However, important efforts were made to disseminate models of elements of system building developed by grantees, and there were notable successes. This subsection discusses some of the diffusion efforts related to transforming nursing.

International diffusion. The contribution of California State University, Dominguez Hills, in Carson to the development of nursing education in Latin America is discussed in an earlier chapter. Here I look at the university's contribution from another perspective: the knowledge base and the skills that the university used to support the Latin American grantees' work.

In an earlier era the Kellogg Foundation had supported the New York State Board of Regents in developing an external degree program in nursing through distance education. This and other learning was carried forward through two grants from 1986 to 1992 to the Center for International Nursing Education to disseminate internationally the lessons learned from Foundation-assisted distance-learning and nursing education projects. (These grants were made when the center was housed in the nursing school at another California State University campus.) In the course of this work the center further developed its skills in the international context by working with a group of nurses seeking to enhance nursing education and practice in their respective countries. When the day came, then, that the Kellogg grantees in Latin America were developing their programs, they already had a solid knowledge base and strong experience in disseminating the knowledge.

Process was just as important as content. The nurses did not go in as "experts" intent on imposing what they "knew" on others. They had an abundance of technical knowledge, but they were flexible and willing to adapt. They were there to listen, and to facilitate. They spoke Spanish. They assured that materials were not only translated into the languages of Latin America but that the content was adapted to the points of view of the people who would be using the materials. They paid attention to and respected where the people they were working with were. In the process we can see the values and principles of both nursing and Kellogg grant-making put into practice.

The Kellogg Foundation in the 1950s supported the piloting of the Associate Degree in Nursing (ADN) in seven community colleges in four states.

National diffusion. An earlier chapter also explored how the Kellogg Foundation in the 1950 supported the piloting of the Associate Degree in Nursing (ADN) in seven community colleges in four states. From the original seven, the number of ADN programs grew to 868 in 1994. By the mid-1990s more than half the number of nursing graduates each year were prepared in ADN programs. The programs continue to provide access to education for traditional and nontraditional students alike. Diffusion and replication were thus so strong that they changed the field. In the mid-1980s to mid-1990s the Foundation made a series of grants that supported a significant change within associate degree nursing education itself. In this case also, diffusion occurred. Community colleges developed and then widely disseminated a model of geriatric-care training for nurses. The grants led to the establishment of the Community College-Nursing Home Partnership program in 200 institutions. The project that established the partnership program was initiated in recognition of the growth in the size of the elderly population. The training enhances the preparation of ADN graduates for work with this population. In the context of practice in nursing homes, the training also prepares graduates to manage and delegate tasks to unlicensed personnel.

Elsewhere in this section I have discussed three projects that deserve to be mentioned again from the perspective of diffusion. One is the Kellogg Nurse Practitioner Fellows project of the Southern Regional Education Board. This project was not just about preparing an additional two dozen nurses to qualify as faculty for advanced-practice training. It was also about diffusing enhanced knowledge and skills into the nursing education institutions in an entire region that comprises close to one-third of the states. We had been very pleased to receive a proposal from a regional board with an affiliated body for nursing education. Networking is a hallmark of Kellogg fellowship training, and one of the particular advantages offered by this project, from the perspective of diffusion, was the periodic convening of the deans of nursing from the fellows' institutions. In other words, the reach of this project went beyond training a certain number of individuals to informing institutional cultures through the development of both deans and faculty associated with 17 institutions across the region and numerous settings of education and practice.

Another is school-based health care project of the College of Nursing of the University of Illinois at Chicago. The Foundation makes a practice of occasionally funding what are explicitly called dissemination projects to increase the impact of grantees' work. The school-based health care project we had funded was included with projects managed by other entities in the Chicago area under a grant we made to the college of nursing for dissemination of application of its primary health care model. The main product of this dissemination project, a video, featured the role in prevention of the Community Health Advocates who worked in the school-based health care project. Other components of the project were a workshop on primary care and a book. Evaluators observed that this dissemination project may have missed an opportunity to educate policymakers, health professionals, community advocates, and academics in public health about primary care policy to the fullest extent possible.

The third is the Parish Nursing project of Advocate Health Care of which the Foundation was an early and continued supporter. The effectiveness of diffusion of the Parish Nurse model will be evident to most readers because the model has been adopted in many places throughout the country. This effectiveness can certainly be attributed to very deliberate dissemination through a resource center that was established in 1985 and used a variety of developmental and diffusion methods to do its work, including print materials and conferences. Ultimately, the International Parish Nurse Resource Center moved to Deaconess Parish Nurse Ministries in St. Louis in 2002. While no longer playing the role of national and international disseminator, Advocate Health Care can be said to have retained a dissemination role within its system. The evaluation team reported, for example, that the documentation system component of Parish Nursing had grown into a pilot project to develop documentation procedures for all Advocate Health Care's service delivery units outside the hospital campus. In addition, in June 1998, while the resource center was still at Advocate Health Care, its staff had, according to the goal evaluators, demonstrated the documentation system at the Cleveland Clinic. The manual with specifications for how to use the system that was produced as part of the documentation project helped make dissemination possible.

Section III.

Evaluating Progress toward the Health Goal

With the board's approval of a goal for comprehensive systems change supported by a plan with nine strategies, we had to be accountable for reporting objectively on the results of putting the vision and theory of change into practice. Achieving the health goal was, in some sense, a single, unified task. But it was also a complex one to be carried out through multiple investments over a period of time. With the dual objectives of accountability and effective management, we asked ourselves some hard questions: (1) How would we know if we were on track and actually doing what we had set out to? (2) How would we know we were making progress? (3) How would we know whether individual grants were each making strategic contributions to achieving the common goal for all grants? (4) How would we get the bird's eye view over all the grants?

The Foundation made a practice of supporting so-called "cluster" evaluations; but these generally applied only to initiatives and told us only about certain groups of projects. There was no established evaluation method to look at field-generated projects from the perspective of the Foundation. There was a solid tradition, instead, of including funds in grants for the grantees to conduct their own evaluations to help them better manage their work. The Foundation did not expect to receive the project-level evaluations. Thus, we had no established evaluation procedure that would enable us to look at the whole.

The Foundation made a practice of supporting so-called "cluster" evaluations; but these generally applied only to initiatives and told us only about certain groups of projects. There was no established evaluation method to look at field-generated projects from the perspective of the Foundation.

We decided to try a new approach—goal-level evaluation. We sharpened our thinking about defining attainment of the health goal in very concrete terms. We searched for suitable evaluators for this new kind of evaluation. We chose The Lewin Group, a nationally known firm. The evaluators began work in March 1996 and used the first six months to prepare a detailed evaluation design. They developed indicators specific to what we were trying to achieve and reflective of the Kellogg Foundation's values and intentions. They learned about our programming in depth from project files, health team meetings, interviews with each program officer, and meetings with the cluster evaluators of our initiatives. We reviewed and accepted the evaluation design and renegotiated the contract for a full-scale evaluation to conclude in 1999.

Once the design was approved the evaluators began field work and, in 1997, began submitting substudies based on site visits around the United States. In all, there were 23 substudies. They covered (1) current or recently ended projects that program staff considered to be exemplary contributors to attaining the health goal, (2) completed projects, which were examined to assess sustainability, institutionalization, policy impact, and replication potential, (3) projects located in the same rural or urban geographic area, although not necessarily related to each other, (4) concepts or phenomena across multiple projects that might, when studied, give a broader understanding than a study of any single project would, and (5) lessons emerging from both direct and indirect efforts to inform policy within both governments and institutions. The evaluators submitted an interim synthesis report early in 1998 and a final synthesis report at the end of 1999. The synthesis reports address progress observed in the field across a wide variety of grant projects—some within initiatives, incorporate findings from initiative evaluations through meta-analysis, and make recommendations on how we may strengthen our grant-making.

Findings. We did not wait until 1999 to begin learning and using the evaluators' findings. Once the field work began, substudies began flowing continuously to us. The consultants made oral presentations on the written reports at our monthly health team meetings. The Lewin Group described our use of evaluation findings in real time in the following way:

> The Health Goal Group reviewed and discussed each of the 23 evaluation substudies and the 1998 Interim Synthesis Report, as well as a discussion draft of this final report. Many of the lessons, and some of the typologies and language generated through the evaluation, have already found their way into the flow of the Health Goal Group's decisions. The Health Goal Group has acted on evaluation findings in the design of new initiatives and in discussions about how to support grantees. The ongoing flow of findings from this Goal Evaluation contributed to the design of the recent Turning Point and Community Voices initiatives, which are the culmination of many years of learning and refining grant-making and management strategies. The Health Goal Group has also used the findings of the evaluation as it continued to develop strategy for future grant making.

The final synthesis report of the health goal evaluation presented findings on 11 subjects. Findings on seven are summarized below:

Access to Care. The Foundation clearly had impact, the evaluators found. Grantees increased access through:

- New or expanded services—most often "seen in medically underserved, isolated communities challenged with extreme poverty and very poor health status"
- Improved acceptability of services to consumers and communities—projects typically strove for culturally acceptable service environments and many contributed to improving fragile economic environments, by hiring local residents
- Removal of logistical barriers related to transportation, child care, and service hours—when the grantee had not foreseen barriers "[t]he Foundation's flexibility in redeploying resources to meet such needs was pivotal"
- Outreach to consumers about services—central to many programs, often stimulating use of new or expanded services

In keeping with their duty to the Health Goal Group, the evaluators rounded out this good news with findings about service improvements that did not last, opportunities that were missed and, overall, ways to translate the lessons about what did and did not work into more productive grant-making with greater impact on access to care:

- A pivotal factor in sustaining improvements was development of organizational and financial infrastructure to support them. Options for better grant-making included pre-grant "sustainability assessments" and, during the grant period, technical assistance and encouragement for the grantee to build infrastructure.
- "Foundation grantees who excelled at providing service environments that attracted and retained consumers" had created "wealth" in the form of strategies and concrete examples that could form a "standard of care" for acceptability of services. The option for better grant-making was to document this standard "to enhance policy debates about consumer rights and quality of care."

- Outreach had impact, the evidence showed. But grantees rarely document how they do outreach. An option for better grant-making was documentation, which is quite valuable to programmatic and policy efforts.
- "The greatest untapped potential" lies in addressing financial barriers to care. The evaluators saw little work on this except in the Comprehensive Community Health Models of Michigan Initiative. The new Community Voices Initiative was taking up the challenge. The option for better grant-making was to deploy technical assistance to build capacity of other community-based grantees to document costs and benefits, since they are typically not expert in operational finance and financing policy. This would boost their ability "to translate a successful program benefit into policymakers' decisions about funding."

Health professions training. Schools of nursing stood out for their success "in shifting curriculum and faculty incentives toward primary care and community-based practice." Further, "[i]n two-year nursing programs, graduates made a detectable impact on nursing shortages in targeted communities." By contrast, medical and public health schools made more limited changes with "a few notable exceptions" attributed to commitment of administrative leadership.

The role of universities in establishing new programs in underserved communities carries an inherent challenge...the return on the Foundation's investment is low, the relative cost of impact high, when programs are not fully institutionalized at the end of the grant period.

Community colleges and tribal colleges stood out as the "settings with the best chance to influence the entry of minority students who were not already college bound." "[A]ccess to mentors, culturally specific adaptation and counseling, and a variety of approaches to developing self esteem and confidence" made an important difference.

The role of universities in establishing new programs in underserved communities carries an inherent challenge. Universities can be both skilled developers and the only willing developers. But "their mission and accountability are not primarily directed at meeting community needs, and they often did not ensure that programs were sustained and turned over to those communities." The return on the Foundation's investment is low, the relative cost of impact high, when programs are not fully institutionalized at the end of the grant period.

Enhanced community participation. Finding that, as a general proposition, community-based organizations used Foundation funds more productively for the benefit of community residents than did mainstream institutions, the goal evaluators focused on building capacity in community-based organizations. Grants not only supported existing community-based organizations, but also starting a number of them and transforming partnerships and informal organizations into community-based organizations having the resources to deliver services.

Community-based organizations had common capacity-building needs and health program directors had some common grants management practices for helping with those needs—permitting funds to be used for technical assistance and referring grantees to consultants or other organizations doing similar work. The evaluators found that a more formal organized approach to information and referral for technical assistance would be warranted. One option would be to support a management services organization that could provide technical assistance regionally, perhaps through a network of consultants. Or a grants-management policy regarding situations in which it would, grantees would be required to use technical assistance.

Effective leadership. Recognizing that the Kellogg Foundation invests heavily in leadership development, the evaluators developed a useful "leadership typology" for us. It showed that leaders are distributed in communities along two continua: positional to non-positional leadership and visible to invisible. While leaders who are both positional and visible may be the ones who more often approach the Foundation or otherwise come to our attention, the success of funded community-based projects may depend more on leaders who are both non-positional and invisible. The evaluators made the following observations about this:

> In many programs, success was linked to the non-traditional leadership roles taken on by community residents. Many individuals who functioned as leaders in their communities held no titled position associated with that role. Many of these leadership roles that prove critical to program success may be "invisible" to people entering a community

While leaders who are both positional and visible may be the ones who more often approach the Foundation or otherwise come to our attention, the success of funded community-based projects may depend more on leaders who are both non-positional and invisible.

> from outside. In assessing the readiness of communities to achieve goals that require community participation, or substantial changes in behavior, it will be important to encourage applicants to describe, value and engage these types of leaders. The Foundation's capacity to convene people is particularly valued by individuals who take on such leadership roles in their communities.

The evaluators also especially noted that their work had "surfaced a wide range of health grants that contributed to community leadership capacity, particularly grants that hired local residents in outreach and program coordinating roles, invested in community participation, and harnessed the networks of local people to support programming."

Institutionalization. The issue of sustainability was very important to us. Grant-makers take chances and expect some failures as part of the search for solutions. But we were interested in combining venturesome grant-making with good rates of success.

The evaluators gave us a valuable distinction between institutionalizing grant-funded projects and ensuring their financial sustainability. These are not identical and both must be given attention.

The Lewin Group defined institutionalization and reported findings that would probably resonate with other grant-makers:

> Institutionalization is the "adoption" of a grant-sponsored function as part of an entity's regular work. We found that may projects were sustained for a period of time after their grant ended with a patchwork of funds, or through aggressive fund-raising, but only a few were truly institutionalized by the organizations that hosted the grant. Some projects were disaggregated, and their functions and capacities were adopted by other participating organizations. Other projects, including some effective ones, disappeared shortly after funding ended, when no organization or group with the capacity to sustain the project work took responsibility for their continuation.

Things we could do to address barriers to institutionalization included: (1) making sure leaders were present with skills not only to innovate solutions but also to implement and institutionalize them; often these skills are not necessarily present in the same person; (2) looking for alignment between projects and the funded institution where, unfortunately, projects sometimes were isolated and invisible; (3) securing formal statements of ongoing commitment from institutions at the outset; and (4) incorporating details in grant agreements that provide the basis for monitoring changes in leadership, shifting organizational conditions, and adhering to commitments.

Financial sustainability. The evaluators observed that early planning for sustainability was being incorporated into our work even as they were conducting the goal evaluation. Thus, we were learning from our own observations and reflections and theirs that "planning for financial sustainability typically did not begin early enough to ensure a smooth transition when grant funds ended." They informed our work with other lessons: grantees found the grant periods too short to allow them to think ahead to sustainability for systems change and other complex endeavors requiring intensive effort; in poor and isolated communities especially there was a lack of expertise and technical assistance for planning financial sustainability.

The Lewin Group also emphasized the important point for us that financial sustainability could hinge on policy change. Their statement reinforced our understanding that evaluation, communications, and policy education should work together to increase impact and support good return on investment.

The Lewin Group also emphasized the important point for us that financial sustainability could hinge on policy change. Their statement that "[t]he 'invisibility' of a great deal of grant-funded work to administrators and policy makers contributes to chronic underfunding" reinforced our understanding that evaluation, communications, and policy education should work together to increase impact and support good return on investment.

Policy relevance and impact. The Lewin Group consultants attended our monthly team meetings during the design phase and then implementation of the goal evaluation. Their analysis of policy relevance and impact was, therefore, informed by watching us strive to incorporate policy targeting into our work and embed policy considerations into our grant-making and grants management, especially in initiatives. At the same, their conversations with us and reports to us informed our thinking.

Their findings recognize our focus on policy work. In suggesting that we might want to develop a pool of "policy coaches," for example, they noted that "[t]his type of coaching is being piloted through the use of a 'Resource Team' of consulting organizations within *Community Voices*." In their field work, which included looking at past projects, they studied both grants with explicit policy goals and grants without them. With respect to the latter grants, the evaluation team "discovered that program staff frequently knew about policies that immediately impacted their work, but were rarely engaged in gaining greater knowledge or in interacting with policy makers." For this tendency to change, the evaluators thought, local grantees might need policy coaching, that is, "technical assistance to identify policy opportunities in different venues, . . . to prioritize opportunities and to link the policy resources of their networks. . . ."

Schools of Nursing as Grantees—A Substudy in the Evaluation

In the health goal evaluation, the magnitude of Kellogg investments in schools of nursing warranted a separate substudy. To prepare this substudy, The Lewin Group drew on eight of the substudies in its goal evaluation that were based on its field work and involved schools of nursing. The firm also drew on selected grant files for the Graduate Medical and Nursing Education Initiative and for newer work the Foundation was undertaking in educational reform and health workforce development.

The substudy had three purposes: "(1) to characterize the factors in the environment in which schools of nursing operate that may influence their effectiveness as grantees; (2) to identify areas in which schools of nursing were most effective and lessons learned from grants to schools of nursing; and (3) to identify implications of findings and lessons learned for future grant-making to nursing organizations."

The evaluators reviewed trends in supply of nurses and the demand for nurses' services in the market (including trends in the locus of practice) and the implications of market forces for schools of nursing. The implications were to (1) adapt training to meet the demand for more highly skilled nurses with bachelor's and master's degrees, (2) revise curriculum in non-hospital based nursing care and establish nursing centers for student training and faculty practice and research, and (3) secure more funding.

From reviewing literature, the evaluators reported that internal institutional factors influence how schools of nursing can respond to market forces. In general, "universities value faculty research and teaching accomplishments over practice, making it difficult for schools of nursing to garner support for increased faculty practice in community-based settings." They concluded that efforts to revise curriculum to respond to market forces would face challenges. Modifying curriculum to shift the locus of practice to outpatient settings in communities or to enhance practice skills related to such settings—finance and reimbursement, data, administration, leadership, public policy, and systems of care—faces institutional barriers. The Kellogg Foundation was helping schools of nursing meet the challenges.

Modifying curriculum to shift the locus of practice to outpatient settings in communities or to enhance practice skills related to such settings—finance and reimbursement, data, administration, leadership, public policy, and systems of care—faces institutional barriers. The Kellogg Foundation was helping schools of nursing meet the challenges.

Our grants produced accomplishments and lessons in overcoming institutional barriers and positioning the nursing profession in the changing market environment. Foundation grants helped some schools or departments of nursing increase their visibility, credibility, and influence among one or more audiences, such as their own and other institutions, other nurse-training programs, boards of nursing, and local communities. Enhanced stature enabled several schools to secure institutional backing for activities the Foundation had initiated and to influence institutional policy change. "Institutional policy changes include the creation of new departments or new faculty positions, modifications to student education and training curriculum, and revisions to faculty tenure requirements."

Nursing schools face another kind of barrier when they try to enter communities in order to organize training for students or establish new health care initiatives for residents. Communities are often suspicious of institutions. The experience of some has been that institutions tried to impose their expertise on the community and then abandoned it once the institutional agenda was met. Our grantees learned how to enter communities. Schools learned three salient lessons: (1) the importance of consulting with communities on their needs, (2) the value of flexibility when the initial project intervention does not work, and (3) the necessity of building relationships with communities based on trust. The evaluators identified the essential ingredients for success community-nursing school partnership as follows:

> Schools of nursing were faced with cultural, ethnic, economic, and geographic barriers to acceptance of their participation in community health work. Working with communities requires a strong project leader who is committed to working with communities, a project staff that values community voice and is culturally sensitive, and an accepting community.

The evaluators noted that various forms of support we provided for the education and practice of advanced-practice nurses backed them as change agents. They cited, for example, training in policy and management as well as technical assistance for billing Medicaid for covered services delivered at school-based health center of a school of nursing.

Involving community residents in delivering services was a very important area in which project leaders in nursing schools were change agents for creating the kind of health systems our programming group envisioned. Formally or informally training and then employing community residents as paraprofessionals supported a pipeline into health care careers, the evaluators observed. There were lasting benefits for some workers from these nurse leaders' initiative: the paraprofessionals often moved into "increasingly responsible health related employment." But the overall message for nursing schools and their funders is less positive. Providing us the advantage of a post-grant look at projects, The Lewin

Group reported that "for the most part, schools of nursing did not sustain training efforts. . . . Individual project investigators . . . were not able to garner department or institutional support for sustainability."

The Lewin Group concluded its report by offering options for strengthening schools of nursing as grantees in the future, suggesting that we might want to:

- Design technical assistance strategies to help nursing school faculty frame community based practice in ways that contribute to promotion and tenure.
- Consider ways to enhance nursing curriculum in the areas of management and policy outside of nursing schools.
- Introduce criteria for schools of nursing and other training institutions to follow that demonstrate their community linkages and their sustainability plans.
- Pursue potential opportunities for schools of nursing to partner with each other and with state nursing associations to influence the legislative agenda.

Section IV.

Conclusions: Alignments and Misalignments, and Their Implications

Serving as coordinator and, later, vice president for health programs at the Kellogg Foundation gave me a chance to be part of a multidisciplinary team that had arrived at a philosophy and vision that were so extraordinarily in alignment with the philosophy and vision of the discipline of nursing. Nursing had developed as a profession, often with Kellogg's help. It had not lost the high vision of helping people help themselves that lived at its origins, just as that vision lived in the Foundation's origins. But nursing had become entangled in contradictions and contrary directions. The result was that the profession had demonstrated

Nursing had developed as a profession, often with Kellogg help. It had not lost the high vision of helping people help themselves that lived at its origins, just as that vision lived in the Foundation's origins.

the ability to articulate statements of lofty principle and vision but not the ability to put them into practice. The principles and vision put the patient first and recognized the patient as a member of a living whole of family, community, society, and social conditions to which caregivers could not be blind if they were to fulfill their noble duty of service. Nursing had competed as it thought it should in a competitive environment where, among other things, women were disadvantaged. It had not lost its heart in the process, but had, in the end, positioned itself where its reach persistently exceeded its grasp. In a way, it was victim to its own success. As I learned more and more about, and participated with the health team, I became increasingly struck by the team's alignment with the vision and principles in nursing. And I came to see the opportunity for aligning the practice of nursing—now too often adrift from its vision—with pursuit of the health goal in a way in which nursing would strengthen pursuit of the goal and pursuit of the goal would strengthen nursing.

Implications of Foundation Experience for Grant-Making to Schools of Nursing

Our learning from Community Partnerships in Health Professions Education, Graduate Medical and Nursing Education, and the goal evaluation showed us both the great promise in nursing schools and the barriers to realizing that promise. The learning demonstrates that the capacity for systems change lives in schools of nursing, and can be fertilized by informed grant-making. It also shows that the barriers are substantial and will not topple easily.

Our work showed what nursing can do, showed the creativity, flexibility, mobility, and force in thinking and action that medicine does not rival and, perhaps, ought to emulate.

Schools of nursing were enabled, especially by the initiatives. We saw that:

(1) Often investments in nursing are sufficient to raise standing and visibility in an institution.

(2) Nursing has demonstrated that it is more likely than other health professions to make changes that will be sustained.

(3) Nursing has the capacity to provide leadership in a community setting; nurses are more willing to enter communities.

(4) Schools of nursing are more willing to work to overcome institutional barriers.

Were the demonstration of educational reform to have another phase of funding, the centerpiece and grantees should be schools of nursing!

Musings on the Implications of How We Did Our Work

Risks and rewards for project leaders. The willingness of nurses to take risks is a resource of great value for producing change. Schools of nursing have demonstrated a significant level of willingness to respond to the visions of risk-takers, to back them in some cases and, even, to reward them from time to time. Take the case of the faculty member from the Northeastern University College of Nursing who headed the multi-institutional Community Partnerships in Health Professions Education consortium in Boston. Northeastern promoted her from the College of Nursing to a senior executive position in the university.

The willingness of nurses to take risks is a resource of great value for producing change. Schools of nursing have demonstrated a significant level of willingness to respond to the visions of risk-takers, to back them in some cases and, even, to reward them from time to time.

The career path of a nurse leader in another Foundation-funded initiative did not take so favorable a course. One of the three projects in the Comprehensive Community Health Models of Michigan Initiative was headed by a nurse. This nurse left the position of chief of nursing at a local hospital in order to lead the project that the Foundation funded through the area's community foundation. This meant she severed her connection with her employer, although the hospital was a participant in the project's convening of consumers, purchasers, and providers and stood to gain or lose by the project's work. A number of results and relationships produced by the project were sustained; a defined project entity was not. The nurse leader who had invested herself professionally and personally in helping her community envision and implement a comprehensive and improved health system was out of a job. No institution, seeing the value of her expertise, experience, and commitment, stepped up to support her.

The career paths for these individuals after they assumed leadership of their respective projects were quite different. The difference brings an issue for organized philanthropy into focus. It happens that both project leaders were nurses, but the issue presented for philanthropy is not specific to nurses. The issue is also not specific to project leaders who sacrifice their institutional connection in order to direct a project: on the one hand, an institution may push out a project director who is not in alignment with its priorities (by not granting tenure, for example); on the other, an institution participating in a broad-based project may, at the conclusion of a project, pick up a project leader with whom it is not directly connected. Philanthropies need capable individuals who are willing to take career and other risks to provide leadership for the testing of innovations. Philanthropies are willing to risk their money on testing innovations. The story of these project directors reminds us that the leaders on whom philanthropies depend to make good with their money may be putting their own careers at risk. Nurses may have particular vulnerabilities because their primacy in assuring that the nation and the world have affordable, high quality health care for all is simply not recognized by the society in general, which still sees nurses as subordinates of doctors—needed helpers but not primary leaders. Sometimes philanthropies stay connected with project personnel after grant periods are over. Funders may, for example, draw on their experience and expertise as resources for consultation in the design and implementation of other projects managed by other grantees. But, obviously, this is not always the case, and the issue of the post-grant career paths of project leaders may deserve more attention.

Are initiatives worth it? Early in this chapter I described how the Foundation transformed how it did its work that was in its early stage when I arrived. Making grants through initiatives and doing work in teams were the main innovations in this transformation. The health team was the first to take on the challenges of initiative grant-making. Investment in developmental work began in 1989 and, from 1991 through 2001, the board of trustees appropriated $179,888,724 for six initiatives in systems change—two in health-professions education with emphasis on medicine and nursing, two in public health, two in comprehensive models of delivering services, financing coverage, and informing policy change. With that much experience and expenditure to go on, the past and present members of the health team should have informed opinions about whether initiatives are worth it.

When initiative grant-making began, the main argument against it was that creative, valuable ideas generated from the field would be squeezed out from funding by the diversion of resources to testing our own ideas of what ought to be. Was it not antithetical to the mission of helping people help themselves for us to think that we knew best? Wisdom from the field has pushed its way through, however, many, many times. This is evident from the advances made in preparing Native American nurses by means of field-generated projects when we had not discerned readiness in the field well enough to think to design an initiative. We might have enabled more development and diffusion if we had done so. Among the many other projects in which this wisdom is evident is the one in which Prairie View A & M University fulfilled a goal recommended 20 years earlier by establishing a program and community-based sites for training nurses in advanced practice and increasing the woefully disproportionate supply of African American and Hispanic advanced-practice nurses. This university did not let the passage of time destroy its dream and was ready to capitalize on the regional Kellogg Nurse Practitioner Fellows program and informal consultation from Foundation staff.

When initiative grant-making began, the main argument against it was that creative, valuable ideas generated from the field would be squeezed out from funding by the diversion of resources to testing our own ideas of what ought to be. Was it not antithetical to the mission of helping people help themselves for us to think that we knew best? Wisdom from the field has pushed its way through, however, many, many times.

That the viability of opportunistic grant-making was repeatedly demonstrated during the decade of health initiatives begs the question: Would the return on investment have been higher if some or all of the millions of dollars devoted to those initiatives had been spent on grantee-initiated projects? My sense is that we did miss opportunities that would have helped us attain the health goal, i.e., demonstrate the feasibility of health systems that meet the needs of people, what they look like, and the logic and means to employ in creating them. Missed opportunities are not only "project grants not made." They are also lost chances to inform policy, to capture and communicate learning, and to disseminate best practices. Grants management for initiatives and grants management for projects are both very difficult to do well and very demanding on staff. Initiatives have the virtue of concentrating the management effort in policy, evaluation, and communications on a defined scope of accumulating work. Achieving the same intensity and continuity of management concentration on these impact services would be considerably harder to achieve for a portfolio of field-generated projects and simply might not be as productive. By commissioning the goal evaluation, we did help ourselves gain more from the blend of initiatives and projects than we might otherwise have. The evaluators recognized the findings and analyses of the various initiative evaluators without duplicating their work; for the field-generated projects in the health portfolio, they did independent field studies. Thus we were able to deploy the whole portfolio to inform our thinking and decisions as a team.

Jump-Starting Rapid Transformation in Nursing and Health Care

During the year preceding my retirement the health team was given an exciting opportunity to advance the development of nursing. The Kellogg Foundation had begun to plan its 75th anniversary celebration in 2005. Each programming team was to develop and submit a plan for use of the extra resources it would receive to fund "legacy" programming of special significance. The programming results from each team's 75th anniversary work would be featured in a celebration in 2005 that would encompass events and publications and shine the light on the capacity of people to help themselves.

The Kellogg Foundation had begun to plan its 75th anniversary celebration in 2005. Each programming team was to develop and submit a plan for use of the extra resources it would receive to fund "legacy" programming of special significance.

The vision of the health team was of a proposal that would intentionally seek cumulative impact from the Foundation's more than 70 years of investments in the development of nursing. Cumulative impact would be sought by lifting up the nursing profession to the central place it merited in overcoming the barriers to access and quality in health care for all. We hoped to achieve results that would be truly a cause for celebration by many. The added resources from the allocation for the celebration were a help, but what was really on our side was timing. We saw the possibility for health systems change to be leveraged through the response to the spreading alarm about the nurse shortage. Hospitals, nurses, and entities that monitor health care quality were among the stakeholders in solving the shortage. Their respective organizations would undoubtedly produce recommendations for reversing the trend in the supply of nurses. Action on the recommendations would probably occur because the gravity of the situation demanded that something be done. The question was whether the "something" would be something truly new. The "same old, same old" approach of working in silos would be more likely. While the awareness in some quarters that there was something different about this shortage—that it was not just a stage in a familiar cycle of supply—could be a force for genuine innovation, such innovation would likely be only at the margin.

As a complex societal problem, the nurse shortage required multidisciplinary solutions and, as previously mentioned, the health team had lived through the process of creating the capacity to invent multidisciplinary solutions—lived through it in our team building and in our grant-making. We had experience with leadership of multidisciplinary approaches and we knew that it is difficult for members of any given discipline to step outside the confines of their discipline to act as—and be perceived as—neutral and trusted conveners. We had some hard-won skill in bringing together people who were not used to being at the same table to actually find new ways to do business. We thought that what we had learned could be applied in a way that would truly be strategic and that would honor and enhance the Foundation's legacy. The press of external events, as the shortage worsened and organizations responded, made prompt action absolutely necessary; the internal expectation for producing results by 2005 created the opportunity in 2001 for seizing the iron while it was hot.

The strategy we proposed to jump-start transformation in nursing and health care would take advantage of crisis to release nursing's enormous potential to improve health care for all. The time for nursing is now, we said. The point is the same for nursing in the hospital and nursing in the neighborhood. In the 1930s, with far-reaching impact on communities and institutions in many places, W. K. Kellogg and the new Foundation had launched the 20-year experiment—the Michigan Community Health Project—that became a learning laboratory that *rapidly* influenced and changed community and institutional practice in education and public health around the country.

In the 1930s, with far-reaching impact on communities and institutions in many places, W. K. Kellogg and the new Foundation had launched the 20-year experiment—the Michigan Community Health Project—that became a learning laboratory that rapidly influenced and changed community and institutional practice in education and public health around the country.

Health care was under enormous pressures that could lead to breakthrough or breakdown. Just as the Big Three automakers chose rapid transformation of production and design of vehicles to survive in the 1980s, when their future was imperiled, health care might now choose, we thought, rapid transformation of how, where, and by whom services are delivered. Action to transform health care delivery must involve, we argued, interest groups with entrenched positions to uphold and territories to protect.

The looming nurse shortage might appear to be only an issue of supply, but it was equally an issue of the utilization of nurses. One source of the angst that was driving practicing nurses out of the profession and blocking prospective nurses from entering was the nurse's place in the system, how the nurse was being used (or abused). Awareness was growing that how nurses were utilized in the system made an enormous difference in recruitment and retention. Nurses were a resource that, if utilized more effectively, could help health care overcome the challenges and perils in which all had a stake of one sort or another.

Our task was to show that the rapid transformation of health care could be jump-started through nursing. Our proposition was that a blueprint for action—with stakeholder buy-in—to reposition nursing in health care delivery and leadership would be a powerful force for such transformation. By 2005 stakeholders would have created the "Blueprint for Action for the Future of Nursing" and actually begun the work of transformation. Concrete demonstrations in communities and institutions would be showing the possibilities for positive change. In demonstration communities, nurse-managed primary-care centers would be delivering cost-effective services where people live. In demonstration institutions, better ways to deliver services to assure quality of care and safety of patients would be tested, documented, and instituted.

We projected that the strategy would have two visible, concrete results:

- *Blueprint for the Future of Nursing*. By its very design, the broad-based process for developing the blueprint would help create the momentum and thrust for action to implement it.

- *Demonstrations in the Field.* Small projects would demonstrate new ways to utilize nurses in leadership, management, and caregiving in order to optimize the benefits of nursing in quality, cost, and access. Results would be documented and would inform development and implementation of the blueprint.

The *Blueprint for the Future of Nursing* would be designed for action. Both those who would take the actions and those who would be affected by them (such as patients) would participate in designing the blueprint. The process for developing the blueprint would bring together the stakeholders, make sure all voices were heard, capitalize on our extensive learning about how to foster collaboration among disparate groups, infuse the best possible information drawn from many quarters, and create a vision of a better future. Key components of the process were to include the following:

- Scan of the field. Determine what already exists of relevance to nursing in health care practice, policy, and finance; professional education; the pipeline to health careers; marketplace incentives; and government regulation. Understand how the varied entities interested in the nurse shortage—as well as larger systemic issues of cost, quality, and access—were planning and organizing for action.

- National Advisory Committee. A national advisory committee that represented a wide range of stakeholders would oversee the scan, guide the development of the blueprint, and spearhead action on the blueprint.

- Convening and engagement. Convening and engagement would (a) bring stakeholders with disparate interests together, and (b) give voice to those unlikely to be heard without sensitive mechanisms for their participation.

- Neutral convener. A nationally respected policy center recognized for its objectivity, expertise, and neutrality would staff the committee, conduct the scan, and support the process.

- Inaugurating action. *Every step* in the process would be calculated to build consensus and commitment for action. The scan would be sensitive to what voices in the field are saying. As a blueprint emerged from the process, it would be treated as a draft and offered to build consensus on solutions.

The *demonstrations in the field* would inform the development and implementation of the blueprint. Small projects would be strategically selected to shed light on and develop two future directions for nursing in the 21st century that we believed could already be glimpsed. The projects would address the following:

- Patterns of utilization of nurses in hospitals. What patterns would enable hospitals to retain nurses on staff while also optimizing the contributions of nursing to patient safety and satisfaction, health outcomes, and cost-effectiveness?
- Nurse-managed primary care in community-based sites. This approach brings high quality, responsive, cost-effective care to accessible locations. Could policy and practice barriers be lowered to enable adoption and sustainability of this alternative model of delivering services?
- Leadership development for nurses in education and practice. In addition to illuminating new directions in hospital and community practice, the small projects would test ways to develop nurse leaders for transformation in all the venues in which it must occur.

We developed the details of the proposal. Major components of these details were funding of the next stage of work (described above) of the Michigan Academic Consortium for Nurse-Managed Primary Care, which was ready to go in 2001, and the impact services of evaluation, communications, and policy education, which were essential to producing the desired results from this jump-start effort.

The transformation we envisioned would take nursing to the next stage of development. The emphasis on practice in neighborhoods would bring Kellogg's special contribution to the development of nursing full circle, linking to the visionary work in the Michigan Community Health Project. The emphasis would bring to the fore the importance of nurses in providing the bridge between communities and the "health care system" in its greatly altered state from the 1930s. We pointed out, for example, that a recent survey by the AARP (formerly the American Association of Retired Persons) found that seniors want more access to care where they live.

The emphasis on safety and quality in inpatient care would also move nursing to the next stage of development. This emphasis would seek to leverage change through response to another kind of growing alarm. In *To Err Is Human,* the Institute of Medicine informed the public that 44,000 to 98,000 patients were dying in hospitals each year because medical error. Just at the time they were trying to respond by improving patient safety, hospitals were battling a severe nursing shortage in some geographic areas and some specialties. At the same time that the dangers hospital patients face had come to light, nurses—the caregivers hospital patients depend on most to have their needs met—were leaving their stressful profession and its difficult working conditions. Too few new nurses were replacing them. No matter what innovative solutions to patient-safety problems hospitals otherwise devised, they could not adequately assure safety with a shortage of the professionals who have the most ongoing contact with patients. Transformative changes to improve the conditions of work could go hand in hand with transformative changes to improve quality of care.

The emphasis on practice in neighborhoods would bring Kellogg's special contribution to the development of nursing full circle, linking to the visionary work in the Michigan Community Health Project. The emphasis would bring to the fore the importance of nurses in providing the bridge between communities and the "health care system" in its greatly altered state from the 1930s.

Nursing in hospitals and nursing in neighborhoods were linked to each other and to the urgent issues around the renewed rise in health care costs matched by a renewed rise in the number of uninsured. Transformative action needed to occur in both venues. Regular primary care helps keep people out of hospitals, where care is much more costly and risks of medical error are higher. *Nurse-managed primary care is a linchpin* in shifting the focus of health care delivery from tertiary care to primary care, prevention, and public health. We pointed to the high barriers to making nurse-managed primary care the norm, and we stressed that the shortage of nurses affected the potential of the health care delivery system to transform into a more cost-effective, responsive, accessible, patient-oriented means for meeting the needs of people.

In sum, we foresaw that the *Blueprint for the Future of Nursing* would lead the profession to realize its central contribution to creation of health systems that meet the needs of all. At this point, the practice of nursing would be in alignment with its own philosophy and vision, as well as with the health goal and strategic plan of the Kellogg Foundation.

Notes and References

The Lewin Group and Brandeis University. (December 1997). *The MEDAC (Access to Medical Careers) Program, Brown University, Providence, Rhode Island, Health Goal Evaluation Substudy Report.* Fairfax, Virginia: The Lewin Group. Used with the permission of the W. K. Kellogg Foundation.

The Lewin Group and Brandeis University. (January 1998). *Sault Ste Marie, MI, Health Goal Evaluation Substudy Report.* Fairfax, Virginia: The Lewin Group. Used with the permission of the W.K. Kellogg Foundation.

The Lewin Group. (November 1998). *W. K. Kellogg Foundation Investments in Chicago, Il, Health Goal Evaluation Substudy Report.* Fairfax, Virginia: The Lewin Group. Used with the permission of the W. K. Kellogg Foundation.

The Lewin Group. (November 1998). *W. K. Kellogg Foundation Investments in the Region of South Dakota, Health Goal Evaluation Substudy Report.* Fairfax, Virginia: The Lewin Group. Used with the permission of the W.K. Kellogg Foundation.

The Lewin Group and Brandeis University. (November 1998). *Rural Health Outreach Program (RHOP), Augusta, Georgia, Health Goal Evaluation Substudy Report.* Fairfax, Virginia: The Lewin Group. Used with the permission of the W. K. Kellogg Foundation.

The Lewin Group. (July 1999). *W. K. Kellogg Foundation Investments, Western Alabama, Health Goal Evaluation Substudy Report.* Fairfax,

Virginia: The Lewin Group. Used with the permission of the W. K. Kellogg Foundation.

The Lewin Group. (August 1999). *Schools of Nursing as Grantees, Health Goal Evaluation Substudy Report.* Fairfax, Virginia: The Lewin Group. Used with the permission of the W.K. Kellogg Foundation.

The Lewin Group. (December 1999). *Health Goal Evaluation: Final Synthesis Report.* Fairfax, Virginia: The Lewin Group. Used with the permission of the W. K. Kellogg Foundation.

Matteson, Peggy S. (Ed.). (1995). *Teaching Nursing in the Neighborhoods.* New York: Springer Publishing Co.

National Association on School-Based Health Care. (2001). *National SBHC Finance and Patient Revenue Study.* National Association on School-Based Health Care. Retrieved August 6, 2004, from http://nasbhc.org/app/finance_survey_overview.htm

U.S. Census Bureau. *Fact Sheet.* Washington, D.C.: U.S. Census Bureau. Retrieved August 6, 2004, from http://factfinder.census.gov/servlet/SAFFFacts?_sse=on

W. K. Kellogg Foundation. *Health (The Five Year Strategic Plan—1994-1999): A Programming Update.* (February 1999). Battle Creek, Michigan: W. K. Kellogg Foundation. Used with the permission of the W. K. Kellogg Foundation.

W. K. Kellogg Foundation. (April 1997). *Community Partnerships in Health Professions Education, Lessons Learned: 1990-96; Graduate Medical and Nursing Education Initiative, GMNE: 1996 and Beyond: Program Initiative Progress Reports.* Battle Creek, Michigan: W. K. Kellogg Foundation. Used with the permission of the W.K. Kellogg Foundation.

World Health Organization. (1978). *Primary Health Care: Report of the International Conference on Primary Health Care, Alma-Ata.* Geneva: World Health Organization.

Part III

Regional Nursing: Latin America and the Caribbean and Southern Africa

Helen K. Grace
Gloria R. Smith
Roseni De Sena
Maria Mercedes Villalobos
Mary Hlalele

Regional Nursing: Latin America and the Caribbean and Southern Africa

Helen K. Grace
Gloria R. Smith

The first parts of this book covered the Kellogg Foundation's long history of work with nursing in the United States. Since 1939, the Foundation has supported nursing abroad—first with Latin America and the Caribbean. In the early eighties, the Foundation supported a major initiative in Australia and New Zealand that provided support for 100 fellowships, primarily in nursing, with some in health services administration. As part of its agenda for the 1980s, the Foundation board made a strategic decision in 1982 to discontinue funding for projects in Europe, which had primarily been in the agricultural domain, for Canada, which had been viewed as an extension of the United States; and in Australia and New Zealand. The board also authorized exploration into opportunities for grant-making in countries in sub-Saharan Africa.

In 1985, Helen Grace was asked by Dr. Mario Chavez, who was then the coordinator for Latin American programming, to visit health projects throughout the region and make recommendations for future work with nursing. As programming began in the southern African region in 1987, Grace was asked to be part of the original team to explore the potential for funding in that area. When Gloria Smith joined the Foundation in 1990 and became the coordinator for health programs, she became responsible for developing, monitoring, and guiding nursing and health projects in all of the regions in which the Foundation did its grant-making. This experience, spanning the years 1985 to 2000, provided an unusual opportunity to observe some of the parallels and contrasts in the development of nursing in three very different settings.

Although events and circumstances differed, the United States and the countries in the other two regions were each declaring a large-scale national health crisis. Health care costs were escalating beyond available resources. The inability to control costs translated into reduced access to larger numbers of people and reduction of types of services available to all. In the Latin American and Caribbean countries and in southern Africa, health care was available to all theoretically; but in reality, a lack of health care resources made this an unrealized ideal. "Health Care for All by the Year 2000" was the mantra throughout the world. But how this goal would be attained was not clear. The predominant approach was to mount a massive effort to train health promoters. While many people were trained as "health promoters," it was not clear how they were to be linked into a comprehensive health care system. Dr. Chavez, in asking Helen Grace to work with him in Latin America and the Caribbean, summarized that "Health Care for All by the Year 2000" would not be achieved unless nursing "plays a major role in linking health promoters to comprehensive health care systems."

Policy analysts and policy makers agreed that one way to create more cost-effective care was to increase access to primary health care services and improve use of primary care. The World Health Organization promoted basic health schemes for primary care to achieve health for all in the developing nations. In southern Africa, as in most African countries, the response was to develop five-year plans that would shift resources from expensive quaternary systems of care to evolving systems of primary health care that more easily matched the needs of young, burgeoning, impoverished populations. In Latin America and the Caribbean, the Pan-American Health Organization worked with its constituent groups to plan strategically for promoting basic health schemes. The United States and other developed nations also began to recognize how to effectively utilize primary health care services. The need for nurses, the largest corps of health professionals, to plan a broader role in a health delivery system with an expanded base in primary health care access, seemed apparent. The big question was how ready was the profession to respond to the challenge.

In the United States, nursing had increased its capacity and demonstrated its effectiveness to deliver primary health care. By 1990, the production of nurses who had advanced practice credentials in primary care had increased and the majority of training programs for nurse practitioners were taught at the master's level. Certification programs for nurse practitioners validated qualifications for advanced practice. So there was a corps of nurses in the United States who had the necessary knowledge and skills. But the nursing profession as a whole remained fixated upon practice in acute-care settings. Despite national recognition of the need and the readiness of nurses to respond, a number of structural and cultural barriers had to be overcome, and to do so would require the united effort within nursing.

The nurses in southern Africa were challenged by the expectations that were being raised in the five-year plans issued by the Ministries of Health. They saw not only challenge but opportunity for the profession to assist in a response to aid the public good. Health care in the southern African countries was delivered largely through a public system administered by a ministry of health. Although the official policies of the ministries acknowledged the need for increased and more effective use of nurse manpower, nurses in southern Africa met resistance that stemmed from structural and cultural barriers. Ministry officials who oversaw nursing services and education recognized that the nursing profession, though willing, was not ready to meet the new challenges. They sought assistance to strengthen the capacity of nursing to manage and deliver primary health care.

Nursing in Latin America and the Caribbean was underdeveloped; education for nursing was uneven. Though a small number of nurses were educated in university programs equal in length to medical training, the majority of "nurses" in practice had little or no systematic training. Many had been trained "on the job" or were political appointees.

The Kellogg Foundation, through its strategic grant-making approach, provided the opportunity for nursing to begin to address its problems in education and practice.

In each context, the status of nursing seemed inextricably intertwined with the status of women. Nurses worldwide were mostly women. Nursing practice was controlled by women. Although large numbers of women had entered the field of medicine, the practice of medicine was still viewed as a field dominated by men. Society had placed a high value on medicine as evidenced by the economics and power of the profession. That seemed reasonable given medicine's importance in saving lives and restoring health. Nursing, though also important to saving lives and healing, was not valued as highly by society. What had rankled, and continues to rankle nursing, is not the fact that it has been valued less than medicine, but that its value has been so highly under-rated and its importance so unheralded. Medicine has been able to wrap itself in the mantle of science as a way of explaining its miraculous powers to heal, thereby keeping its mysteries and formulas to itself. Thus, one can only imagine the great wealth of knowledge needed to direct healing. Fear of losing access to medical "healers" has kept public dissent at a minimum. Medicine, the discipline of men, is viewed as science. Nursing, the discipline of women, is viewed as intuitive and an extension of mothering needing no scientific base. The nursing profession in the United States has embarked on a mission to prove its worthiness as a "learned profession" and a "science." The course for doing this does not necessarily parallel the way to prove the profession's value to the public. Nursing's desire to be valued equally with medicine creates huge dilemmas for the profession. That desire makes it difficult for the profession to break away from medicine's lead to assume its own place as a leader in healing and promotion of health, as compared with the disease-orientation of medicine.

The Kellogg-funded initiative Community Partnerships in Health Professions Education provided a unique laboratory in each of the three regions for the discipline to further demonstrate its efficacy in delivering primary health care. The initiative also offered an opportunity to determine the profession's next steps in enlarging its consumer base. The more nursing demonstrates its effectiveness, the greater the likelihood that medicine will perceive its activities as an economic threat and oppose its advances. This means that the nursing profession will have to focus its efforts to protect its development until nursing is firmly rooted in policy, structure, and practice. Nurses, whether they are primary-care practitioners or work in other specialties, whether they are hospital employees or work in other settings must vest in the vision of a different future for the profession—a future of agreements with consumers and partnerships with communities rather than one of service under physicians and through hospitals.

Nurses around the world need to unite in this vision for the future. Much is to be learned from nurses in other countries of the world—particularly in developing countries. With the United States population reaching approximately one third of European origin, one third of African origin and one third Hispanic, there is an urgent need for nursing to become culturally attuned to this changing populace. The health care problems in the United States for underserved communities are remarkably similar to those in Latin America and the Caribbean, and in southern Africa. The nurses within these regions have much to teach us, if only we will look carefully at the lessons to be learned. This part of the book tells the international story from the perspective of nurse leaders in Latin America/the Caribbean and in southern Africa, and from the perspective of the Kellogg Foundation. It is hoped that the story of nurses in these regions will be instrumental in bringing nurses together from around the world in a shared vision for the future.

11 Nursing in Latin America and the Caribbean

Helen K. Grace and Gloria R. Smith

The relationship between Latin America and the Caribbean and the Kellogg Foundation has been ongoing since 1939. That year, at the request of the Rockefeller Foundation and in consultation with the State Department, the Kellogg Foundation hosted in Battle Creek a delegation of 28 representatives from 26 countries of South and Central America and the Caribbean. The underlying interest was to build alliances with our neighbors to the south at a time when the war clouds were building over Europe. Since this initial meeting until today, the Kellogg Foundation has maintained its commitment to working with these countries to improve their quality of life. During the early years, the major emphasis was upon the health professions, and through the years nursing has received ongoing support for its integral role in providing quality health care for people.

1940–1985

The Foundation's early work in the field of nursing abroad fell under the supervision of the Nursing Advisory Committee. Mildred Tuttle and members of the Committee made numerous trips to the South and Central American countries to assess the situation and to recommend funding initiatives. The work in nursing focused primarily upon

The Kellogg Foundation has maintained its commitment to working with these countries to improve their quality of life. During the early years, the major emphasis was upon the health professions, and through the years nursing has received ongoing support for its integral role in providing quality health care for people.

collegiate-based nursing education without addressing larger issues surrounding the nursing field. In these early years, the focus was upon building educational programs through developing library materials, translating textbooks into Spanish, and through providing fellowships to nurses to receive graduate education, primarily in the United States. As fellows returned to their home countries, they continued to build and strengthen educational programs. In the 1970s, educational technology centers were built in strategic locations across the region to continue improving nursing education. The funding of projects related to nursing paralleled that of funding in other health professions such as medicine and dentistry. But there was little interaction between the professions in these early years.

1985–1994

Shortly after joining the Kellogg Foundation, Helen Grace was approached by Dr. Mario Chavez, the program director for Latin America and the Caribbean, who requested that she visit countries in Latin America. Dr. Chavez was concerned about the World Health Organization's approach in working toward the goal of "Health for All by the Year 2000." In many areas of the world, including Latin America, they were seeking to achieve this goal through preparing community health workers, but in many instances they were not linking these workers to nurses. It was Dr. Chavez's belief that unless it made this important link to nursing, "Health for All" would not be achieved. At his invitation, Helen Grace visited four countries (Argentina, Brazil, Colombia, and Ecuador) in 1984. She was accompanied by Dr. Georgina de Carillo, a nurse who served as a consultant to the Foundation and had been involved directly in the nursing technology centers. This was a crash course in learning about nursing in a very different context than that of the United States. The important lessons related primarily to the context in which nursing was practiced. The educational programs within the universities were amazingly similar to those in the United States. As faculty described the "vertical" and "horizontal" strands of their curricular design in Spanish, it was easy to understand what they meant; Grace recalled very similar discussions among the faculty at the College of Nursing at the University of Illinois, where she was dean before joining the Foundation. The nursing consultants and the fellows who had returned from education in the United States had transferred much of the content of baccalaureate nursing education into the framework of university programs in Latin America. But as one looked where nursing was actually practiced, the context was dramatically different. Unlike in the United States, where nurses accounted for the majority of health professionals, in Latin America licensed nurses were proportionally few. There, the ratio of physicians to nurses was 5:l. The university programs preparing nurses and physicians were nearly the same in length; nursing educational programs based in universities were four years, with a year of specialization required post-graduation. The only licensure for nurses came through completion of a baccalaureate program in nursing at a university. Preparation of physicians required completion of a five-year program of study within the university and then postgraduate specialization. Upon completion of their educational programs, the salaries for physicians were four times that for nurses. Yet despite this disparity in numbers of nurses and in the salaries they received, the few nurses there available supervised the primary care networks. For example, in an area outside of Sao Paolo, Brazil, a community health center made up of a clinic and a 100-bed hospital served as a hub of a system of smaller clinics in outlying villages and a third level of very small clinics in remote rural areas. The clinic in a small remote area was staffed by a lay midwife. The nurse within this system supervised patient care in the hospital and also in the clinics. She made weekly visits out to the rural areas and supervised the care, provided primarily by auxiliary nurses and nurse aides. The physicians came only to the hub, and patients were referred in to them—physicians did not go out into the outlying communities. Although they were required to provide services in the clinic as "payback" for their medical education, most worked only a few hours per day, spending the rest of their time in their private practices. It was clear that no matter how many physicians might be practicing, nurses were the "glue" for the system. It was also clear that in order to address issues in nursing education and practice, the large numbers of auxiliary nurses and nurses aides would need to be included

Unlike in the United States, where nurses accounted for the majority of health professionals, in Latin America licensed nurses were proportionally few. There, the ratio of physicians to nurses was 5:l.

It was clear that no matter how many physicians might be practicing, nurses were the "glue" for the system. It was also clear that in order to address issues in nursing education and practice, the large numbers of auxiliary nurses and nurses aides would need to be included as part of an overall approach.

as part of an overall approach. As Chompre and Villalobos noted in the journal *Nursing and Health Care*, "in terms of personnel to develop primary health care (PHC), state and government authorities preferred quantity to quality and engaged a large number of people who had no real training and only minimal preparation at hospitals or community institutions for health care" (*Nursing and Health Care*, p. 193). For example, in Argentina, Eva Peron simply declared people to be nurses as a reward for their political support. Linking jobs, such as nursing, to the political process continues to this day in some of the countries within the region.

As an outgrowth of this original visit, the Kellogg Foundation convened a group of nursing leaders who had gained visibility through their work with previously funded projects. This meeting, in Caracas, Venezuela, in March 1985 led to formation of a base for a blueprint for action that has evolved to this day. In the Caracas meeting, participants identified five basic premises to guide the nursing profession:

- Health problems in Latin America demand immediate action to ensure the well-being of the population;
- Rapid social changes in Latin American countries are producing changes in health services that imply scientific, ethical, and technical responsibilities for health professionals;
- The problems of nursing are interdependent and can be understood only if analyzed as part of the multidisciplinary universe of the health team;
- Improving care and service implies developing health professions and the leadership capacity of their members;
- Cooperative work—indispensable in today's world—requires interaction among the different modes of nursing (teaching, service and research) and with other health professionals at local, regional, and international levels (Conference Proceedings of the Meeting for Nursing Leaders, 1985, p. 171).

It is important to note that in developing a blueprint for action, the focus was on the health problems of people and the responsibility of nursing to be an active participant in addressing the problems, rather than on the profession per se. It was clearly understood that nursing would gain stature and recognition through being a responsible participant in working toward improvement of health care for people and in being part of a multidisciplinary team.

It was clearly understood that nursing would gain stature and recognition through being a responsible participant in working toward improvement of health care for people and in being part of a multidisciplinary team.

Development of human resources in nursing involved a number of proposed strategies:

- Train nursing professionals appropriate to the situation in Latin America, building on a knowledge base common to other health professionals, particularly in the sciences and technology;
- Prepare nursing leaders in order to encourage advancement of the profession;
- Train nurses who will work to train others, such as nursing assistants and auxiliaries;
- Create an international network to promote cooperation and information exchange between educational institutions in Latin America;
- Prepare teachers—through different types of continuing education—to spearhead curriculum change;
- Replicate innovative, practical experiences that lead to curriculum changes in graduate study and that address the realities of Latin American life;
- Emphasize the need to focus on primary health care without neglecting other levels of care;
- Stimulate scientific work of nurses and disseminate findings (Conference Proceedings of the Meeting for Nursing Leaders, 1985, p. 172-177).
- Four specific courses of action were to be undertaken:
- Develop education oriented toward community needs;
- Develop new technology for undergraduate and postgraduate studies;
- Develop documentation and information systems;
- Develop systems for dissemination (Conference Proceedings of the Meeting for Nursing Leaders, 1985, p. 181).

At this time, opportunities were very limited for postgraduate study in the Latin American countries; most nurses seeking graduate education to prepare them for teaching or leadership roles within the profession came to the United States.

The Caracas meeting was followed by a number of subsequent meetings of the core leadership group and a variety of shared experiences, including traveling seminars to the United States. A follow-up meeting of the original group was held in Guanajuato, Mexico, in 1986. At that meeting the Federal University of Minas Gerais in Brazil proposed a major project, "Support for the Development of Postgraduate Studies in Latin America." This was intended to establish postgraduate courses at the master's level with an emphasis on primary health care. At this time, opportunities were very limited for postgraduate study in the Latin American countries; most nurses seeking graduate education to prepare them for teaching or leadership roles within the profession came to the United States.

One early goal established by the leadership group was to build the capacity of nursing to prepare its basic educators and leaders in the Latin American context. A 1987 meeting in Barbacena, Brazil, provided opportunity for the leadership group to critique the proposal. Consultants from the United States participated in this meeting, which became a very important turning point for the future. The most significant aspect of this meeting was that the nursing leaders from Latin America "took charge" of the initiative, recognizing that the approaches that needed to be taken were appropriate to nursing in the region, rather than adopting U.S. models that might not fit. For example, one discussion centered on the need for a more adequate information base for nursing practice in the region. Thelma Schorr, then editor of the *American Journal of Nursing,* presented a proposal that had been developed with Maricel Manfredi from the Pan American Health Organization to translate the *American Journal of Nursing* into Spanish and Portuguese. In the discussion, the nurses described a culture in which women were not encouraged to read. Many of the nurses serving as auxiliaries or nurses aides were not literate, and there were few licensed (university-educated) nurses in practice. Both Thelma Schorr and the nurses from Latin America concluded that this was not an appropriate avenue for addressing the problem, and that they needed to develop their own base of information and research. There was in place an amazingly effective information system that had been developed under funding from the Pan American Health Organization to provide information for physicians from throughout Latin American.

In the discussion, the nurses described a culture in which women were not encouraged to read. Many of the nurses serving as auxiliaries or nurses aides were not literate, and there were few licensed (university-educated) nurses in practice.

The organization that disseminated the information, with the acronym BIREME, based in Sao Paolo, Brazil, agreed to expand its network to address nursing needs. An initial emphasis was placed on trying to capture nursing data and information and build upon that base. The computer-based system allows nurses from throughout the area to submit queries about research or information and to access all available literature. Through the years, research received more attention, and now there are at least three scholarly journals reporting research findings. Previously, textbooks in nursing had been translated from English into Spanish and Portuguese. Now a number of textbooks have been written by authors in Latin America, and nursing education is increasingly more relevant to the realities of health care in these countries.

Within the original plan for expanding graduate education, it was proposed that the institutions that were best equipped to do so be identified and supported. As part of the discussions at the 1987 Barbacena meeting, it was also proposed that the focus should be expanded to include other aspects of work that had been identified as needed. The concept of Poles of Development was advanced. After considerable debate about the terminology—"Development Centers" vs. "Poles of Development"—the consensus was that the "Centers" concept connoted everyone coming to a central place, while "Poles of Development" connoted a reaching out of specific institutions to a broader field. The initiative became known as the Latin American Program for Development of Nursing Education (PRODEN).

Six proposals from institutions wishing to become a "Pole of Development" were submitted to the Kellogg Foundation as an outgrowth of this meeting, and in 1989 four schools were selected:

- The School of Nursing at the Federal University of Minas Gerais in Brazil was supported to "organize a development pole centered on the following: postgraduate and undergraduate education; continuing education and, particularly, retraining for nursing personnel working in nursing at intermediate and elementary levels; development of nursing research and nursing technology; and support for data processing." This development pole was to work with the entire state of Minas Gerais.

- The nursing school of the University of Nuevo Leone in Monterrey, Mexico, was funded to develop nursing in the northeast area of the country to provide opportunity for professional improvement through professional practice and education.
- The school of nursing at the Catholic University in Santiago, Chile, was funded to prepare nursing leaders and to take responsibility for a database support program and to establish an information system that would support nursing throughout the Latin American countries through the BIREME education and research system.
- The nursing project of the health school at the University of del Valle in Cali, Colombia, was funded for a distance-learning program to strengthen undergraduate and graduate education as well as continuing education (Polos de desenvolvimento de enfermagen America Latina, 1989).

One of the basic principles adopted at the 1985 Caracas meeting was that in all of the health projects funded by the Foundation, nurses should be in significant leadership roles.

In addition to funding for the PRODEN, the Kellogg Foundation funded numerous other projects in support of nursing (See Appendix A). One of the basic principles adopted at the 1985 Caracas meeting was that in all of the health projects funded by the Foundation, nurses should be in significant leadership roles. In the general funding for health projects, the integration of teaching and service was emphasized, and in a number of projects nurses took leadership. Here are some examples:

- In a comprehensive health care project to serve women during pregnancy, delivery, and immediately after childbirth at the Federal University of Ceara, Brazil, appropriate technology was developed and the knowledge and experience of lay midwives was incorporated in health care delivery for rural and marginal urban areas. In this model, nurses educate and supervise lay midwives and traditional healers, teach them what to look for as they work with expectant mothers and young children, and what to do if they see something problematic. Most of this teaching is through pictures, as the lay midwives and traditional healers are mostly illiterate. The lay midwives, with support from nurses, organized a three-level system of care. At the first level, expectant mothers would go to the lay midwives' home to deliver their children, using a hammock suspended on a porch. If any problems arose, or if the mother was an adolescent, she was sent to a mini-clinic, the second level of care. If problems developed in the delivery process, transportation was available to the university hospital—the third level of care. The infant mortality rate in this northeastern state in Brazil dropped dramatically from very high levels to those paralleling or falling below the infant mortality rates in the United States.

- In a rural area of Mexico, a project funded to the University of Guanajuato uses a comprehensive nursing model for primary health care to improve living conditions of a peasant population. Working with underserved rural populations, comprehensive primary health care is offered with particular attention to women and children. The School of Nursing at Irapuato has had the lead role in this effort, which includes participation of the schools of engineering, nutrition, and agriculture, among others.
- At the Federal University of Minas Gerais, Brazil, a program for community action was developed. In this multidisciplinary project, including participation from the schools of medicine, dentistry, psychology, and nursing, nursing assumed a leadership role in a primary health care model targeted to children, pregnant women, and adults.
- A nurse leads a health self-care education project at the Catholic University of Chile in Santiago. This multidisciplinary health care model has been recognized by the Ministry of Health of Chile for helping to improve the quality of primary health care services throughout the country.
- In Bogota, Colombia, nurses lead a multidisciplinary community project that emphasizes mother-child care in a marginal community in the southeastern section of Bogota.

As nursing work evolved throughout Latin America, the importance of upgrading the entire workforce took on greater significance. A three-pronged strategy developed: (1) using distance education to improve the educational preparation of nurses; (2) developing support structures to maintain, accelerate, and extend the process of change throughout the entire region, including the Caribbean; and (3) enhancing leadership development.

Building on prior Kellogg Foundation funding to develop a non-traditional baccalaureate nursing program within the University of California system, the University of California at Dominquez Hills provided important resources to help make nursing education accessible in the most remote areas of the country.

Building on prior Kellogg Foundation funding to develop a non-traditional baccalaureate nursing program within the University of California system, the University of California at Dominquez Hills provided important resources to help make nursing education accessible in the most remote areas of the country. A number of projects were funded to develop distance-education programs, and technical assistance was provided by faculty from the University of California-Dominquez Hills. These faculty made it their business to understand the context in which these programs were working and to help develop appropriate educational programs and assure their quality.

As for structures to support nursing development, an Association of Schools and Faculties of Nursing built up a network in nursing education, and another network organization, the Latin American Nursing Network (REAL), worked to establish communication networks. REAL is now in place as a self-sustaining, permanent organization. Additionally, the International Council of Nurses (ICN) headed two significant projects that have strengthened the entire effort. The first was a review of regulations governing nursing practice in the varying countries, which led to a systematic approach to updating regulations through the political process. As a result of this effort, nurse-practice regulations have been upgraded throughout the region. A second ICN effort was leadership development throughout the region. A series of workshops were conducted, and nursing leaders from throughout the region have become a coherent group. Along with the ICN leadership development and the individual projects, a number of fellowships continued to be supported during this period. While the opportunity remained for nurses to pursue graduate education in the United States, as the capacity of the nursing educational programs has increased over time, many nurses now choose to seek out graduate education in Spanish or Portuguese-speaking universities. A number of universities now offer doctoral education that combines study in areas outside of nursing with available graduate education in nursing. This has enabled more nurses to gain advanced education without having to leave their responsibilities as wives and mothers in a culture in which women's roles are still somewhat circumscribed, and without having to master English as a prerequisite.

While most of the original work was centered in the countries of Brazil, Colombia, Mexico, and Chile, a systematic approach has extended it to all regions. The REAL network has taken responsibility for convening groups in all regions of Latin America and the Caribbean to work toward upgrading nursing education and practice. Some areas lag far behind others, particularly in the Caribbean. But the opportunity for nursing leaders to meet with one another from different areas enables them to learn approaches that have worked in other countries and adapt them to their own realities.

While the opportunity remained for nurses to pursue graduate education in the United States, as the capacity of the nursing educational programs has increased over time, many nurses now choose to seek out graduate education in Spanish or Portuguese-speaking universities.

One important characteristic of the work in Latin America is that it has not focused solely upon nursing as an isolated profession, but also on achieving recognition of nurses as part of the multidisciplinary health team. Before 1985, nurses were rarely included in any forums related to health care. But the conscious push for inclusion in multidisciplinary forums, coupled with nursing leaders who are articulate and adept in negotiation, has led to nursing's recognition and consideration as a full partner in health care delivery to communities throughout Latin America. Dr. Marcos Kisil, who succeeded Dr. Mario Chavez as coordinator of Latin American programming, was highly instrumental in seeing that nursing was valued and supported. And nursing also has its unsung heroes. Marcos's mother-in-law was a nurse—she worked with her husband, a physician in a clinic in a rural area of Brazil. Marcos would report that every time he visited her, she would ask, "What have you done for nursing?" He didn't dare visit without having something to report.

In recent years, a major funding initiative throughout Latin America has aimed to build partnerships between health-professions education and community health services. This effort was patterned after the U.S. Community Partnerships/Health Professions Education initiative. In this endeavor, known as UNI, nursing has played a key leadership role. Nurses in the initiative have provided leadership to many of the activities of UNI, such as the organization of networking meetings and in the individual projects (see Appendix B). At the same time, nurses have maintained their active role in spreading the work that has been done in nursing to all areas of Latin America and the Caribbean.

The Latin American and Caribbean experiences illustrate the tremendous gains that can be achieved when philanthropy and a field such as nursing work hand-in-hand.

The Latin American and Caribbean experiences illustrate the tremendous gains that can be achieved when philanthropy and a field such as nursing work hand-in-hand. While in the early years the Kellogg Foundation's approaches may have been prescriptive (i.e., books, libraries, learning technology centers, and the like), in the latter years, the efforts were directed more by a group of committed nursing leaders who work together to achieve agreed-upon goals. These nurses provided leadership not only to the nursing profession but also to their counterparts in medicine to make dramatic changes in health care throughout the region.

The Foundation's approach in Latin America was framed by a strategic plan specific to the region. Nurses in Latin America demonstrated the capacity to plan a major social movement. The nursing leadership had coalesced over a 10-year period, and was positioned to play a larger leadership role in the UNI initiative that supported community-based health approaches advanced by reform-minded medical schools. Many of the nurses became managers and project directors in UNI projects, thus enabling the move toward community-based primary care. These institutions were in the forefront of the community-based medicine movement. A number of projects had shown the impact of partnerships between medical schools and communities in raising community health status and overall quality of life. Individuals and clusters of faculty immersed themselves in literature on radical social change. They designed projects that infused health services with strategies to empower communities such as neighborhood purchasing and cooperatives. A psychiatrist who taught students in a maternal and child health clinic organized female patients in groups that developed long-term relationships and became an avenue for peer health instruction, counseling, and support. The women not only learned parenting skills, but they were encouraged to develop entrepreneurial skills as well as learned how to preserving their mental health in a *machismo* culture.

Nursing leaders in Latin America transformed nursing significantly, focusing nursing education on community-based primary health care and upgrading the entire nursing workforce through career mobility educational programs. The leadership was focused not on the nursing profession per se, but upon the health care needs of people. Their vision for the nursing future was revolutionary; their base broadened continuously, and their knowledge and tools deepened through networking and education. At the heart of the work was "new thought" about people, their value, their capacity, and their potential.

Notes and References

Conference Proceedings of the Meeting for Nursing Leaders. 1985. Caracas: FEPAFEM. *Nursing and Health Care, 16, no. 4.* 1995.

Polos de desenvolvimento de enfermagen America Latina. 1989.

Appendix A: Fellowships: 1984-2000

Individual Fellowships:

Pilar Amaya de Pena	Bogota, Colombia	$82,500
Avila Rincón Luz Alcira	Bogota, Colombia	73,000
Rosemary Bacacas	Quito, Ecuador	40,000
Celina Camilo de Oliveira	Belo Horizonte, Brazil	75,000
Margarita Cerna Barba	Cajamarca, Peru	99,500
Roseni Chompre	Belo Horizonte, Brazil	65,000
Andrea Correa de Oliveira	Belo Horizonte, Brazil	75,000
Bertha Enders	Natal, Brazil	5,500
Marina Estrada	Cajamarca, Peru	89,500
Esther Gallegos	Monterrey, Mexico	78,000
Sonnia Gomez	Guayaquil, Ecuador	33,122
Sunny Gonzalez	San Jose, Costa Rica	45,000
Marlene Montes	Cali, Colombia	77,000
Luz Munoz	Valdivia, Chile	110,000
Aura Perez Escalante	Cali, Colombia	50,000
Francisca Perez Saavedra	Lima, Peru	85,000
Bertha Salazar	Monterrey, Mexico	80,000
Maria Sensevey	Rosario, Argentina	50,000
Jose Silva	Itajuba, Brazil	81,000
Gloria Consuela Fernandez	Bogota, Colombia	62,309
Rosa Vaiz	Mexico, Peru	43,000
Nora Agnes Vega Villalobos	San Jose, Costa Rica	57,000
Doris Valasquez	Lima, Peru	80,000
Ruth Vilchez	Iquitos, Peru	50,000
Maritza Villanueva	Iquitos, Peru	45,000

Group Fellowships:

Catholic University of Chile	Santiago, Chile	85,000
Cayetano Heredia University	Lima, Peru	30,000
Asociación Ecuatoriana de y Facultades de Enfermeria	Escuelas Quito, Ecuador	47,011

Appendix B: Nurse-Related Project Funding

Goal: Community Based-Health Services
Strategy: Comprehensive Services
Cluster: Infant/Child

Ministry of Health of the Province of Cordoba, Argentina $737,360

Strategy: Health Professions Education
Cluster: Nursing

Ecuadorian Association of Schools and Faculties of Nursing		94,960
Escuela Superior de Enfermeria	Rio Negro, Argentina	209,200
Fundacao de Amparo a Pesquisa E Extensao Universitaria	Florianópolis, Brazil	1,180,041
Fundacao de Desenvolvimento de Pesquina	Belo Horizonte, Brazil	157,953
Pan American Health Organization		422,000
Pan American Health Organization		62,000
Pontifica Universidad Catolica de Chile	Santiago, Chile	385,383
Universidad de la Republica	Montevideo, Uruguay	202,000
Universidad del Valle	Cali, Colombia	1.038,720
Universidad del Zulia	Maracaibo, Venzuela	356,285
Universidad Autonoma de Nuevo Leone	Monterrey, Mexico	924,333
Universidad Nacional de Colombia	Bogota, Colombia	758,210
Universidad Nacional de Colombia	Bogota, Colombia	64,466
Universidad Nacional de Rosario	Rosario, Argentina	488,527
Universidad Nacional de Tucaman	Tucaman, Argentina	509,690
Universidad Nacional Autonoma de Honduras	Tegucigalpa, Honduras	200,000
Universidade de Sao Paolo	Sao Paolo, Brazil	38,725
Universidade Federal de Minas Gerais	Belo Horizonte, Brazil	795,796
Universidad Nacional de Rosario	Rosario, Argentina	168,360

University of Puerto Rico	San Juan, Puerto Rico	78,796
University of Puerto Rico	San Juan, Puerto Rico	129,917

Strategy: Educational Reform
Cluster: Nursing Education

California State University – Domínguez Hills	Carson, California	111,500
Sistema Provincial de Salud	Rawson, Argentina	225,788
Universidad del Valle	Cali, Colombia	346,300
Universidad del Zulia	Zulia, Venezuela	120,035
Universidad Autonoma de Nuevo Leone	Monterrey, Mexico	186,993
Universidad Nacional de Colombia	Bogota, Colombia	400,700
Universidade Federal de Minas Gerais	Belo Horizonte	561,048
Universidade Federal de Mato Grosso	Cuiaba, Brazil	372,430
Pan American Health and Education Foundation	Washington, D.C.	361,093

Strategy: Leadership Development
Cluster: Leadership Development in Health

International Council of Nurses	Geneva, Switzerland	749,930

Appendix C: Multi-Disciplinary Projects

Strategy: Health Professions Education
Cluster: Community Partnerships Health Professions Education

UNI		
Universidade de la Frontera	Brazil	$273,000
Universidade Federal de Bahia	Brazil	750,000
Fundacion Universidad Del Norte		726,600
University of Illinois at Chicago	Illinois, USA	349,560
Universidad del Zulia	Argentina	663,876
Universidad Autonoma de Yucatan	Mexico	411,214
Universidade Estadual de Londrina	Brazil	817,000
Universidad Centro-Occidental "Lisandro Alvarado"	Venzuela	487,000
Fundacao Municipal de Ensino Superior de Marilla	Brazil	800,000

Universidad del Valle	Colombia	600,000
Universidade Estadual de Londrina	Brazil	500,000
Universidade Federal de Rio Grande do Norte	Brazil	200,000
Universidad Nacional Autonoma de Nicaragua	Nicaragua	4,273
Fundacao Universitaria Do ABC	Brazil	100,000
Universidad Federal de Bahia	Brazil	50,759
Universidade Estadual de Londrina	Brazil	40,000
Universidade Federal do Rio Grande do Norte	Brazil	400,000
Pan American Health Organization	Washington, D.C.	70,000
Universidade Federal de Minas Gerais	Brazil	69,000
Universidade Federal de Bahia	Brazil	10,288
Universidad Nacional de Trujilo		610,907
Universidade de Brasilia	Brazil	301,550
Fundacao Municipal de Ensino Superior de Marilia	Brazil	703,416
Universidad Nacional Autonoma De Nicaragua	Nicaragua	852,554
Universidad de la Frontera		811,004
Universidad de Antioquia	Colombia	1,121,170
Universidade Estadual Paulista	Brazil	954,322
Universidad de Colima	Colombia	553,634
Universidad Nacional de Tucaman	Argentina	700,000
Universidad Autonoma de Nuevo Leone	Mexico	769,800
Universidad Nacional de Tucaman	Argentina	76,700
Network of Community-Oriented Educational Institutes for Health Sciences		189,000

12 Outcomes of Nursing Development in Latin America

Roseni de Sena and
María Mercedes Durán de Villalobos

Organized support from the W. K. Kellogg Foundation for nursing in Latin America came at a critical social and political time for the countries of the region. There was considerable inequality in nursing education and in the development of nursing services, along with common regional problems, such as the establishment of different types of health systems and the use of nursing aids and auxiliary personnel as a low-cost strategy for expanding coverage, even within existing economic constraints.

Countries lacked coherent policies on formal training and preparation for non-professional personnel, and the governments showed little interest in establishing policies on including non-professional personnel in the health care labor force. This led to highly dubious results in the type of service provided to users of health care services at all levels.

Also, professional nursing associations lacked the necessary presence at political levels and, therefore, were neither representative nor able spokesmen for advocating a solution to health problems, or to the problems originating within the profession itself (Grace, 1985).

Countries lacked coherent policies on formal training and preparation for non-professional personnel, and the governments showed little interest in establishing policies on including non-professional personnel in the health care labor force. This led to highly dubious results in the type of service provided to users of health care services at all levels.

As adverse as the situation was, some conditions favored establishing a process for change, with groups of personnel at operational and decision-making levels who were able to encourage and mobilize to develop problem-solving strategies.

Camilleri (2000) noted that during the diagnostic process undertaken by the Foundation, "there was appreciation for the need to prepare leadership groups to respond to the difficulties. With this in mind," Camilleri explained, "nurses who had shown initiative in creating and developing health care programs as part of health services were identified. Most of them had demonstrated a capacity for leadership and were situated in nursing schools. Also, they were interested in promoting the primary health care model, with a community-based approach, and wanted to incorporate these experiences into educational programs."

These nurses had the task of helping the nursing profession in Latin America to develop to the point where it could help improve the quality of health in the countries of the region. As Camilleri noted, "the group carried out its work through a series of interrelated, innovative projects to provide service at different levels of care through students. Another criterion was that the nursing schools had to be able to sustain the programs developed with the support of the Foundation, once its financial backing was no longer available."

The strategy to support and develop the nursing profession was based on a global initiative known as the Latin American Program for the Development of Nursing Education (PRODEN) and focused on the following aspects:

- Support for the nursing development poles (PRODEN) at the following universities: Autónoma de Nuevo León, Mexico; del Valle, Colombia; Católica, Chile; and Federal de Minas Gerais, Brazil;

- Support for a second phase of PRODEN, emphasizing work with universities in southern Brazil, Argentina, Venezuela, and Colombia;
- Support for the Latin American Nursing Network;
- Collateral support for the Latin American projects, through advisory assistance from the Center for International Nursing Education (CINE) at California State University, Domínguez Hills, Carson; and
- Support for the Nursing Development Program for Latin America and the Caribbean (PRODEC), which emerged later (Project to Support the Development of Postgraduate Nursing Programs in Latin America, Barbacena, 1987, and Proposals for Innovative Projects in Education, 1996 and 1997).

An important feature of this support was that it encouraged activities parallel to the primary concern of the different projects. These activities promoted networking, coordination among various institutions at the national and international level, and the production of reference material. This encouraged participant interest in systematization and production, as well as publications.

The emphasis was to be on the search for alternatives to increase coverage, to facilitate access to educational programs, to upgrade care, to consolidate research into nursing practice, and to facilitate progress toward political, technical, and cultural organization of the profession and nursing professionals themselves.

In the wake of the diagnosis done in 1985 (Grace, 1985) and the Barbacena meeting (1987), consideration was given to the possibility of coordinating a nursing program that would respond to the uncertainties and difficulties facing the profession. The emphasis was to be on the search for alternatives to increase coverage, to facilitate access to educational programs, to upgrade care, to consolidate research into nursing practice, and to facilitate progress toward political, technical, and cultural organization of the profession and nursing professionals themselves (De Sena and De Villalobos, 2000).

A Force for Strategic Inter-institutional Action

The Caracas meeting in 1985 served as a point of reference for systematizing the support provided by the Foundation. As described in the previous chapter, the diagnosis, based on visits to the countries of the region and expert analysis, was intended to lead to strategies that would enable nursing to construct its own development within the context of health care delivery as a whole.

The primary reason for strategic action was the stock of experience acquired by certain nurses through non-traditional work. There was also the willingness of institutions to intervene in processes that implied responsibility and a commitment to the future. And, there were nurses who were committed to the new proposals and to coordination of projects in their institutions.

The nursing development poles (PRODEN) were seen as programs with the capacity to generate their own growth and that of their institutions. In turn, they were intended to serve as models for developing nursing in their regions and in the areas they served. The following were the criteria proposed to define the PRODEN:

The nursing development poles (PRODEN) were seen as programs with the capacity to generate their own growth and that of their institutions. In turn, they were intended to serve as models for developing nursing in their regions and in the areas they served.

- Development of innovative experiences based on a strategy that combines teaching with care;
- Emphasis on postgraduate programs oriented toward primary health care;
- Development of programs for continuing education;
- Specialization and refresher training with an eye toward reorganizing basic care in the health systems; and
- Production and dissemination of innovative technology for health education and education in services (Proyecto de Apoyo, 1987).

The poles had other features as well: institutional leadership capacity, capacity for interdisciplinary work, and potential for better teaching. Four poles with these characteristics were established initially. Here is a glimpse of each:

(1) The Nursing School at the Universidad Federal de Minas Gerais in Belo Horizonte (Brazil) emphasized undergraduate and graduate education, and particularly training for intermediate and basic personnel, in addition to developing its own research and technology for these processes.

(2) The Nursing School at the Universidad Católica de Santiago (Chile) became a development pole to promote the teaching care strategy. It also disseminated conventional and non-conventional literature, and sought to modify the quality of people's health through community centers where undergraduate and graduate students underwent practical training.

(3) The Nursing School at the Universidad del Valle Health Sciences School in Cali (Colombia) developed distance education aimed at reinforcing undergraduate and graduate programs, as well as for continuing education. The teaching care model proposed as the basis for support focused on "nursing care in local health systems." The distance education model would be applied in different parts of Colombia and Latin America.

(4) The School of Nursing at the Universidad Autónoma de Nuevo León, in Monterrey, Mexico worked on the development of teaching care activities and also proposed a pioneering taxonomic project that classified community nursing and primary health care activities. In addition, it reinforced the college degree program in nursing, which served as a curriculum model for many nursing schools in Mexico.

Three other initiatives were developed to support the PRODEN. They included: (1) programs for conventional and non-conventional nursing databases; (2) the creation of a nursing information sub-area in conjunction with the Regional Health Sciences Library (BIREME), with nursing schools and poles in Latin America acting as affiliate centers to generate data on nursing education, leadership, and the development of nursing care and management; and (3) encouragement for continuing education to train leaders in primary health care (Polos de Desenvolvimento de Enfermagen na América Latina, 1989).

The strategies for the poles were applied through innovative programs for nursing education that centered on three aspects: (1) adopting educational programs based on reflexive critical theories, (2) creating alliances between the service sector, the professional sector and the teaching sector, and (3) maintaining a supply of classroom programs and semi-classroom programs, based on the distance teaching method.

The results of these experiences are varied and relate to the proposals of each program, which are framed by their own contextual, socioeconomic and cultural reality. However, in a general way, the following innovations are evident:

- Changes in undergraduate and graduate curriculum strategies to reinforce leadership development, with an emphasis on the strategy for primary health care management (PHC).
- Improvement in existing postgraduate programs and the creation of new cooperative programs at strategic sites in Latin America, together with the development of programs for continuing education directed to teaching nurses as well as in-service nurses.
- Development of primary health care (PHC) programs that maintained the tie between teaching and nursing care. These models included an evaluation of appropriate health care technology in every area: promotion, prevention, rehabilitation, and even community participation.
- A build-up in the leadership required to manage health and nursing programs.
- Stronger national and international networks to facilitate and reinforce the work generated by the schools and the poles (De Sena and De Villalobos, 2000).

Several entities were responsible for developing the leadership program, including teaching institutions, the educational technology centers of the Pan American Health Organization (PAHO), the PRODEN and, in general, the institutions that had an opportunity to produce innovative projects associated with health education and health services. In addition, scholarships were available for postgraduate programs and non-formal education through in-service training and roving seminars (De Sena and De Villalobos, 2000).

Based on the program proposals, the Foundation supported a series of strategic projects that reinforced the central endeavor through initiatives in other countries. These included educational projects with an emphasis on non-traditional models (Argentina and Uruguay); service based on the primary health care strategy (PHC); projects that combined teaching care with community-based initiatives (Brazil, Chile, Colombia and Mexico), projects to improve maternal-infant health (Bolivia, Ecuador, Honduras, Paraguay, the Dominican Republic and Venezuela), and initiatives to develop conventional and non-conventional databases (Brazil, Chile and Colombia) (Chompré and De Villalobos, 1995).

The Foundation also provided the Latin American Association of Nursing Schools and Faculties (ALADEFE) and the School of Nursing at the Universidad de San Pablo, Riberon Preto, with support to publish the *Revista Panamericana de Enfermería (Latin American Nursing Journal).*

A second phase saw the possibility of continued support for the four PRODEN and for additional projects in Maracaibo, Venezuela; Patagonia, Argentina; and Santa Catarina, Brazil. On the whole, these projects were intended to develop educational programs for nursing personnel at every level. The School of Nursing at the Universidad Nacional de Colombia in Bogotá was also included in this second stage, through the development of a postgraduate program. During this time, the programs received continuous support from CINE.

At this stage, the emphasis of the programs in southern Brazil was on converting nurse's aids into nursing auxiliaries and nursing auxiliaries into technicians. The basic premise of the conversion programs is "that better preparation leads to better professional performance on the part of those who receive it and is one of the strongest pillars for progress in nursing education. Better preparation for nursing personnel has a direct impact on two areas: better care for those who use nursing services and increased professionalization" (Gallegos de Hernández, 2000).

The use of empirical and auxiliary personnel as human resources in the field of nursing is a reality in most countries of the region. It represented a challenge for the projects and compelled educational institutions to revise their priorities, given the commitment to these innovative programs.

The use of empirical and auxiliary personnel as human resources in the field of nursing is a reality in most countries of the region. It represented a challenge for the projects and compelled educational institutions to revise their priorities, given the commitment to these innovative programs. The following general features were common to all the programs developed at this level of instruction:

- Study-work demands. It was difficult to negotiate agreeable work-study ratios with the institutions that employ students and because of the many roles students play when they are taking nursing courses.

- Curriculum design. Given the traditional curricula, it was impossible to respond to the students' need for comprehensive training in cognitive, psycho-motor and emotional-behavioral areas.
- Motivation. Teaching and service personnel, as well as trained workers, continually fulfilled a variety of roles to achieve their objectives. Working students remained interested in improving their personal skills and job performance, in spite of the aforementioned difficulties. Nursing professors and service personnel became the driving force of innovation, with benefits for their own appreciation, performance and the quality of nursing care (Arraigada, et al., 2000).

The programs were characterized by their large number of students and by the development of strategies used to train the teachers who would play an active role in them. The involvement of teachers from nursing schools and nurses from health services allowed these individuals to apply the teaching care strategy in a more objective, concrete way.

There are accomplishments to be emphasized, such as those in Patagonia and Rosario (Argentina), with spectacular results that changed the makeup of the nursing workforce in those regions.

There are accomplishments to be emphasized, such as those in Patagonia and Rosario (Argentina), with spectacular results that changed the makeup of the nursing workforce in those regions (Blanco and Lloyd, 2000).

An integrated, non-traditional curriculum was used in the program at the Universidad Federal de Minas Gerais Nursing School in Belo Horizonte. It featured participatory methods and involved service institutions and the university itself. The result was a pioneering initiative in the field of nursing education in Brazil (Bastos and Magalhaes, 2000).

Programs were expanded for students who live and work far from school campuses or educational centers. These programs were the only quality alternative for access to a good education. They enabled the participating universities to fulfill their social role by developing methods and proposing teaching models in keeping with the real needs of the community (Ceballos and De Sena, 2000).

The conversion of nursing technicians into college graduates with nursing degrees occurred in Mexico and Argentina as a way to overcome the deficit of licensed nurses in those countries. Throughout the process, there was concern about how to grasp, in a more precise way, the disciplinary and multi- or interdisciplinary knowledge that is required to support the emerging roles of new personnel with college degrees. Their prior training had already given them a command of the technical and mechanical components of some of these roles.

Although this process is also necessary in regular courses for a college degree, its relevance is not as clear as when the student is not a newcomer to the nursing profession.

In that case, the focus is on managing teaching and learning methods that emphasize sound judgment and decision making. Including relevant disciplinary content will accomplish little if new professionals do not apply it intelligently. Finally, there is the process of changing the role of the professor when working with adults and the role of students when they change their subordinate role to become independent professionals.

Including relevant disciplinary content will accomplish little if new professionals do not apply it intelligently.

Accordingly, the programs designed to transform technicians into professionals with college degrees represented a step forward and a challenge that led to the following results:

- Application of teaching-learning methods to facilitate acquisition of the intellectual skills and capacity for sound judgment and decision making.
- Reconstruction of the teacher's role in college degree programs, so the teacher is training not only adults with nursing experience but also students who must perform as professionals.
- Development and use of active learning methods that center on the student and require the teacher, the student and the in-service nurse to change their values and attitudes with respect to the student's learning experience.

Postgraduate programs for higher diplomas or master's degrees were offered at the Universidad del Valle Nursing School (Cali, Colombia), the Universidad Federal de Minas Gerais Nursing School (Belo Horizonte,

Brazil), the Universidad de Santa Catarina Department of Nursing, and the Universidad Nacional de Colombia Nursing School in Bogotá. These programs were strategically important in developing proposals that entailed a commitment to ensuring broader coverage in their respective areas, besides offering flexibility and academic excellence. They were derived from the effects of reform in the health sector, and incorporated technology as a response to the need for care and attention of those who use nursing services (De Sena and De Villalobos, 2000).

Here are some examples of what these programs accomplished:

- The development of methods that allow for a relationship between theory and practice, and that focus on solutions to real problems in nursing practice.
- Proposals for combining theory and practice that went beyond formal academic propositions.
- Research incorporated and used as a tool to modify the way knowledge is acquired. This, in turn, reinforced intervention on the part of nursing students to change the quality of care in their working environment.
- Signature of specific agreements between universities and service institutions, and with professional associations.
- Programs placed in the context of continuing education, to ensure the learning process is tied into health care and health service management.
- Definition of competence in each field, with the necessary detail depending on to the required expertise.

The master's programs were very successful: They represented a break with tradition and orthodox mechanisms for curriculum management at the universities, without jeopardizing academic excellence.

The master's programs were very successful: They represented a break with tradition and orthodox mechanisms for curriculum management at the universities, without jeopardizing academic excellence. They lent priority to the researcher's role and the purpose of the research and encouraged professional disciplinary development through rigorous examination processes that earned recognition from the nursing scientific community in and outside Latin America (Canaval and collaborators, 2000).

Benefits and Challenges

Once projects were completed, the following benefits were determined:

- The use of semi-classroom programs provided access to education for nurses who are far from centers of higher learning.
- The use of tools for distance education helps facilitate access and coverage. It also establishes a connection between work and teaching by enabling students to learn by doing.
- The adoption of approaches that respond to the problems students face at work allows for prompt and adequate intervention, not only in terms of curriculum development, but also by changing the way health services operate.
- The establishment of new regulations for semi-classroom programs and distance education at universities was perhaps the most outstanding administrative accomplishment. To a certain extent, it obliged nursing schools to negotiate a change in university bylaws, in order to legalize the innovations. The doubts and concerns expressed by traditional universities were resolved by the positive outcome of various programs derived from the projects. These were adopted as models and eventually used by the universities to develop new educational options.
- Most administrators and professors were motivated by the various changes. This enhanced their desire to learn, to innovate, to change and to implement a new concept of problem-based teaching in their everyday work. It also revitalized teaching and lent unprecedented relevance to nurses.
- Those who played a leading role developed a critical view of the educational process, which enabled them to look at education from a different perspective in order to change in the way programs are taught and to judge new proposals in a positive way.

Almost every program used the concept of continuing education to change teaching practices, direct care and management. The programs were executed to respond to the demands of providing quality care, to facilitate access to care and to increase coverage (Canaval and collaborators, 2000).

Almost every program used the concept of continuing education to change teaching practices, direct care and management. The programs were executed to respond to the demands of providing quality care, to facilitate access to care and to increase coverage.

The projects also faced a number of challenges, including the need for programs to be directed to nursing personnel as a whole. Given the social and technical divisions in the work of nursing, this was important in eliminating discrimination and improving health care in general. Another challenge was to ensure that the requirements for quality service throughout the progressive network of nursing care, which includes non-traditional areas such as institutions for palliative care, home care, elderly care, non-governmental organizations, and research fields.

The projects and development strategies to support nursing education never deviated from the unifying philosophy that guided the work from the beginning. Although the participating institutions worked on different projects, they pursued an ideology intent on changing the design of education. This was part of an effort to eliminate traditional rigid nursing structures and propose flexible teaching methods in their place. There was increased give-and-take between teaching and nursing care, as part of an effort to put the learning process in context and make the curriculum more relevant. Communication technology, including data processing, played an extensive role, depending on the needs of each institution (De Sena and De Villalobos, 2000).

The Latin American Nursing Network (REAL)

In 1992, after analyzing the evolution and outcome of the Latin American Program for the Development of Nursing Education, networking was emphasized as a way to exchange experiences, circulate results, and develop databases. This initiative was the culmination of action derived from the Caracas meeting. It was based on the potential for cooperative work and on the need to eliminate the differences that still exist in Latin America.

The mission of the Latin American Nursing Network (REAL) is to "reinforce nursing through contact among individuals, groups and institutions that operate as part of a network." As such, they can make a more rational contribution to improving people's health and, consequently, to raising the standard of living in Latin America (Chompré and De Villalobos, 1995 and *Boletín REAL*, No. 1, 1993).

With its horizontal structure and lines of action, and by including groups of students, REAL has facilitated communication within the nursing profession in Latin America and has created ties with international nursing organizations. In doing so, it has generated a series of opportunities to strengthen the presence of nursing in health systems.

With its horizontal structure and lines of action, and by including groups of students, REAL has facilitated communication within the nursing profession in Latin America and has created ties with international nursing organizations.

Building Bridges between Regions and Continents

Interaction among the countries was possible through concerted activities among individuals and universities. The Foundation played an important role in defining strategic actions that were linked to the projects and designed to strengthen the bonds and exchanges needed to facilitate the processes, given the institutional and regional disparities that existed.

With the initial activity in the 1980s, when the diagnosis of nursing education was conducted in more than 12 countries of the region, three working groups were established, each with the participation of two nurses from Latin America and one from the United States. The U.S. representatives were from three universities: the University of Texas at Galveston; the University of Illinois at Chicago; and the University of North Carolina at Chapel Hill.

The results of the intense efforts of these groups were presented at the meeting in Barbacena, Brazil. The diagnosis was fundamental to establishing a university exchange program that allowed teachers from Colombia and Mexico to pursue doctoral studies, with the support of the Foundation.

Another positive exchange activity was developed with California State University, through a program with the Center for International Nursing Education (CINE). It involved continuous consulting through four so-called institutes, which were developed in support of the Nursing

Education Development Program. The first institute was conducted in July 1995 at California State University. Its objective was (1) to familiarize 20 participating teachers (from Argentina, Brazil, Colombia, Mexico, and Venezuela) with state-of-the-art technology for distance and semi-classroom programs in nursing education; (2) to conduct brainstorming sessions that would facilitate the development of proposals and projects; and (3) to draft work plans and strategies for each country and for groups of countries.

Strategies were applied for an ongoing exchange among the projects and teachers involved in the CINE initiative. The teachers conducted seminars and workshops in their countries, according to the particular needs of the projects, thereby helping to strengthen bilateral relations and exchanges.

The institute eventually prepared five projects—one per country. They featured cooperative work and optimum use of resources. Each project was negotiated in the respective country and became the second phase of support provided by the Foundation for the development of nursing in Latin America. Based on the California institute, strategies were applied for an ongoing exchange among the projects and teachers involved in the CINE initiative. The teachers conducted seminars and workshops in their countries, according to the particular needs of the projects, thereby helping to strengthen bilateral relations and exchanges.

The second institute was held at the Universidad Nacional de Colombia School of Nursing in Bogotá in January 1997. With 40 participants, its goals included working on strategic project administration and management of human and financial resources. Difficulties related to incorporating and using technology for management and financial control were analyzed, and strategies to facilitate project administration were discussed.

The third institute occurred at the Universidad Nacional San Juan Bosco in Comodoro Rivadavia, which is in the Patagonia region of Argentina. The historic event brought together more than 60 nurses from South and Central America to discuss teaching-learning methodologies. The most important aspect of the third institute was the opportunity it provided for a significant number of nursing teachers, in-service nurses, and nursing associations from various parts of Argentina to take part in an international event. It lent visibility to the work performed by nurses in the region and was an unquestionable source of support for the development of nurses involved in the different projects. For example, by the time the Foundation initiated the second phase of its support for projects in South

America, Patagonia had three nurses with college degrees and 43 nursing technologists (Blanco and Lloyd, 2000).

The fourth and final institute was the responsibility of the Nursing School at the Universidad Federal de Minas Gerais in Belo Horizonte. Held in August 2000, its goal was to analyze the production and use of educational resources and materials. In doing so, it developed a way to organize and classify experiences using educational materials from previous projects. More than 150 nurses took part, including nursing teachers, in-service nurses and representatives of professional associations.

The institutes created a curriculum to prepare project participants in a more relevant way. They also pointed out several aspects of the work that needed to be upgraded and examined in more detail. Moreover, a culture of trust was built up through an exchange among the projects. This facilitated a flow of information and materials, which allowed for optimum use of individual resources.

A culture of trust was built up through an exchange among the projects. This facilitated a flow of information and materials, which allowed for optimum use of individual resources.

The materials produced and applied for consultation were collected and later used as reference materials for the new projects. Consulting and an exchange of advice among the projects also reinforced the capacity for regional cooperation (De Sena and De Villalobos, 2000).

The innovations transcended the educational programs; there was a demand for a revision of nursing practice. This led to concern about the quality of care and a clear awareness of the need to expand coverage. The projects also encouraged a new approach to research that emphasized a solution to practical problems. Teaching used resources that played a fundamental role in developing future generations of nurses at all levels, as is still the case.

One element discussed during the institutes was the development of research projects based on common problems. No conclusion was reached in this respect, mainly because project participants were responding to their own academic and administrative demands. This was unfortunate. It would have been extremely valuable for research on nursing education to benefit not only from a narrative of the results, but also from scientifically structured evidence on the advantages of the curricular and methodological experiences being developed.

Support for Nursing in Central America and the Caribbean (PRODECC)

The situation of nursing in Central America was discussed, in detail, in 1998. After a careful analysis, and given the improvement in political conditions in most of the countries, it was suggested that an effort be made to close the gap between this region and the rest of the continent. There had been no projects to provide sustained support for nursing in Central America, and the disparity in relation to the other countries was notorious, even in spite of important support for Central American nursing from the Pan American Health Organization, various European agencies and the Japanese. These international agencies played a significant role, as they still do. However, each operates in keeping with its own interests. There is no attempt to strengthen cooperation initiatives through common goals, nor are they interrelated. The nursing institutions that receive this support have yet to encourage proposals that reinforced multiple efforts.

The Program for the Development of Nursing in Central America and the Caribbean originated when the Foundation invited Costa Rica, El Salvador, Guatemala, Honduras, Nicaragua, Panama, and the Dominican Republic to develop a regional proposal for reinforcing the link between teaching, service and professional associations. The idea was to improve the quality of nursing care and the visibility of the nursing profession in these countries. Specific needs were to be determined through a regional diagnosis and country surveys, based on the same philosophical and strategic principles applied to the set of projects for South America and Mexico (PRODECC, 1999).

The Program for the Development of Nursing in Central America and the Caribbean originated when the Foundation invited Costa Rica, El Salvador, Guatemala, Honduras, Nicaragua, Panama, and the Dominican Republic to develop a regional proposal for reinforcing the link between teaching, service and professional associations.

Without losing initiative or ignoring the general goals of PRODECC, and with the experience acquired through the earlier projects, the Central American and Caribbean initiative was organized as a program with three stages:

1. Support for the individual proposals of each country.
2. Support for postgraduate education in nursing (master's degrees for 33 nurses) to strengthen the general development of nursing and to create a critical mass of leaders to promote general and institutional progress.
3. Encouragement of fostering inter-institutional and cross-sector development projects led by nursing.

Each country produced a specific proposal. Most of them, with the exception of the Costa Rican proposal, which focused on continuing education for in-service nurses, involved a wide range of educational initiatives, such as programs in Honduras and the Dominican Republic to enable nurse's aids to earn a college degree, advanced degree programs for technicians in El Salvador, Guatemala, and Panama, and college degree programs in Nicaragua with an emphasis on maternal-infant health and pediatrics.

PRODECC was accompanied by a series of workshops for strategic planning, project evaluation, project management and evaluation of materials. Each country sought individual assistance as well. In most cases, this involved innovative educational methods, the design of materials or teaching aids, and curriculum development (Cluster evaluation reports, 2000 and 2001).

According to the visits, project evaluations, and assessments, the workshops created the conditions necessary to develop and move the projects forward. In each case, the parties had to learn to collaborate.

According to the visits, project evaluations, and assessments, the workshops created the conditions necessary to develop and move the projects forward. In each case, the parties had to learn to collaborate. However, the support activities were the backbone of effort to promote project institutionalization, insertion, and prospective planning (De Sena and De Villalobos, 2000). The type of structured support envisioned for the project made two things possible: networking and appreciation for the experience of others. This helped to accelerate processes and to avoid unnecessary wear and tear.

During these activities, it became evident that nursing teachers and in-service nurses had to receive postgraduate training if they were to become the driving force of change in their countries and in the region. Being conscious of the experience acquired through the Innovar Project at the Universidad Nacional de Colombia, the Foundation provided the support required for 33 nurses to earn master's degrees in nursing. This initiative also included a comprehensive plan for additional development through work with the respective countries. The plan was carried out with professors from the postgraduate program at the Universidad Nacional de Colombia.

The proposal to train nurses concluded with two research forums. The first was held in San Salvador; the second was scheduled for Guatemala City. These events are used to present the students' dissertations and to encourage research in the area. In addition to the students, nurses from all the countries, including the host country, are invited to take part.

The hope is that, in the short term and with what has been learned by those involved in educational processes and from the participating countries, an academic community can be formed that is representative of the Central American region.

The forums are designed not only as an academic arena where students can defended their dissertations, but also as an opportunity for professors from the Universidad Nacional de Colombia to promote international work, to defend proposals and the rational behind the way they conduct research, and to help circulate research results and other materials of interest.

Because an invitation to the forums was extended to the directors of the health institutions and schools where the students were developing their work, results are seen in appreciation and clarification of roles. Expectations are being met, although incipiently, of making dissemination and openness to research a custom. However, the hope is that, in the short term and with what has been learned by those involved in educational processes and from the participating countries, an academic community can be formed that is representative of the Central American region (Evaluation Reports on the I Central American Research Forum, 2003).

Although objective figures are still not available, one can assume the Central American experiences are fruitful. Projects to upgrade personnel have increased the number of college graduates in nursing, which should improve the quality of nursing care. This is an important accomplishment, particularly since the coverage of health care at country level is almost entirely in the hands of auxiliary personnel. The students involved in the master's program are performing leadership roles and have become motivators and generators of change. They have succeeded in unleashing action for change, and this has produced a cascade effect at the institutions where they work. However, strong ties have yet to develop between students from the different countries. The most striking aspect of this group of nurses is its capacity to apply, in practice, the proposals developed through the master's program. Perhaps this is because the program tried put into context each nurse's learning.

A major shortcoming of PRODECC was not having proposed a systematic effort for auxiliary nursing personnel. Giving them access to university programs is not enough. Because they are a fundamental part of the nursing team, working strategies should have facilitated dialogue among the various groups of personnel. Although there were timid attempts, they were not able to bring nursing personnel closer together and allow for the possibility for joint work. With the experience acquired by the countries, not only in Central America, but also in Mexico and South America, it is hoped that nursing as a whole will be able to devise proposals for joint development. It is also hoped continuing education will be geared toward improving the professional performance of all types of nursing personnel.

With the experience acquired by the countries, not only in Central America, but also in Mexico and South America, it is hoped that nursing as a whole will be able to devise proposals for joint development.

The third stage of PRODECC involved proposing a working hypothesis to nursing teachers, in-service nurses, and professional associations in each country that went beyond the uni-disciplinary work of nursing. The suggestion was that, "if nurses work in conjunction with other professions, in an inter-institutional, cross-sector and participatory way, with a defined territorial base, nursing would be able to influence local public policy."

With this utopia in mind, the countries are trying to start projects that go beyond nursing work to involve other agents from health institutions and the social sector who are applying the same philosophy to overcome health problems in their areas. The goal is for sustainable action that can be institutionalized, monitored and evaluated (Reports on the workshops on participatory project management, 2002).

Besides the people who directly benefited from the projects' innovative opportunities, the indirect beneficiaries were the educational programs, the conventional institutions and, to a lesser degree, the service institutions. It is difficult to know what the repercussions were for professional organizations and nursing associations. At any rate, the integrating effort must continue, because changes in education ultimately affect the quality of care provided. This is the real interest behind an effort of the magnitude implied by the "support program of the Foundation."

The Foundation financed a number of projects that were not conceived originally as major contributions of the Latin American program but that clearly helped to reinforce nursing in the region.

Independently, and through direct contacts with the International Board of Nurses (CIE) and its local chapters, the Foundation financed a number of projects that were not conceived originally as major contributions of the Latin American program but that clearly helped to reinforce nursing in the region. These initiatives include: (1) the Project for International Classification of Diagnostics, Interventions or Situations in Nursing Practice—CIPE (Brazil, Chile, Colombia and Mexico), (2) the Leadership for Change project, which was nearly continental in scope, and (3) the Leadership for Negotiation project, which also included most of the countries. They have all used reproductive methodologies and, although not directly related to education, have had an important impact on nursing groups and associations.

In recent years, particularly through the PRODECC projects, there has been an attempt to coordinate these projects and those supported through the International Council of Nurses. However, there has been little success. This should prompt nursing to stop and think. Had it adopted a more holistic view, these experiences might have had a broader impact.

Clearly, only a few, rather privileged institutions have had this experience, which is not the common denominator in Latin America and the Caribbean. The fact that the lessons can be replicated in other educational institutions and programs for nursing is a source of satisfaction. It is not necessary to live each step over again. And this prevents us from making the same mistakes and from investing valuable time to correct them.

The knowledge and technology that were accumulated in the course of more than 15 years is certain to continue in the future. Also, other international institutions have come to appreciate the value of the Foundation's regional work (De Sena and De Villalobos, 2000).

Going It Alone

The developments and advances with the support of the Foundation led to accomplishments and lessons that can be used to address the new and changing nursing situations. They also seem to indicate that the future will not entail the difficulties observed in the past, since we now have different people and institutions.

The countries have made a political decision to convert empirical nursing personnel into auxiliary personnel. They are also trying to transform auxiliary personnel into a technological staff. This means institutions will have to start educational programs that consider a student body with experience and expectations different from those of students who are involved in conventional programs. It also means professional and work regulations will have to be established within the realm of the policies of the sectors involved; that is, health, social security, education and labor (Arraigada and De Villalobos, 2000). Financial support for the courses is a constant challenge. The guarantee that human resources will be trained is fragile at best, given the changes in social policy and in the government agencies that fund educational programs. With this scenario, those responsible for managing the programs must identify and establish new alliances with funding agencies and consolidate recognition and responsibility for human resource development on the part of the agents involved (education, health services, and profession associations).

The guarantee that human resources will be trained is fragile at best, given the changes in social policy and in the government agencies that fund educational programs. With this scenario, those responsible for managing the programs must identify and establish new alliances.

As demonstrated by the results up to now, preliminary as they may be, schools that offer a course based on a combination of teaching and work experience assume a social commitment from which there is no turning back. In addition to doing what is necessary to provide an educational program consistent with the adopted teaching approach, the school agrees to furnish health services for the community. This requires reaching agreements with health institutions to update health professionals, to give them the tools for research and to increase the production of knowledge, all of which contribute to quality nursing care. However, doing so also requires new objectives, such as the development of partnerships with health services, the assessment of impact indicators, and the establishment of strategies to guarantee financial support and ensure joint responsibility for intervention that will change health services through an approach based on social ethics (Bastos and Magalhaes, 2000).

Both regular programs and those for continuing education face the challenge of educating nursing professionals who have a sense of critical awareness. They must be capable of identifying solutions to problems particular to their profession. If they are to operate properly, these educational programs require a decided policy for institutional support. They also require an efficient and effective effort on the part of the teaching care team, which must be familiar with and dedicated to the characteristics of the model. Consequently, for conventional universities, proposals that are truly innovative represent a challenge to institutional policy (Canaval and collaborators, 2000).

The supply of innovative programs must be able to make extensive use of technological resources for distance and semi-classroom education. Training for teachers and service personnel involves changing the work culture, of the values underlying nursing practice, and of personal enrichment for the participants.

The supply of innovative programs must be able to make extensive use of technological resources for distance and semi-classroom education. Training for teachers and service personnel involves changing the work culture, of the values underlying nursing practice, and of personal enrichment for the participants. These dynamics will be improved through an educational project that not only contemplates academic knowledge, but also the individual group know-how that comes with experience (Davini, 1995).

Nonetheless, one of the aspects to consider and reinforce, as a priority for promoting nursing programs and moving them forward, is the theoretical disciplinary basis on which nursing care and curricula are founded. This goes hand-in-hand with other technological developments and with give-and-take. In other words, "we are talking about knowledge that moves away from the prescriptive field of medicine, without overlooking professional nursing intervention, and leads to decision making and nursing action that is independent and autonomous" (Gallegos de Hernández, 2000).

Nursing urgently needs to take into account the different the levels of nursing education and their impact on teamwork. This applies to the entire range, from the technical level up to the highest postgraduate degree, and is important, even if an institution does not offer all of the programs.

But it is important to note that an expansion in pertinent models and venues for nursing care requires more than curriculum changes. It requires rethinking professional practice for greater clarity, and includes adopting the principles of fairness, universality, quality, and cost-effectiveness as its own.

Nursing in Latin America has grown, matured and become sophisticated during the past 25 years, thanks to experience and the continuous and persistent efforts of a large group of nursing professionals who have helped to develop the Kellogg projects and other initiatives. Maturity has given nurses access to numerous opportunities for professional performance, and nursing practice has been enriched in a way that makes it possible to see new goals, propose action for professional reinforcement and progress, and improve living conditions.

Nursing in Latin America has grown, matured and become sophisticated during the past 25 years, thanks to experience and the continuous and persistent efforts of a large group of nursing professionals who have helped to develop the Kellogg projects and other initiatives.

"At the dawn of this new century," Garzón wrote in 2000,

> the talents of nursing face the challenge of transforming nursing practice and reducing the gap between practice and education. We must remove practice from alienating, instrumental and personal routines and place it in the realm of interpersonal processes for humane and individualized care that is the product of scientifically based analysis and decision making. Although nursing education may be at the forefront of scientific and technological progress, it should not leave practice behind.

We must face the challenge of educating nursing personnel, not only for the specific work involved in nursing care, but also as citizens with social values.

"There is clearly a sustained demand for better education that allows health professionals to confront a more competitive and uncertain job

We must face the challenge of educating nursing personnel, not only for the specific work involved in nursing care, but also as citizens with social values.

market. Consequently, maintaining the status quo is not enough, even if the quality is high. What must be maintained is innovation and prospective" (Manfredi, 2000).

The Kellogg Foundation worked with sectors of nursing in Latin America to address complex challenges through a new educational proposal. This was done by integrating partners and, more recently, by laying the basis for a prominent interdisciplinary, inter-institutional effort, engaging governmental and non-governmental entities and the population to help resolve the health problems of specific groups. The responsibility is, therefore, in the hands of the profession, which has no choice in the matter. Processes cannot be turned back. On the contrary, they exert pressure to move ahead and afford careful thought to the future.

Notes and References

Arraigada, J.; & De Villalobos, M. M. (2000). El caso mas común: Formación de auxiliares a técnicos. In J. Arraigada, G. E. Canaval, M. E. Cevallos, R. De Sena, & M. M. De Villalobos. *Recursos de Enfermería: Contribuciones al Proceso de Desarrollo.* Bogotá: Unibiblos. p. 60–61.

Bastos, M. A. R.; & Magalhaes, Z. R. (2000). Experiencia de la escuela de Enfermería de Minas Gerais. In J. Arraigada, G. E. Canaval, M. E. Cevallos, R. De Sena, & M. M. De Villalobos. *Recursos de Enfermería: Contribuciones al Proceso de Desarrollo.* Bogotá: Unibiblos. p. 42–54.

Blanco, L.; & Lloyd, V. (2000). Programa alternativo para la formación de enfermeras. In J. Arraigada, G. E. Canaval, M. E. Cevallos, R. De Sena, and M. M. De Villalobos. *Recursos de Enfermería: Contribuciones al Proceso de Desarrollo.* Bogotá: Unibiblos. p. 62–69.

Camilleri, D. (2000). Creando y alimentando un sueño americano. In J. Arraigada, G. E. Canaval, M. E. Cevallos, R. De Sena, & M. M. De Villalobos. *Recursos de Enfermería: Contribuciones al Proceso de Desarrollo.* Bogotá: Unibiblos. p. 166–169.

Canaval, G. E.; Gonzalez, M. C.; Acosta, C. M.; Santesteban, C.; Sena, R. R.; & Sesevy, M. G. Maestría otra oportunidad de formación. In J. Arraigada, G. E. Canaval, M. E. Cevallos, R. De Sena, & M. M. De

Villalobos. *Recursos de Enfermería: Contribuciones al Proceso de Desarrollo*. Bogotá: Unibiblos. p. 124–137.

Canaval, G. E.; Cevallos, M. E.; Mesa, I.; & Montes, M. (2000). Formación universitaria del auxiliar de enfermería: Una experiencia. In J. Arraigada, G. E. Canaval, M. E. Cevallos, R. De Sena, & M. M. De Villalobos. *Recursos de Enfermería: Contribuciones al Proceso de Desarrollo*. Bogotá: Unibiblos. p. 73–80.

Ceballos, M. E.; De Sena, R.R. (2000). Oportunidades para Ayudantes de enfermería. In J. Arraigada, G. E. Canaval, M E. Cevallos, R. De Sena, & M. M. De Villalobos. *Recursos de Enfermería: Contribuciones al Proceso de Desarrollo*. Bogotá: Unibiblos. p. 41.

Chompré, R. R.; & Villalobos, M. M. (1995) Nursing Leadership Development and the W. K. Kellogg Foundation. N&HC: Perspectives on Community. 16:4, p. 192–198.

Davini, M. C. (1995). Educación permanente en salud. PALTEX Series for Health Program Teachers. No. 38, Washington.

De Sena, R.; & De Villalobos, M. M. (2000). La especialización: estrategia para la excelencia de la práctica. In J. Arraigada, G. E. Canaval, M. E. Cevallos, R. De Sena, & M. M. De Villalobos. *Recursos de Enfermería: Contribuciones al Proceso de Desarrollo*. Bogotá: Unibiblos. p. 96–98.

De Sena, R.; & De Villalobos, M. M. (2000). Introducción. In J. Arraigada, G. E. Canaval, M. E. Cevallos, R. De Sena, & M. M. De Villalobos. *Recursos de Enfermería: Contribuciones al Proceso de Desarrollo*. Bogotá: Unibiblos.p. 15–26.

Gallegos de Heranndez, Esther. (2000). Formación de Técnicos a Licenciados. In J. Arraigada, G. E. Canaval, M. E. Cevallos, R. De Sena, & M. M. De Villalobos. *Recursos de Enfermería: Contribuciones al Proceso de Desarrollo*. Bogotá: Unibiblos. p. 81–83.

Gallegos de Hernández, Esther. (2000). Formación de Recursos de enfermería: una estrategia clave. In J. Arraigada, G. E. Canaval, M. E. Cevallos, R. De Sena, & M. M. De Villalobos. *Recursos de Enfermería: Contribuciones al Proceso de Desarrollo*. Bogotá: Unibiblos. p. 27–34.

Garzón, N. (2000). Un punto de vista sobre el desarrollo de los recursos humanos ligado a la práctica de enfermería. In J. Arraigada, G. E. Canaval, M. E. Cevallos, R. De Sena, & M. M. De Villalobos.

Recursos de Enfermería: Contribuciones al Proceso de Desarrollo. Bogotá: Unibiblos. p. 154–157.

Grace, H. (1985). The Kellogg Foundation Focus on Nursing. Nursing in Latin America: Strategies for Development. In *Proceedings of the Meeting for Nursing Leaders.* Caracas: FEPAFEM, 9. p. 50–181.

Informe Evauativo del I Foro De Investigación En Centroamérica. (2003). Confidential document prepared by the Nursing School at the Universidad Nacional de Colombia, Bogotá.

Manfredi, M. (2000). Un comentario de la experiencia de la Organización Panamericana de la Salud. In J. Arraigada, G. E. Canaval, M. E. Cevallos, R. De Sena, & M. M. De Villalobos. *Recursos de Enfermería: Contribuciones al Proceso de Desarrollo.* Bogotá: Unibiblos. p. 158–160.

Memorias de la reunión de Barbacena. (1987). Project to Support the Development of Postgraduate Nursing Programs in Latin America. Minas Gerais. p. 34–36.

Memorias de los Talleres de Gestión de Proyectos Participativos. (2002). W. K. Kellogg Foundation.

Nájera, R.M. (2000). La visión de l a Asociación Latinoamericana de Facultades y Escuelas de Enfermería. In J. Arraigada, G. E. Canaval, M. E. Cevallos, R. De Sena, & M. M. De Villalobos. *Recursos de Enfermería: Contribuciones al Proceso de Desarrollo.* Bogotá: Unibiblos. p. 161–162.

Propuestas de Proyectos Innovadores de Educación. (1996 and 1997). A document for internal use by the W. K. Kellogg Foundation.

Proyecto de Apoyo para el Desarrollo de Portgrados de Enfermería en América Latina. (1987). Proceedings of the meeting in Barbacena, Minas Gerais, Brazil. p. 5–92.

Proyecto de Desarrollo de Enfermería en Centro América y El Caribe—PRODECC. (1998 and 1999). PRODECC internal document. Panamá.

Reports on Evaluation of the Cluster of Innovative Educational Projects. (1999 and 2000). W. K. Kellogg Foundation documents.

13 Lessons from Development in Latin America

Roseni de Sena and
Maria Mercedes de Villalobos

An assessment of the Kellogg Foundation's 60 years of support for the development of nursing in Latin America highlights the capacity to respond to the new challenges generated by the way nursing and health systems have changed. These dynamics call for solutions that take into account the capabilities of human resources and facilities.

The Foundation has also displayed confidence in the way that projects are managed financially. This has allowed recipient institutions and individuals to focus efforts on important academic processes (the development and management working proposals), instead of spending time and energy on budgets and financial management.

But financial support, which has been significant and indispensable to achieving the results outlined in the previous chapter, is not the only element of vital importance. The strategies that emerged at the same time, whether planned or spontaneous, made a difference in processes for change.

Another positive outcome was the series of lessons and multiple capabilities—both personal and institutional. They are part of a uniquely rich endowment that has left an important mark, not only on nursing, but also on the people who benefited indirectly from the Foundation's support.

An assessment of the Kellogg Foundation's 60 years of support for the development of nursing in Latin America highlights the capacity to respond to the new challenges generated by the way nursing and health systems have changed.

This wealth is expressed in the projects, institution building, the intellectual and technical advancement of those involved, as well as in nursing's ability to define its own policies and those of the health sector.

In taking stock of the Foundation's work on behalf of nursing in Latin America, and in an effort to explain our experiences with the Foundation, we have grouped our observations in a way that reflects the thinking of those who collaborated in the process and proposed solutions to some of the problems. Complementing the Foundation's financial assistance, the essential elements of the Foundation's support include: (1) respect for the social, cultural, historic, and institutional context; (2) the role of instigator and promoter of ideas, individuals, and institutions; (3) development of strategic action for sustainability and institutionalization; (4) systems for monitoring and evaluation; (5) output of knowledge; (6) support for people; and (7) continuity of program and project managers. Following is a description of each.

Respect for the Social, Cultural, Historic, and Institutional Context

This feature of the Foundation's work allowed for support and development of the nurses and people involved in training, care, management, project planning, and related activities.

The ability to understand and respect both general and specific contexts enabled us to address concrete problems and to play an active role in planning, development, and evaluation. But above all, it helped to foster the ability for self-management and leadership among individuals and groups. As a reflection of this, when selecting consultants and support personnel, the Foundation preferred to lend its backing to the requests submitted by the projects. Outside technical assistance was used only when warranted.

The Foundation also insisted that those who proposed projects for financial support be capable of dealing with the context. This requirement ensured the proposal would be viable (administrative and institutional insertion), and it guaranteed ethical management by the individuals and communities involved.

Initially, the experience gained in the work provided another intangible dividend: it guaranteed the continuity of the many actions generated by the projects. This helped to encourage institutionalizing these actions in local and general circumstances.

The Role of Instigator and Promoter of Ideas, Individuals, and Institutions

The Foundation played an important role in developing ideas throughout the health sector in Latin America in general, and in nursing in particular. Its initial support during the 1940s helped a significant number of nurses obtain training through postgraduate programs and short courses at universities in the United States. Upon returning to their countries (Brazil, Colombia, Costa Rica, Puerto Rico, and others), they took on responsibility for organizing hospitals and public health services, and for creating new nursing schools.

Since that time, the Foundation has been concerned that nurses receive a sound education in order to exercise leadership in their fields, particularly at the heart of their institutions.

Nursing training set the stage for development of the conditions under which work has been carried out for the last 20 years, not only within the scope of structured nursing programs, but for a leading role in other grass-roots projects for material-infant care, maternal-perinatal care, etc. (Kisil, 1992).

The Foundation played an extremely important role in disseminating ideas on the need for nurses to organize in order to play a political role in shaping public policy and nursing policy in education and service.

The Foundation played an extremely important role in disseminating ideas on the need for nurses to organize in order to play a political role in shaping public policy and nursing policy in education and service.

Education for nursing personnel was a decisive factor in guiding the primary health care policy in Latin American countries. Moreover, the work developed in support of the health sector was crucial to defining a number of public policies, some of which became programs for care at the grass-roots level and were highly consistent with services created by the health systems (Mobilizacao de Desenvolvimento de Liderancas, 1989).

The Foundation's capacity as an instigator was also reflected in its effort to support and circulate ideas by encouraging cooperation between project agents whose interests coincided with these initiatives and by suggesting the importance of developing work based on related topics and projects.

One important aspect of the regional work, which provided support for some of the more recent developmental proposals, was the effort to encourage a transfer of knowledge and technology that leads to developmental action and support for people. The bid to support individuals helped broaden and deepen nursing in projects for institutional development at universities, in health services, and within the community. The continued emphasis on specific projects favored the development of individual and group capabilities in institutions and education.

The Foundation's strategy of working with the three I's—Ideas, Individuals, and Institutions—further advanced the work that was already underway.

The Foundation's strategy of working with the three I's—Ideas, Individuals, and Institutions—further advanced the work that was already underway. This can be regarded as a major accomplishment by a community dedicated to philanthropic work; it made intervention through groups and consultants much easier.

We learned that ideas must be appropriated by the people involved in order to be relevant. Individuals, in their particular environments, can encourage the development of new ideas that depend on support from others, institutional conditions, the creation of mechanisms for future sustainability, plans for institutionalization, and processes for evaluation.

The Foundation's support for institutional development showed that communities dedicated to philanthropic work must focus on this issue as part of a strategy to ensure projects that feature new and challenging ideas. This is particularly true for those who want to continue to improve their own projects and themselves.

Therefore, by showing concern for people and institutions when developing a project, and once financial support was over, the Foundation helped to foster the belief among participants and consultants that "when an idea is not self-sustaining over time" it is essentially because of flaws in strategic planning. In other words, a critical analysis of the situation and contextual and strategic proposals for a solution is required for everyday work (UNI Project Cluster Evaluation Reports, 1995, 1996, and 1997).

By insisting that those responsible for project management and execution receive training on key aspects of strategic planning, we have been able to use resources efficiently and effectively.

Moreover, there are no strategic plans or programs without proper context. By insisting that those responsible for project management and execution receive training on key aspects of strategic planning, along with reiterating the need for cost-effective actions, we have been able to use resources efficiently and effectively.

The Foundation's emphasis on combining respect for people and institutions, with the concept of work based on a critical analysis of situations, and on defining problems in the search for relevant solutions, created a culture that offered support for the development and evaluation of proposals and projects. Eventually, it also helped to shape managerial and administrative elements at institutions and with respect to curricula. The development of technologies related to the projects was encouraged. These were transferred to other situations when warranted.

Development of Strategic Action for Sustainability and Institutionalization

The Foundation's insistence that individuals and institutions develop a sustainability plan emphasizing finance, individual training, and institutional development was extremely important. This must be a consideration when a project is being designed. In a number of cases in which the Foundation provided cooperation, the lack of a plan for sustainability was one of the factors that made it difficult to continue the project, or to advance ideas. This is why it is critical that organizations dedicated to philanthropic activities be able to help individuals and institutions prepare

In a number of cases in which the Foundation provided cooperation, the lack of a plan for sustainability was one of the factors that made it difficult to continue the project, or to advance ideas.

projects to face the future, particularly for the time when outside support is no longer available (UNI Project Cluster Evaluation Reports, 1995, 1996, and 1997).

In the nineties, when the Foundation began to implement the idea of working in "project clusters"—such as the UNI and UNIR programs—participating universities, services, and communities were required to provide sustainability plans in order to receive support. Sustainability plans were also necessary for project development and evaluation. In the case of nursing, which was part of the UNI projects, this was an opportunity to learn to work strategically with multiprofessional groups, whether on institutional development or on plans and projects at nursing schools.

Monitoring and Evaluation Systems

An extremely important development was the ability of the Foundation to induce processes for monitoring, systematizing, and evaluating project activities. Few institutions and individuals in Latin America had much experience in these areas. Efforts to institute evaluation to show progress and compare it with other projects taught us to acquire skill for this procedure. It also required people within their institutions to be able to respond to the organizational needs of the projects.

Evaluation seminars allowed for growth as projects progressed. When faced with the difficulty of evaluation and the strategy required for a solution particular to each context, the Foundation itself, the consultants, and the people involved in the projects learned to provide responses.

The evaluation seminars became forums for exchanging experiences and information in order to define strategies to help overcome weaknesses, both in the projects and in the evaluation systems themselves (UNI Project Cluster Evaluation Reports, 1995, 1996, and 1997).

A community dedicated to philanthropic work could learn much from this experience. In many service institutions and universities, the development fostered by the projects was a source of knowledge and methods that circulated in a number of academic areas and became one of the most important resources for institutional evaluation processes.

By the same token, the development of ways and means to monitor and systematize processes enabled the projects and institutions involved to set up monitoring systems of their own. These activities helped in the shift from a culture with scant use of methods for monitoring and systematization to one based on the certainty that these methods make a difference in verifying the progress and difficulties of change. The processes generally are complex and have to be mediated through dialogue and consensus. These, in turn, generate proposals for improvements.

Production of Knowledge

The Foundation also lent important financial support for publications and other ways to disseminate experiences. This was something new for communities dedicated to philanthropic activities in the region. What was novel was the production of literature based on the results of project experiences. This process broke with the traditional pattern of support for literary production by individuals or groups and became a mechanism for production by project groups, as part of their management model.

Support for specific project publications in the three languages in the region (Spanish, Portuguese, and English) eliminated the language barrier in the Americas. This made it possible to record history and to share experiences. It also helped to narrow the gap in literary production in the field of nursing, which is still a problem in Latin America. For those who helped to develop the projects, the consultants, and the nursing community in general, the assistance provided to produce and circulate literature has been an important form of support. The publications, regardless of their type, are reference materials for work in the different areas of nursing and health care. (See the bulletins and newsletters published by the Latin American Nursing Network; the series *Nursing Resources: Contribution to the Development Process*, available in Spanish, Portuguese, and English; the publications derived from development of the UNI program; and *Nursing Education in Latin America* by Chompré Editores. These are but a few examples.)

Support for specific project publications in the three languages in the region (Spanish, Portuguese, and English) eliminated the language barrier in the Americas… it also helped to narrow the gap in literary production in the field of nursing, which is still a problem in Latin America.

Support for Individuals

Support for individuals or groups of people who have the potential to take charge of certain activities, then to encourage change, was clearly of tremendous value to the work developed in the region.

> *The possibility of accompanying the project from the proposal and startup phase to the final evaluation generated in-depth knowledge and prompted individual initiatives that would have been impossible to achieve under other circumstances.*

One important aspect is the individual training we participants received to become familiar with the policies of the Kellogg Foundation and the way it operates. The possibility of accompanying the project from the proposal and startup phase to the final evaluation was significant as well. These activities generated in-depth knowledge and prompted individual initiatives that would have been impossible to achieve under other circumstances. The possibility of the consultants working as a team, to establish guidelines and conceptual frames of reference, made it easier to clarify the theoretical components and to propose methodologies that could be easily transferred to similar situations.

One of the most important experiences, besides being part of a unique project, was our work as consultants to the UNI and UNIR projects, and later with the South American Development Project and the Project for the Development of Nursing Resources in Central America and the Caribbean (PRODECC).

As mentioned earlier, being part of these work teams has been a source of additional encouragement and personal growth. The Foundation urged the consultants and its own staff to work together. This produced a sense of identification with the endeavor and a bilateral understanding of what the projects accomplished and the challenges they faced. A culture of continuing education began to emerge. Because it stemmed from genuine needs on the part of both the projects and the consultants, it led to personal development as well. This even extended to institutions that were not directly involved with the projects (Kisil, 1992; Project to Support the Development of Nursing Postgraduate Programs in Latin America, 1987; Evaluation of the UNI Project Cluster, 1997, 1998, 1999; Arraigada & collaborators, 2000).

Continuity in Program and Project Managers

Having project directors in the Latin American countries made it easier to accompany and initiate ideas and institutional development. The fact that the Foundation had offices in the region, initially in Rio de Janeiro (Brazil) and later in the different sub-regions (Mexico, Chile, Argentina, and the Dominican Republic), guaranteed a close relationship between the Foundation's staff and the individuals and directors at the institutions financed by Kellogg. This factor was also conducive to consultation and facilitated direct access to the consultants.

The fact that the Foundation had offices in the region, initially in Rio de Janeiro (Brazil) and later in the different sub-regions (Mexico, Chile, Argentina, and the Dominican Republic), guaranteed a close relationship between the Foundation's staff and the individuals and directors at the institutions financed by Kellogg.

What would we do differently if the experience were to repeat itself? How would we accelerate efforts to guide and accompany it? We acknowledge that experiences are unique and cannot be repeated. Still, experience changes perspective and provides wisdom to address problems. Following are several observations in this respect.

- **Better coordination or integration between project partners.** Those involved in teaching, health care, and professional processes should be encouraged to play a more active role. This capacity for integration emerged almost spontaneously during the course of the projects. However, had it been addressed from the start as a strategy, its fruits would have been brought to bear sooner. For instance, initiatives required to train assistants and auxiliary personnel could have occurred without so many individual changes in personnel and services. This also might have given the partners and other interested parties a broader understanding of the importance of the proposals.

- **Broadened working strategies.** Because most of the projects were the responsibility of nursing units at the universities, the development of leadership among in-service nurses was more limited, compared with the leadership achieved by teaching personnel. It would have been more appropriate, at the crucial point of development in many projects, to have involved different areas of nursing without ignoring the importance of primary health care. These strategies would have to be based on surveys of the proposals.

Strategies should be proposed to overcome the structural weaknesses of nursing in Latin America. These include the limited number of people with the ability to lead processes for change, the limited capacity for participation in processes involving institutional policies, and the virtual lack of influence on policy decisions at the regional and national level.

- **Support for overcoming structural weaknesses.** It is important to remember the disparities throughout regions as well as countries. Keeping this in mind when analyzing the results will help to explain the delays. In order to generate more harmonious processes for development, strategies should be proposed to overcome the structural weaknesses of nursing in Latin America. These include the limited number of people with the ability to lead processes for change, the limited capacity for participation in processes involving institutional policies, and the virtual lack of influence on policy decisions at the regional and national level.

- **Emphasis on students.** Students are the future of nursing. As such, it is fundamental for them to be directly involved at every stage of project proposal and development. If they are mere recipients of activities, we cannot expect them to contribute ideas that support innovation. We are not being fair to these people, who are so important and make so many contributions to the success of proposals, if we regard them only as producers of action or targets of educational projects. Their creativity makes them indispensable to the success of goals that would be impossible to achieve without them, or would be a long time in coming. However, the most important thing is for students to incorporate and support these ideas for change in their daily work. We believe the experience students would gain by becoming fully involved in the work of a project and in decision making would not only make a contribution to their personal education, but would also make them creative and enterprising. In other words, the student becomes a generator of ideas.

- **Holistic view of nursing.** There were two elements for nursing development that were not emphasized enough when promoting initiatives for ideas and projects. They are the need to view the quality of nursing care from the standpoint of "the work of nursing" and to support continuing education for in-service nurses as well as for teaching nurses. This would help to achieve the prospect of sustained professional quality. These two elements go hand in hand. To interpret this proposal, it is important to understand that the nursing team in Latin America, which includes professional personnel, technical people, and nurse's aides, is a constant feature in the history of the profession and has been determined almost exclusively by economic factors. It has created a culture that emphasizes a horizontal division of multiprofessional work and a vertical position that reinforces the capacity for work, but could affect relations among workers in nursing and their roles in providing health care. As part of this make-up, the role of nurse's aides has been particularly significant. They largely outnumber professional personnel at all levels of care, from primary to high-tech services. Occasionally, they even replace nurses in service management activities and in care that implies a great deal of responsibility. Given the impossibility of changing this situation in the near future, and considering the fact that nurse's aides are strategic to health services, it is unfortunate that labor policies neglect to facilitate access to continuing education for them. Continuing education for nurse's aides surely would alter the results of nursing care in terms of quality and cost effectiveness. This situation is the result of inadequate planning and inappropriate use of resources. This problem was mentioned in virtually every proposal. There was no understanding, early on, that the answer was to work with nursing, not with nurses, and that continuing education was the way to help solve many of the difficulties of nursing personnel, including the teachers in undergraduate and graduate programs, provided it is applied in a way that encourages a constant move towards professional progress and achievement. This means taking into account the realm of labor and educational capabilities.

Education for nurse's aides surely would alter the results of nursing care in terms of quality and cost effectiveness. This situation is the result of inadequate planning and inappropriate use of resources. This problem was mentioned in virtually every proposal.

- **Coordinated efforts.** Finally, we believe the Foundation and the teams that propose projects must learn to work with other cooperation agencies that provide assistance simultaneously to institutions in different countries. Multiple funding is a common phenomenon in the region, given the interest of funding agencies in promoting their particular proposals. Most cooperation agencies finance projects with highly specific objectives and do not deviate from the predefined goals. This means that, at a given point in time, an institution might be receiving assistance from a variety of collaborators with philosophies that are different and a bit dispersed. If cooperation agencies were to view their actions from an integral standpoint, the results could be better. This is a lesson for philanthropic institutions and consultants, as well as those who participate in proposals and projects. The Kellogg Foundation, with its characteristic flexibility in accepting proposals, might consider looking for strategies to remedy this common problem and thereby improve the cost-benefit of its activities.

The inequalities in the region have yet to be overcome, and may not be for quite some time. Nonetheless, we are in an irreversible period of change and motivation.

The opportunity to work with the Foundation has provided a valuable lesson to the people involved in the nursing program, particularly as of 1985. During the course of this effort, regions incorporated experience that had not been considered initially in the proposals, but certainly warranted attention and cooperation to embark on the path toward sustainable development.

The inequalities in the region have yet to be overcome, and may not be for quite some time. Nonetheless, we are in an irreversible period of change and motivation. Institutions and groups of individuals have gained strength in the last 18 years. We now possess methodologies of our own and experiences that can be replicated. Technology has created opportunities that offer the possibility for collaboration and articulation, and will accelerate change. We have networks, but, above all, we have the will to work together, not only for nursing, but also for equitable health care throughout Latin America.

Notes and References

Arraigada, J.; Canaval, G. E.; Cevallos, M. E.; De Sena, R.; Duran de Villalobos, M. M. (Eds.). (2000). Recurso de Enfermería: Contribuciones al Proceso de Desarrollo. Bogotá: Unibiblos.

Evaluación de Cluster de los Proyectos UNI. (1995, 1996, 1997). Documentos de la Fundación W. K. Kellogg.

Evaluación de Cluster de Proyectos de Educación Innovadora. (1999). W. K. Kellogg Foundation document.

Kisil, M. (1992). Acoes para o Desenvolvimento de Enfermagem na América Latina. Reunión de desarrollo de enfermería de América Latina. (Coordination: Rosení Chompré). PRODEM: Universidad Federal de Minas Gerais. p. 12-13.

Movilizacao de Desenvolvimento de Liderancas em Enfermagen para Actuarem en Atencao Primaria. (1989). Proceedings of the meeting held in Contagem, Minas Gerais, Brazil. p. 86-88.

Proyecto de Apoyo para el Desarrollo de Postgrados en Enfermería en América Latina. (1987). Proceedings of the meeting in Barbacena, Minas Gerais, Brazil. p. 5, 91-92.

14 Nursing in Southern Africa: 1987–1994

Helen K. Grace and Gloria R. Smith

As the resources of the Kellogg Foundation grew in the mid-1980s, the board made a strategic decision to shift its project funding from Canada, Europe, and Australia to sub-Saharan Africa. The decision was based on the results of an earlier investigation into the feasibility of extending grant-making to this part of the world. This was at a time when the international community was becoming increasingly concerned about the apartheid system in South Africa, and businesses were divesting from the region as a way to create pressure for change. Recognizing that change was inevitable, and intending to prepare for the changes that would occur, the Foundation decided that one way to contribute to the future of the area would be to provide scholarship support for black African students studying in universities in the region. Full scholarship support was provided through the universities for black Africans studying in the fields of agriculture, education, health professions, and business. Support was also provided to the universities to develop programs to support the students. More than 2,000 students received assistance in the first five years of the Foundation's involvement in the region. The majority of students receiving scholarships were in South African universities, and a number of scholarships went to nursing students.

The Foundation decided that one way to contribute to the future of the area would be to provide scholarship support for black African students studying in the fields of agriculture, education, health professions, and business.

Building upon this, the Kellogg program staff was assigned to assess the potential for making project grants within the overall framework that had been established to guide programming across the Foundation. In March 1987 Helen Grace made an initial visit to Botswana, Swaziland, and Lesotho. As part of this tour, she explored the ministries of health and the nursing educational programs in each of these countries. At the same time, another program staff member from the Kellogg Foundation visited Zimbabwe and South Africa. From these initial visits, the staff gained an understanding of the enormous need for support that guided the thinking about future programming. It was clear that nurses managed most of the primary health care within these countries, with little backup medical support. For example, in Swaziland there were 700 registered nurses in the country and 100 physicians. All but one of the physicians were expatriates from other countries. The registered nurses were predominantly Swazis. Nurses ran the local health clinics and provided all of the care, with physician consultation to consult on complicated cases. In Botswana, Helen Grace visited a clinic in the rural area, and her field notes recorded the following observations:

> I visited a rural health center—within the clinic there were a large number of mothers and children waiting for services. All children are weighed and an assessment of their nutritional status made. Then the nurse sees all mothers and children coming for care. The mothers bring the health record with them as they bring their children. Unlike many clinics throughout the world, where one day is "injection day" and another for "well baby care," they saw patients as they came, and took care of all of their needs at one time. The mothers were responsible for bringing the child's health record with them to the clinic, and the nurse recorded the visit and what had been done for the child, and then returned the record to the mother for its safekeeping. The nurse typically saw 200 patients each morning. Afternoons were spent making home visits and visits to schools. The nurse lived at the clinic, and mothers came to the clinic for delivery of their children 24 hours a day,

seven days a week. A physician visited the clinic twice a week for consultation on difficult cases.

One conclusion reached in the health area was that if the health care of people within the region was to be improved, nurses were the backbone of the system.

An advisory group for the overall programming of the Kellogg Foundation was convened in Botswana in April 1987. This group was drawn from the five English-speaking countries within the region—Botswana, Lesotho, South Africa, Swaziland, and Zimbabwe—and represented the program areas in which the Foundation funded projects: agriculture, education, and health. In the health area, the major area of concern was for primary health care for the majority of the people. Ndiki Ncongo, Botswana's undersecretary for health manpower in the Ministry of Health, was a representative from nursing on this advisory group. One conclusion reached in the health area was that if the health care of people within the region was to be improved, nurses were the backbone of the system.

In September 1987, Helen Grace traveled again to southern Africa, this time to South Africa and Zimbabwe. The issues related to nursing were in some ways similar to those in the other countries, but in others very different. In South Africa, nursing was clearly linked to the apartheid system. The majority of nurses practicing, particularly in rural areas, were black South Africans who had been trained in colleges of nursing which, in their system, are similar to diploma schools of nursing in the United States. With primary health care provided predominately by nurses, nursing in South Africa was not too different from the other countries. But in the political structure, the power was in the hands of university-educated white nurses. In the cities and in the sophisticated tertiary care hospitals of South Africa, the leadership was almost exclusively white. The professional association was governed by whites. In health care as in all other domains, it was not feasible to work through formal channels if nursing was to be upgraded.

Zimbabwe was similar to Botswana, Swaziland, and Lesotho in many aspects, but differed in the educational structure for nursing. While most practicing nurses had been prepared in the equivalent of diploma schools of nursing, the University of Zimbabwe had a fledgling medical school, and was in the process of trying to establish a nursing educational program as a department within the school. With this very preliminary assessment completed, the next challenge was developing a strategy for working to improve nursing education and practice in the region.

Nursing education in southern Africa had been established largely through the work of missionaries in the region. As Vermak notes,

> In the latter half of the 19th century groups of trained nurses came to South Africa from European countries. Most of them were associated with religious orders based in Europe. The Bishop of Bloemfontein realized that if "women's work," including teaching and nursing, was to become effective, it would have to be done on a permanent basis and the organization would have to be done locally and not by remote control from mother houses in home countries. The Anglican Sisterhood, which was founded in Bloemfontein, was responsible for the establishment of professional nursing in South Africa. . . . Sister Henrietta obtained state registration for nurses in the Cape Colony before this was achieved in any other country in the world (Loots & Vermaak, 1975).

Ironically, although South Africa was a leader in achieving state registration for nurses, professional nursing was restricted to whites only. As Kupe quotes a letter from Dyke to Jamieson in 1933, the policy regarding education of nurses in all of southern Africa was summarized as follows:

> We need to establish such training as would more appropriately meet the needs of Natives in rural areas. We have got to produce a type of girl with a moderate amount of theory but well-trained practically—and one who has not become sophisticated and detribalized by a long course of training. Native girls must be given a particular training more in keeping with the present stage of development of natives in rural areas (Selelo-Kupe, 1995).

A program was developed to prepare subordinate African nurses who would work under the supervision of white nurses and were restricted to practice in rural areas. Most nurses were prepared in programs similar to diploma nursing education in the United States. Although licensure of nurses was possible, it was available only for professional nurses. And there were many restrictions that limited the licensure of nurses. For example, in Rhodesia in 1930 the medical council specified that "a general trained nurse cannot be registered unless she produces a certificate of training in a general hospital with not less than 50 European beds." Since all services were segregated on the basis of race, it was impossible for a black African nurse to meet this requirement (Selelo-Kupe, 1995, p. 205). Although independence had been attained in Botswana and Swaziland in 1966, in Lesotho in 1967, and in Zimbabwe in 1980, these countries had small numbers of poorly trained nurses, with their education mainly in curative hospitals.

There were many restrictions that limited the licensure of nurses. For example, in Rhodesia in 1930 the medical council specified that "a general trained nurse cannot be registered unless she produces a certificate of training in a general hospital with not less than 50 European beds." Since all services were segregated on the basis of race, it was impossible for a black African nurse to meet this requirement.

In South Africa the major universities had established baccalaureate nursing education programs, but these programs educated white nurses only. In South Africa, nursing and nursing education was embedded within the apartheid system, with white nurses enrolled in university programs and African nurses in "colleges" of nursing. Because of the apartheid system in South Africa, a strategic decision was made to focus upon the four other English-speaking countries in the region that had achieved independence. This was part of an effort to improve primary health care through support of nursing education and practice in these countries. Support from the Foundation for changes in nursing coincided with the adoption of public policies focusing on primary health care for all and the need to change the nursing educational systems.

Southern Africa offered the refreshing example of nurses mounting a major reorganization of education and practice through governmental policy change.

Southern Africa offered the refreshing example of nurses mounting a major reorganization of education and practice through governmental policy change. In Botswana, for example, nurses organized a process that convened a series of open meetings for nurses, nurse assistants, consumers, and local and national policy and decision makers. The meetings were held in villages, townships, and cities across the country. It is a small, rather affluent nation of 1.6 million people. Botswana is one of the few African nations that has little debt and sufficient reserves to support its own development. The teams of nurses who traveled across the country were representative of available power: educators from the university and the Ministry of Health training programs, directors and staff from the nursing service, representatives from the professional association, and staff from the board of registration. It was necessary in this case for the leaders of the reform to convince nurses, the local authorities, and parliament that reform was in the public's best interest. The Botswana were a conservative people who were prudent purchasers. Health care delivery and training in Botswana were largely public expenditures. Making a case for upgrading the nursing profession and lengthening the course of study necessitated an emphasis on the health care needs of people. Nurses in Botswana were already involved in primary care delivery; they felt, however, that they needed more preparation in physical assessment, critical thinking, problem solving, leadership, and management in order to function more effectively in settings where often they were the only, or most readily available, personnel. Though the example of Botswana can hardly be applied in large nations with more complex structures, it is useful for other small nations and their governments. Botswana also offered an example of a national nursing association providing programs and projects to improve health status of communities that were not just intermittent exercises. The Botswana Nurses' Association owned its own headquarters building in which it established a daycare program for children. It secured funding from private foundations and village governments to establish feeding programs for children and the elderly. It worked with the agriculture ministry to promote and assist the development of community gardens to feed families and produce cash crops. A community clothes closet was established in the headquarters building; it was supported by an annual drive for resources and clothing that engaged nurses across the country.

The Nursing Initiative

In all of the countries, nursing leaders were interested in upgrading nursing education by linking the educational programs to universities within the countries, changing the focus of nursing education to primary health care, and moving nursing education from the ministries of health into university-based education. In Botswana, they had made considerable progress in this area and had established a baccalaureate nursing educational program within the university that allowed registered nurses to complete a degree modeled after U.S. degree completion programs. But by combining the two educational models—the British with the United States—the program took 10 years from beginning to end to complete a baccalaureate degree.

A first step in building a nursing initiative for the region was to follow an example from the work in Latin America and the Caribbean and convene a small group of nurses from the four African countries of Botswana, Lesotho, Swaziland, and Zimbabwe. The International Council of Nurses was working with the Latin American countries to assess the regulations governing nursing licensure and practice. A series of workshops were convened as part of this project. In January 1988, a workshop was planned in Tobago, and nurses from Africa were invited. Two representatives from each of the countries, with the exception of Zimbabwe, participated in the conference.

A first step in building a nursing initiative for the region was to follow an example from the work in Latin America and the Caribbean and convene a small group of nurses from the four African countries of Botswana, Lesotho, Swaziland, and Zimbabwe.

The setting of Tobago was particularly significant for the Africans in that most Tobagoans had African roots, and for the nurses from Africa, it was their first opportunity to visit a country outside of their region. Each of the countries joined with representatives from countries throughout Latin America and the Caribbean and participated in reviewing the regulations that governed their education and practice.

The African countries also met to develop and recommend a strategy for working to upgrade nursing education and practice in the African region. The recommendations of this group were to convene a meeting within the region with representatives from four sectors: (1) ministries of health; (2) nursing education; (3) universities; and (4) nursing associations. The meeting was to be held in Botswana in March 1988. Marilyn Edmondson, a consultant from the Academy for Educational Development that was part of the faculty at the University of Botswana, assisted in the planning of this meeting.

On March 28, 1988, a meeting of representatives of the four countries met in Gabarone, Botswana. Botswana, Lesotho, and Swaziland each had four representatives from the ministries of health, the basic nursing education programs, universities, and nursing associations. Zimbabwe was represented by two people—one from the nursing education sector and the second from the Ministry of Health. At this conference each country made a report on the status of nursing and their particular needs. After each of these reports the meeting then focused on developing common objectives for the group as a whole. These objectives were:

- Develop continuing education to upgrade the skills of nurses in practice and to provide for educational mobility;
- Develop linkages between basic programs and universities;
- Prepare leaders in nursing and allied health management and practice;
- Strengthen nursing practice through a review of the regulatory system that influences participation in primary health care;
- Develop networking activities and strengthen professional nursing associations within the countries.

The strategies for achieving these objectives were to be a combination of in-country and cross-country activities: "It was agreed initially that it would be best to explore ways of addressing these issues at a country level, recognizing that over time regional approaches may develop" (What Have We Learned?). Each country was to develop a proposal for work within their country. To assist them in this effort they requested consultation for a counterpart from the United States who would work with them. They also requested support for a country representative to participate in workshops sponsored by the International Council of Nurses on the regulation of nursing education and practice that were to be conducted in Africa.

As a follow-up to this meeting, Dr. Elnora Daniel, dean of the College of Nursing at Hampton University, was contracted as a consultant to visit each of the four countries to assess the status of nursing education and practice. In October 1988, Dr. Daniel provided her consultation report and summarized the needs of the countries in four categories: (1) consultation; (2) fellowships to support the training of nurses for education and leadership; (3) provision of personnel to replace faculty on leave to complete higher education; and (4) technical assistance in the form of workshops and short courses. A meeting on October 24 with Dr. Daniel and Dr. Treadwell from the Kellogg Foundation produced a recommendation that a core group of consultants be identified, and that individual consultants would work with individual countries over a period of time. Two fellows from the Kellogg National Fellowship program were identified: Drs. Ora Strickland and Faye Gary-Harris. Two additional consultants—Dr. Dorothy Powell, dean of the School of Nursing at Howard University and Dr. Lucille Davis, dean of the School of Nursing at Northwestern University—were invited to serve as consultants.

In the January 1989 meeting in Harare, Zimbabwe, the U. S. consultants were introduced and began work to address issues in each of the countries. Dr. Strickland was paired with Botswana, Dr. Gary with Lesotho, Dr. Powell with Swaziland, and Dr. Davis with Zimbabwe. The January 1990 meeting held in Swaziland continued this work. At this meeting participants agreed that the focus for the next meeting would shift from planning to reporting on activities that had been undertaken in each of the countries over the three-year period.

The next meeting, scheduled for January 1991, was postponed because of the unrest in the area related to the Gulf War. It was rescheduled for May, but again was cancelled because of the political situation in Lesotho and the degree of unrest. Finally, the fourth meeting was convened in January 1992 in Maseru, Lesotho. Participation in this meeting was expanded to include the regional director from the World Health Organization based in the Congo; the Secretary of the Eastern Central Southern African Colleges of Nursing (ECSACON), which was a newly forming regional organization; and a representative from South Africa, Carolyn Ntoane, director of the School of Nursing at the University of the Western Cape. U.S. representatives included Hattie Bessent, deputy director of the American Nurses Association; Joan Cobin, project director, International Health Manpower, University of California System; and Jean Kearns, executive director of the Western Institute of Nursing (WICHEN). In a summary paper, Helen Grace reviewed the work to that point: "To this point approximately $4-4.5 million has been invested in strengthening nursing within the region. About 25 nurses have been provided support for study outside of the region. The average level of support for one student is $80,000; about $2 million has been directed toward this support" (What Have We Learned?). As part of this "support," faculty from the United States were filling spots vacated by nurses from Botswana and from Swaziland who were studying for advanced degrees outside of the region. The consultants had made repeated visits to the countries to work with them around their priorities and to conduct workshops that were a part of the overall plan. While South Africa had not been included in the original work because of the political situation within that country, the participation of a South African was a first step toward building relationships between nurses in the region. A second major factor was the participation of ECSACON. The stage was set for the next phase of development. As noted in the summary presentation by Helen Grace:

While South Africa had not been included in the original work because of the political situation within that country, the participation of a South African was a first step toward building relationships between nurses in the region.

> This meeting constitutes a significant turning point from a number of perspectives. First, there is a transition in leadership from the Kellogg Foundation. Gloria Smith has assumed responsibility for the leadership of this effort, while my role has shifted to include some broader responsibilities within the Foundation. Secondly, this is a time for transition in the pattern of our interaction. In the initial phase, the Foundation has taken the initiative in starting the work to strengthen nursing; from this point on, it is important that you, within the region, assume responsibility for

> setting direction and establishing priorities. Our goal has been, and remains, to build capacity with the region—to encourage self-direction and increase independence. To achieve this end, leadership and future direction must come from within the region (What Have We Learned?).

As an outgrowth of this meeting, a steering committee was formed that assumed responsibility for planning further activities within the region.

The fifth Southern African Nursing Network Conference was convened in Gabarone, Botswana, in January 1993. The focus of this meeting was "Bridging the Gap between Nursing Education and Nursing Practice," and participants at this conference included consultants from the United States, and inclusion of South African nurses. Representatives of each of the countries presented formal papers; consultants from the United States and the United Kingdom provided papers on topics such as nursing leadership, regulation of nursing practice, and distance education. In this meeting a framework for regional cooperation was developed. The objectives for regional cooperation included the following:

- To improve the quality of primary health care services by strengthening the capacity of nurses;
- To promote a uniform level of minimum standards of education and practice to ensure quality of service throughout the region;
- To develop the competence of nurses as managers, participants, and leaders of multidisciplinary teams of PHC practitioners;
- To promote networking in the region to help develop nursing services and nurses.

While in the initial meetings in 1987 the participant countries had adopted a policy of each working independently, by 1993 they had taken the first tentative steps toward regional collaboration.

While in the initial meetings in 1987 the participant countries had adopted a policy of each working independently, by 1993 they had taken the first tentative steps toward regional collaboration.

The sixth Southern African Nurse Networking Conference was convened in January 1994 at Victoria Falls, Zimbabwe. The theme of the conference was "Economics, Power, Politics, and Leadership in Health Care Management." As Clara Mufuka-Rinomhota, director of Nursing Services in the Zimbabwean Ministry of Health, summarized,

> The message came out loud and clear that if nurses were to make a difference in health care, they must break from the Florence Nightingale tradition of a "good," unquestioning woman. This must be achieved by being more assertive and educationally sound. It is the management of change and accommodation to change that makes nursing what we want it to be. No one is going to give nurses power to manage their own affairs, but it is now time for nurses to say, "We harnessed the past, help us secure the future" (Proceedings of the Sixth Southern African Nursing Networking Conference).

In this conference, the proceedings show increased progress in each of the countries in achieving the original goals. The sophistication of the papers presented indicated the progress that had been made, and the commitment to moving forward to achieve the primary goal of improvement of health care for all through improved nursing education and practice.

By 1994, significant progress toward reaching the original goals could be seen. In particular, the entire educational system had been upgraded throughout the region. Zimbabwe and Botswana had increased their baccalaureate programs.

By 1994, significant progress toward reaching the original goals could be seen. In particular, the entire educational system had been upgraded throughout the region. Zimbabwe and Botswana had increased their baccalaureate programs. With assistance from Case Western Reserve University, these programs integrated distance education as a way to reach out to the large number of nurses prepared in the diploma programs. Under assistance from the University of South Africa, doctoral education was available for nurses within the region. This enabled many nurses who could not uproot themselves to leave the area to get doctoral education in other countries. Some, for example, held responsible positions within the ministries of health, and the availability of the UNISA program made it possible for them to maintain their positions while gaining additional preparation.

While changes within the Kellogg Foundation refocused funding efforts within the region and de-emphasized nursing, they at least led to the establishment of an infrastructure for nursing development within the region. In particular, the numbers of nursing fellowships that were supported during this time period significantly contributed to a strengthened leadership base within the profession. Significant changes had been achieved through other support strategies such as the development of distance education, which contributed significantly to upgrading the nursing workforce. Options for graduate education were developed throughout the region, and new alliances were built between the educational and practice sectors throughout the region. Regulations governing nursing education and practice were strengthened, and the role of nursing in primary health care was more clearly delineated. And perhaps most importantly, the ministries of health had reached a new understanding of the contributions of nursing to primary health care in their countries. As Kupe notes, "The success in six years is unprecedented. It is now possible to prepare B.S. and M.S. nurses within the region, and a number of teachers and administrators, nurse specialists and researchers have been prepared. . . . The Kellogg Foundation introduced the idea of collaboration or regional networking. . . . It has not been an easy marriage, but sharing ideas is bringing us closer together for the benefit of nursing and the people of the region" (Selelo-Kupe, 1995, p. 208). Appendix A summarizes the funding that was directed specifically to the nursing initiative. Appendix B provides a listing of the fellowships that were awarded to nurses in the region.

While the dream was to develop long-range cohesion and coherence in nursing in southern Africa, what was demonstrated, instead, was fierce national identity and loyalty. Regional cooperation was lost to competition between nations. Nurses from each country were driven to excel, and in some instances, engaged in intense national rivalry.

The Southern African Nurses Network, during its seven years of existence, achieved many successes. The management and oversight of the Network was in the hands of a steering committee whose members had been selected by country-specific processes. The steering committee engaged in strategic planning, program development, evaluation, and monitoring. Within each country, it devised processes to develop and implement plans that helped move the regional agenda. Annual meetings and conferences included broader representation of nursing communities from each country. The annual meeting provided the opportunity to

There was consensus on the fact that the countries were in desperate need of a better prepared nursing workforce; resolve to upgrade the nursing workforce; and recognition that the faculty workforce would have to be expanded and better qualified.

report on progress (or lack thereof), receive input, and critique, negotiate, and resolve. There was consensus on the fact that the countries were in desperate need of a better prepared nursing workforce; resolve to upgrade the nursing workforce; and recognition that the faculty workforce would have to be expanded and better qualified, as would also be true for nurse administrators and clinical leaders.

The regional plan that emerged supported establishing master's programs at the University of Botswana and the University of Zimbabwe, where bachelor's programs in nursing already existed. Lesotho and Swaziland determined that they would develop and institute programs for nurses at the university level. These goals were supported. The new vision included entry-level university programs in nursing for new high school graduates to complement existing programs for registered nurses; the University of Botswana led this development. Practicing nurses indicated the need for options that were more consistent with the realities of their lives. They sought distance-learning strategies and access to learning sites closer to their homes in remote villages and towns. The Botswana Ministry of Health was more responsive to this request than any other; it successfully launched a massive upgrade program using a variety of distance-learning strategies. Creating access to graduate education programs was not just a matter of creating new programs. There were many negotiations over administrative and governmental policies. Nonresident student fees were almost prohibitive for nurses from poor nations like Lesotho and Swaziland. University admission requirements were rigid and exclusionary for most nurses. But despite these impediments, regional planning supported reforms in education that led to better education for nurses and improved access to higher education for others in the region.

Focus on Southern Africa

The approach to working with nursing in South Africa took a far different direction than that in the other four English-speaking countries in the region. Nursing in South Africa was embedded in the apartheid system of government, and strategies for strengthening nursing in primary health care had to be different. Nursing education in South Africa was separate and unequal. The majority of nurses of African descent were trained in "colleges of nursing," programs equivalent to diploma schools of nursing in the United States. White South African nurses were educated in university programs of nursing equivalent to our baccalaureate programs. The majority of nurses providing primary health care to black South Africans were graduates of colleges. In seeking ways to improve primary health care, the efforts to strengthen nursing used primary health care projects in various regions of the country.

In community-based health services, nurses provided leadership in primary health care to rural populations and to squatter communities that developed around the major population centers. The Ithuseng project was located in a largely rural area in northern South Africa. There a comprehensive community development project had developed through the years providing primary health care to the region, and early childhood education and economic development activities in the agricultural area. This project was headed by a nurse, and Kellogg Foundation support helped them expand their daycare activities and improve their agricultural practices.

In community-based health services, nurses provided leadership in primary health care to rural populations and to squatter communities that developed around the major population centers.

The Refenghotso project was also headed by nurses who had established a clinic in a community outside of Johannesburg. Rural South Africans migrated to the larger cities in a hope of finding some form of employment. They built make-shift houses out of whatever building material they could find or afford. Usually their houses were made of sheet metal and built upon land that they could find, usually in large settlement areas housing thousands of people on a small parcel of land. The Refenkghotso project supported the work of nurses who had established a primary health clinic in the area.

Rural South Africans migrated to the larger cities in a hope of finding some form of employment. They built make-shift houses out of whatever building material they could find or afford. Usually their houses were made of sheet metal and built upon land that they could find, usually in large settlement areas housing thousands of people on a small parcel of land.

The Kwa Mashu Christian Care Society had been developed by African nurses in a densely populated area outside of Durban. The initial concern was for the large numbers of elderly who were living alone and in destitute conditions in the community. Traditionally, African families care for their elders, but because many more families migrated to seek a better life, elders often were separated from them. In the Kwa Mashu community, many elders and disabled individuals tried to subsist in their homes. Three nurses had sought out money from businesses in Durban to build a "nursing home" for those who needed ongoing nursing care. They also did an extensive outreach into the community with a program of home visits meal delivery. Grocery stores in the Durban area donated food that was near the limits of being useable. The program received support to expand facilities and services, and for the addition of a child-care component. The program engaged elders to provide child care for working mothers.

In health-professions education, a number of projects that involved nursing were funded. For example, in rural areas of South Africa, lay midwives attended many of the baby deliveries. A particularly successful way to improve primary health care for this population was to upgrade the education of nurses so that they in turn would work with midwives within their region to improve the quality of care. Two projects were funded to support this effort—one in the northern rural communities in South Africa, and another in the Durban area. From these projects, a group of nurses traveled to Brazil and worked with nurses there who had established a highly effective program of working with lay midwives and developing a system of care that had markedly reduced the infant mortality rates in their region.

Another avenue for strengthening nursing was through the Community Partnerships/Health Professions Education effort, patterned after the U.S. endeavor. The South African CP/HPE effort involved nurses in every phase, starting with the original advisory committee and including full participation in each of the projects that was ultimately funded (Appendix C). A total of $25 million was invested in the Community Partnerships/Health Professions Education in South Africa. For the first time nurses of African descent were recognized as nursing leaders and participated fully with their counterparts in other health professions.

Summary and Analysis

Nursing initiative. In both southern Africa and in Latin America and the Caribbean, nursing has been recognized as an essential profession in the provision of primary health care. Support from the Kellogg Foundation has been important in strengthening the role of nursing and increasing nurses' effectiveness in providing important primary health care, particularly to underserved populations. The Latin America/Caribbean experience differs in many important aspects from that in southern Africa. Unlike in Latin America, where the effort to strengthen nursing stemmed from 50 years of Kellogg Foundation support in the region, the southern Africa experience in many ways was the Foundation's introduction to that region. This was problematic because the ministries of health in the four countries of Botswana, Lesotho, Swaziland, and Zimbabwe had other priorities. For example, Zimbabwe was developing a medical school, and Botswana sought to create its own. But the students enrolled in the medical school in Zimbabwe were mainly expatriates and Europeans who had lived in the area for some time. The medical school was often used as the offshore schools in the Caribbean, where young people who could not get admitted to medical schools in the U.K. or in other European countries sought admittance. Their plans were to return to their home countries upon completion of their medical education, and they did little to improve the status of health care within their countries. Support for nursing and nursing education brought increased visibility to the role that nursing could play in improved primary health care for people, and over time the ministries of health began to recognize it as important as, although less prestigious than, support for medicine.

Unlike in Latin America, where the effort to strengthen nursing stemmed from 50 years of Kellogg Foundation support in the region, the southern Africa experience in many ways was the Foundation's introduction to that region.

The point of entry to support of nursing within the region was also very different. In Latin America and the Caribbean, the leadership group for nursing development grew out of projects that the Kellogg Foundation had funded in the past. In southern Africa, the approach was to engage titular leaders—the representative for nursing in the ministries of health, the director of the nursing educational program, and the director of the professional nursing association. Also, all four countries in the region were represented by these titular leaders. In Latin America/Caribbean, there was no attempt in the beginning to include all of the countries, although through time this has occurred. By working with all of the countries in the southern African region with the exception of South Africa, and by following the recommendations of the initial group to focus upon the individual countries rather than upon a more collaborative approach, the political process grew complex. There had been little history of collaborative work between the countries, with the exception of the common licensure examination developed between Botswana, Lesotho, and Swaziland. Zimbabwe had recently gained its independence but was enmeshed in the process of moving from a colonial system to a more democratic society. Over the six years of work with the four countries, considerable progress was made toward greater collaboration, particularly in the area of education. Distance education made it possible for nurses who had been trained in the old system to link to university-based educational programs and gain bachelor's and master's degrees and gain status as professionals. By linking to the University of South Africa and its distance education programs, nurses could also gain doctoral preparation within the region. This was important for women who, by tradition, were limited in their mobility by their culture.

There had been little history of collaborative work between the countries, with the exception of the common licensure examination developed between Botswana, Lesotho, and Swaziland.

Perhaps the most outstanding contribution of the nursing initiative was the number of fellowships granted for advanced study. Fifty nurses from the four countries received fellowships, mainly for advanced degrees in nursing, and are now back in these countries providing leadership at all levels of nursing education and practice.

In the mid-nineties, within the Kellogg Foundation, programming priorities shifted to a focus on integrated development. Within this framework, specific support for the nursing initiative dissipated and sustainability wasn't achieved. For example, in Lesotho and Swaziland, the smallest countries in the region, support was terminated; although officials had gone through all of the legislative processes to move nursing education into the university, they could not get the programs established without financial support. Consequently, the final step of securing a place for nursing education within the universities was not achieved.

South Africa. In South Africa, the Kellogg Foundation's major contribution in support of nursing was to gain recognition of the contributions nursing could make toward improved primary health care in the country. As the apartheid system was breaking down, the need to address the pervasive issues of separate but unequal systems of health care was of utmost importance. Kellogg's support of the Community Partnerships/Health Professions Education initiative and the insistence of full participation of nurses made an important statement, especially in a country in which the profession of medicine was so predominant. African nurses demonstrated their abilities to be strong and forceful leaders and to have important links to communities that were essential in developing comprehensive primary health care throughout the country.

As the apartheid system was breaking down, the need to address the pervasive issues of separate but unequal systems of health care was of utmost importance.

In summary, the funding of the Kellogg Foundation to nursing in the southern African region extended over a 10-year period of time in which the seeds were planted for development of the nursing profession. The following chapter tells of these efforts from the perspective of nurses within the region. With this leadership in place, perhaps those seeds will nurture and develop in the years ahead.

Notes and References

Loots, Idalia; & Vermaak, Mollie. (1975). Preface. In P. J. de Villiers, *Pioneers of Professional Nursing in South Africa.*

What Have We Learned? *Fourth Southern African Network Nursing Conference,* p. 68.

Proceedings of the Sixth Southern African Nursing Networking Conference, p. iv.

Selelo-Kupe, Serara. (June 1995). Strengthening the Nursing Profession for Leadership. *Nursing and Health Care: Perspectives on Community,* 16, no. l.

Appendix A: Grants in Support of the Nursing Initiative in Botswana, Lesotho, Swaziland, and Zimbabwe: Community-Based Health Services

Comprehensive Health Services

Date	Grantee	Amount
2/91	Ministry of Health-Zimbabwe	$1.314.647
6/92	Nurses Association of Botswana	399,286
7/94	Maluti Adventist Hospital	9,123
9/93	Scott Hospital	98,596

Health Professions Education

Date	Grantee	Amount
10/89	Ministry of Health-Botswana	75,753
1/93	Ministry of Health Botswana	223,271
4/92	University of Botswana	238,145
8/94	Ministry of Health-Botswana	800,000
1/94	University of Botswana	1.162,000
1/89	Ministry of Health-Zimbabwe	169,588
2/91	Ministry of Health-Zimbabwe	1.314,647

Health Professions Education

Date	Grantee	Amount
7/92	Ministry of Health-Zimbabwe	129,780
6/92	Ministry of Health-Zimbabwe	199,649
10/90	International Council of Nurses	18,578
12/89	Ministry of Health-Lesotho	22,825
1/91	Ministry of Health-Lesotho	116,945
1/92	Ministry of Health-Lesotho	117,200
6/93	Ministry of Health-Lesotho	582,500
4/94	Ministry of Health-Lesotho	70,497
12/89	Ministry of Health-Swaziland	32,348
8/91	Ministry of Health-Swaziland	98,049
7/92	University of Florida-Gainesville	45,987
7/88	University of Illinois at Chicago	359,958
9/91	University of Illinois at Chicago	256.613
7/91	World Health Organization	96,100
7/90	World Health Organization	124,200
7/91	World Health Organization	96,100
12/92	Case Western University	184,018
1/94	Case Western Reserve University	1,833,163
	TOTAL	$10,360,511

Appendix B: Fellowships Made to Develop Nursing Leadership in Botswana, Lesotho, Swaziland, Zimbabwe, and South Africa:

Botswana

1997	Miriam Sebego	$136,939
1994	Koona Keapoletswe	95,000
1990	Onalenna Lemo	89,000
1990	Florah Lepodisi	97,000
1991	Mponah Manthe	115,000
1992	Keamogetse Masupologo	120,000
1989	Keitshokile Mogobe	82,800
1994	Keitshokile Mogobe	145,000
1992	Mary Moseki	100,000
1993	Ephraim Ncube	53,000
1992	Patricia Ncube	124,006
1992	Nthbiseng Phalade	306,000
1992	Thandia Phindela	144,000
1990	Sarah Rathedi	101,613
1991	Sinah Relu	110,000
1992	Motshedisi Sebone	81,000
1992	Miriam Sebego	110,000
1991	Mmapula Lucy Sechele	120,000
1993	Esther Seloilwe	142,000
1991	Sheila Shaibu	232,000
	TOTAL	$2,307,358

Lesotho

1991	Malechaba Hoeane	$35,000
1994	Katleho Lelosa	80,000
1991	Adeline Lesaoana	35,000
1990	Elsie Makoa	80,000
1992	Tiisetso Moiloa	140,300
1991	Alina Ramokoena	102,000
1994	Makhabiso Ramphoms	307,000
1992	Eliza Sethinyane	110,000
1991	Victoria Thelejane	95,000
1993	Isabel Yako	268,000
	TOTAL	$1,252,300

Swaziland

1995	Beauty Makhubela		$152,000
1990	Winnie Nhlengethwa		100,000
1991	Lena Radebe		45,000
1993	Nonhlanhla Sukati		147,000
1990	Bertha Vilakati		126,000
1990	Isabel Zwane		101,000
		TOTAL	$671,000

Zimbabwe

1992	Rosemary Bakasa		$162,200
1993	Eileen Chinyadza		129,421
1991	Doreen Chonoma		32,000
1993	Ursula Dengedza		121,000
1993	Helen Gundani		158,000
1994	Margo Mapanga		45,000
1991	Margo Mapanga		120,000
1991	Anne Maruja		33,000
1996	Regina Matambanadzo		137,000
1994	Elizabeth Moyo		116,000
1992	Auxilia Munodawafa		157,000
1995	Sylvia Mupepi		175,000
1994	John F.K.Mutikani		312,000
1993	Maria Ncube		101,500
1994	Florence Ndhlovu		96,000
1994	Sithokozile Simba		149,289
		TOTAL	$2,044,410

South Africa

1994	Rhosta Gcaba		$17,000
1993	Rachel Gumbi		15,000
1997	Elizabeth Malomane		117,000
1991	Roselyn Mazibuko		55,000
1992	Lindiwi Mhlanga		110,000
1995	Mankuba Ramalepe		141,050
1990	Amelia Ranotsi		85,000
1994	Amelia Ranotsi		96,000
1995	Hilda Selepe		145,000
			$781,050
		GRAND TOTAL	$7,056,118

Appendix C: Grants made in South Africa in which Nurses were Project Leaders

Community-Based Health Services

4/91	Refengkgotso Community Project	$344,243
3/93	Ithuseng Community Association	25,375
6/90	Ithuseng Community Association	67,692
12/95	Ithusheng Community Association	88,400
8/90	Kwa Mashu Christian Care Society	365,225
	TOTAL	$890,935

Health Professions Education

10/91	Department of Health and Social Welfare-Gazankulu-Giyani	$4.000
6/94	Gompo Welfare Organization	126.023
8/94	KaNgwane Government	5,228
12/92	University of the Western Cape	15,596
8/89	Natal University Development Foundation	83,320
2/92	University of Natal	17,665
8/92	University of Natal	133.125
7/94	University of Natal	63,887
5/94	University of South Africa	36.725
10/94	University of South Africa	673.946
10/88	University of Cape Town	55,467
	TOTAL	$1,214,382

Community Partnerships/ Health Professions Education

9/93	University of the Orange Free State	$463,269
7/94	University of the Orange Free State	3,285,691
3/93	University of the Western Cape	256,265
7/94	University of the Western Cape	3,181,094
3/93	University of the Witswatersrand	432,701
3/93	University of the Witswatersrand	171,653
7/94	University of the Witswatersrand	2,339,002
7/94	University of the Witswatersrand	3.542.247
3/95	University of the Witswatersrand	380,635
11/94	University of the Witswatersrand	670,020

3/94	University of Durban-Westville	6,575
4/96	University of Durban-Westville	1,796,325
8/93	University of Fort Hare	118,830
7/94	University of Fort Hare	1,196,409
8/89	Natal University Development Foundation	83,320
	Community Partnerships/ Health Professions Education	
4/90	University of Natal	$62,250
2/91	University of Natal	50,000
10/92	Natal University Development Foundation	507,229
8/91	University of Natal	953,315
3/93	University of Natal	144,359
7/94	University of Natal	991,884
3/93	University of Transkei	486,529
7/94	University of Transkei	1,898,684
9/94	University of Transkei	47,000
6/95	University of the Transkei	96,000
1/91	University of Capetown	55,467
4/95	University of Cape Town	87,233
12/98	University of South Africa	1,069,765
	TOTAL	$24,374,271

15 Nursing in Southern Africa: A Regional Perspective

Mary Malehloka Hlalele

During the 1980s, the vast majority of countries in the developing world were facing the imposition of the World Bank's structural adjustment policies. These policies, from the World Bank's perspective, were intended to reduce and eliminate unsustainable external and internal imbalances in economies. The World Bank also saw this as a way to increase economic flexibility in order to respond to change and enable affected nations to use resources more efficiently. Contrary to this viewpoint, the structural adjustment policies were a source of increasing impoverishment for the "adjusting" countries.

In the countries where structural adjustment programs were imposed, real incomes declined, costs of living rose, and the government reduced spending on social services. Soon it was clear that the living standards of the majority of people were in decline. As governments intensified the processes of privatization, retrenchment of workers led to severe cuts in public health services, and areas of social services that had been offered for free now came at a price.

In 1990, the UN General Assembly on International Economic Cooperation concluded that the structural adjustment programs in many instances had led to exacerbation of social inequity without restoring economic growth and/or development, while also threatening political stability. The negative effects of the structural adjustment

In the countries where structural adjustment programs were imposed, real incomes declined, costs of living rose, and the government reduced spending on social services. Soon it was clear that the living standards of the majority of people were in decline.

programs could be well illustrated through an example of Zimbabwe, which, under the auspices of the expanded structural adjustment programs, had to drop its health expenditure by 20%. In Zimbabwe, the effects of the expanded structural adjustment programs, compounded by the scourge of HIV/AIDS, has prompted a rapid reversal of the impressive gains in areas like life expectancy and literacy rates that Zimbabwe made in the seventies and early eighties.

The people's realities and their roots. For the majority of people in southern Africa one of the most pressing and overarching challenges has remained crushing poverty, with the vast majority of poor people living in rural areas or in crowded peri-urban slums. There, basic human needs for survival continue to be problematic, and long distances continue to be the main barriers to people accessing basic essential services. As is common across the sub-region, most infrastructure and employment opportunities are concentrated in the urban centers. Inadequate and impure water supply, lack of adequate food for sound nutrition, and sub-standard housing all contribute to the ever-increasing incidence of infectious diseases.

These problems are rooted in the region's history. Medical services were introduced into southern Africa by the colonialists, well over a hundred years ago, solely to preserve the health of the European community, and to keep the native labor force in good working condition while controlling epidemics. In South Africa, the oppressive system of apartheid was introduced primarily to exclude black populations from having access to any facilities whatsoever. In the health sector this situation was exacerbated by the fragmentation of the health department through the creation of homelands. The problems were further compounded by educational systems that were constructed around racial divisions, with health sciences institutions designed as separate and unequal.

To this day, years after the collapse of the colonial era, health care services in southern Africa still perpetuate the old tradition of expensive, high-tech curative care in large hospitals, with Western-trained doctors as the main providers of medical care, almost all confined to the towns. This colonial model, inherited by all southern African governments and maintained over many decades, was tightly controlled with centralized funding and management systems. Today a real challenge to the fragile democratic governments in southern Africa, all of which face the daunting responsibility of providing equitable health care services to their citizens, is managing the progressively deteriorating health services in the face of declining economies.

Today a real challenge to the fragile democratic governments in southern Africa, all of which face the daunting responsibility of providing equitable health care services to their citizens, is managing the progressively deteriorating health services in the face of declining economies.

In the early 1980s, following the Primary Health Care conference in Alma Ata (1978), all participating United Nations member states made firm commitments to adopt and implement primary health care as an integral part of national socioeconomic development—a revolutionary step in addressing the concerns about inequalities in the health status of people across the Third World nations. It was during this same period that the W. K. Kellogg Foundation took a stand to begin its programming in southern Africa; a time when many governments were grappling with the practical application of strategies that would address the ambitious goal of "Health for All by the Year 2000." Many African governments during this era were working on shrinking national budgets, and were struggling to redirect resources for use within their national health care systems. This sad state of affairs, compounded by the economic recessions, exacerbated the problems of financing the health sector in many countries.

So in 1986, the Kellogg Foundation entered a sub-region fraught with political turmoil, with several of the nations falling in and out of unprecedented civil unrests. Aligning itself to the health priorities of the five southern African countries where it was to establish partnerships, the Foundation prioritized its health programming around strengthening primary health care systems in the sub-region, through linking health providers and communities, and improving the education and practice of health professionals. Three key strategies focusing on health care services, health worker education, and public health partnerships drove the Foundation's focus.

Health Care Services

- Support emerging national comprehensive health care systems that build on primary health care structures and include family-centered, community-based health services, foster collaboration

The Foundation prioritized its health programming around strengthening primary health care systems in the sub-region, through linking health providers and communities, and improving the education and practice of health professionals.

among multiple sectors, and engage communities and institutions as partners.

- Support efforts of traditional health care provider institutions to become more responsive by incorporating community perspective in governance and delivery.

Health Personnel Education

- Foster development of qualified students from communities to enter educational programs for careers in community-based health services.
- Foster institutional change to redirect health professions higher education, create new models for training health personnel, and encourage multidisciplinary approaches to meeting the health needs of people.
- Foster development of leadership for integrated, comprehensive health services and education systems.

Community-Institution Partnerships in Health

- Foster the improvement of health services as a result of community-education provider partnerships that link public health and preventive and primary health care to tertiary care.
- Encourage the introduction of curricular changes in health professional education programs that reorient training toward community.
- Foster development of mechanisms that bring community perspectives into the design and delivery of health services and health personnel training.

In spite of its strong focus in southern Africa, the Kellogg agenda faced many challenges. Following are the frustrating realities of health care service delivery that have been expressed by practitioners across the five southern Africa countries of Botswana, Lesotho, Swaziland, South Africa, and Zimbabwe, where the Kellogg Foundation was to start programming in 1986.

- **Access to health care services:** In a good number of the countries, community health facilities were inaccessible to the communities they were supposed to serve.

- **Staffing:** There was a chronic shortage of staff in the health service area, and inadequately trained personnel were constantly deployed in the outlying health centers.
- **Malnutrition:** This has remained a real problem and a threat to human development. It is fueled by chronic poverty and high levels of illiteracy, and is compounded by natural disasters and the high unemployment rates in the sub-region.
- **Community-based health care:** Communities continue to be inundated with diseases of squalor and poverty, all of which are preventable; for example: diarrheal illnesses, malaria, and tuberculosis. These are, however, still the greatest cause of morbidity and mortality among those under five years of age.
- **Staffing in the schools of health sciences.** With the intent and desire to embrace a multidisciplinary, community-based approach to health personnel education, it became imperative to improve the status of staffing in the nursing schools. Following are examples of how those realities were manifested in two countries.

Botswana health indicators over 15 to 20 years. Tables 1 and 2 provide some indication of trends in the changing health status in Botswana over the years.

Year	1971	1991	
Infant mortality rate	100	71	45
Under 5 mortality	147	109	56

Table 1. Infant and child mortality rates (per 1,000 live births).

Year	1987	1990	1994
Full immunization at right interval	66%	67%	57%

Table 2: Immunization coverage.

Year	1982	1988	1994
Underweight in %	27.5	15.5	14.8

Table 3: Nutritional Status of Under Fives. Source: NDP 7, 1991- 97, NDP 8, 1997-98—2002-03).

Though the full immunization coverage rates illustrated in Table 2 show a dip in 1994, other concerns were being addressed more successfully. For instance, the issue of access to health care facilities improved remarkably over the years. In 1989, 86% of the total population of Botswana was within 15 kilometers of a health facility, while 73% was within 8 kilometers of a health facility. The rest of the population is reached through a network of mobile services.

In Lesotho, access to remote communities has always posed major challenges for any type of services

Lesotho: nurses in remote locations. In Lesotho, access to remote communities has always posed major challenges for any type of services, given the country's impenetrable terrain. The general nurse with midwifery was for many years the sole health service provider for the majority of people, particularly those living in the most remote areas. It was the same double-qualified nurses who were plucked from the clinical areas to serve as tutors in the nursing schools.

It was not until their serendipitous meeting with Dr. Helen Grace at the Kellogg Foundation that nurses in Lesotho began to visualize quality nursing services, run and managed by competent nurses with tertiary level academic preparation. They imagined a situation in which nurses would have the capacity to provide quality health care services, through effective leadership. At this stage, the notion of identifying nurses for university-level preparation became a matter of serious consideration. The reality, however, was that a good number of nurses in active practice did not have the requisite high-school grades which would permit them to access university-level programs. So when the Kellogg Foundation formed a partnership with the nurses of Lesotho, it invested in programming directed at improving the high-school grades of enrolled nurses and nursing assistants.

Mobilized through the Lesotho Nurses Association, a good number of nurses took advantage of this opportunity and enrolled, with a view to completing, and in some cases improving, their high-school grades. Nurses also attended a series of empowerment workshops across the country. Results of these two activities are captured in Tables 4 and 5.

Targeted Candidates	Dropped Courses	Examined	Passed	Outcome
73	17	56	3	Failure· Project terminated (analysis of the LNA leadership)

Table 4: Upgrading of Enrolled Nurses and Nursing Assistants—Lesotho.

Targeted Candidates	Actual per Health Service Area	Total trained in country
120 Nurses	1 per Health Service Area	60 Nurses

Table 5: In-country workshops—Lesotho.

While the nationwide empowerment workshops had been successful overall, the upgrade program was regarded as a dismal failure, given that only 4% of the registered candidates passed. This experience was a valuable eye-opener that informed future programming plans and decisions that were later supported by the Kellogg Foundation.

The status of nurses and nursing in the sub-region. In spite of the universal acknowledgement that nurses form the backbone of health care services of any nation, many policy makers questioned the expressed need for the academic advancement of nurses. Where nurses saw the need to improve their knowledge to provide quality services to the majority rural poor, others attached no added value to the desires of nurses for further education. In some instances, policy makers were quoted as having questioned the need for nurses to engage in tertiary education programs. In some instances, senior government officials are on record as being opposed to the notion of higher education for nurses. Among documented objections are quotes like, "Nurses don't need a degree to change a patient's bed pan, or to take child's temperature."

The upgrade program was regarded as a dismal failure, given that only 4% of the registered candidates passed. This experience was a valuable eye-opener that informed future programming plans and decisions that were later supported by the Kellogg Foundation.

This narrow view was upheld by many policy makers in strategic positions, in spite of the complex nature of the universally accepted approach necessary to design and develop comprehensive primary health care services for the poor.

The Foundation's Entry into Southern Africa

The Foundation recognized the complexity of community problems and the need to bring many disciplines and sectors together in order to link the resources of the countries to the needs of the people.

Consistent with its historic mission of "helping people help themselves," the Kellogg Foundation, upon entry into southern Africa, focused its programming efforts on individual, institutional, and community capacity building, while actively fostering development of leadership skills. With this as its primary approach, the Foundation recognized the complexity of community problems and the need to bring many disciplines and sectors together in order to link the resources of the countries to the needs of the people.

It took a while for potential grantees in southern Africa to embrace the Foundation's style of operation, in which all parties were regarded as partners, with valuable assets to bring to the table. This empowering, albeit unusual and thus overwhelming tradition of the Foundation was later recognized and appreciated by the partners as a means, a measurement, and an integral part of total community development, which is the very tenet of primary health care. As most nations of southern Africa were already proponents of the primary health care concept, this somewhat helped them accept the unfamiliar notion of partnerships.

Soon all stakeholders and partners in the initiatives recognized that the concept of primary health care was never meant as cosmetic surgery for existing health care delivery systems. In fact, they saw that a total reorientation of existing systems was crucial; that they *could not* address health as a disease problem, but as a problem of poverty, of social justice, and equity, and most importantly, a problem of people with different concerns.

At this point, it is important to recognize a real limitation in this review of the impact of Foundation interventions in the countries of Botswana, Lesotho, Swaziland, South Africa, and Zimbabwe. Unfortunately, no baseline studies were conducted at entry point to determine more scientifically the true nature of the status of the health and social services in the southern African countries in which the Foundation planned programming. However, without such baseline studies, and only guided by the programming priorities of the Foundation, it can be assumed that the grants that were made were to a large extent a fair reflection of the real gaps that existed, requiring attention in the countries' health care needs and development priorities.

Some of the central initiatives that were supported by the W. K. Kellogg Foundation in southern Africa included nursing education, community partnerships in health personnel education, service delivery, and community-based health care development projects. Each is described below:

Nursing education in southern Africa. Some of the projects funded by the Kellogg Foundation were in response to southern Africa's critical shortage of nurses, who happen to be the principal providers of primary health care but who were often inadequately prepared academically for the tasks they were expected to perform, and expected to relocate and live in remote and barely accessible areas in order to reach sparsely populated rural communities.

Some of the projects funded by the Kellogg Foundation were in response to southern Africa's critical shortage of nurses, who happen to be the principal providers of primary health care but who were often inadequately prepared academically for the tasks they were expected to perform.

Kellogg Foundation support was therefore aimed at the design and development of effective and efficient health service delivery for the majority of southern Africans. Over the years, the process would include working with communities in the sub-region, and developing health education strategies likely to promote positive changes in risky behaviors at the individual and family level of health care and development. Major investments were focused on education programs for health personnel, particularly the training of nurses in primary health care to address the needs of rural communities.

Efforts to improve health service delivery can be tracked through many projects supported by the Kellogg Foundation in the early days of programming in southern Africa. These included:

- Nationwide accelerated in-service training workshops for nurses in Lesotho.
- Support for nurses to undergo high-school completion courses—a pilot project offered at two sites in Lesotho.

- Provision of funding to enable nurses to hold nationwide conferences to explore critical unifying elements for the profession in South Africa, a Concerned Nurses of South Africa-led process.
- A nationwide effort to upgrade the enrolled nursing program through both residential and distance-learning methods in Botswana.
- Building of classrooms at Mpilo Hospital, in Zimbabwe, to quickly develop a critical mass of nurses who could be deployed to rural areas.
- Offers of study grant opportunities for nurses to pursue undergraduate and graduate studies within the region and abroad.
- Assistance in establishing nursing departments within national universities. New degree programs are now in place in the national universities of Lesotho and Swaziland, major revisions and improvements were introduced into the University of Botswana, and residential and distance-education courses were started at the master's level in Zimbabwe (the Zimbabwe program was established to serve the entire sub-region). A master's program was introduced in Botswana, and a PhD leadership program was offered through the University of South Africa.

Table 6 features some examples from Botswana, Lesotho, and South Africa showing achievements in human resource development efforts supported by the Kellogg Foundation, with specific focus on nursing education.

Health Training Institute	Previous General Nursing Annual Output	Current General Nursing (2000) Annual outpout
Gaborone	80	80
Francistown	40	40-80
Lobatse	Enrolled Nurses 40	40-60
Molepolole	40	40
Serowe	Enrolled Nurses upgrading	30
Ramotswa	Enrolled Nurses upgrading	36
Mochudi	Enrolled Nurses upgrading	30
Kanye	30	50
Total	190	340

Table 6. Output of the nationwide general nurses upgrading program in Botswana.

To date, Botswana is the only country in Africa that has completed the enrolled nursing upgrade program.

The Bachelor of Education Nursing program at the University of Botswana was phased out, and a generic Bachelor of Science Degree in Nursing (BSN) was started. Table 7 shows progress and achievements made in this transition.

BEd. Nursing Education/BSN Output	**1981-1990**	**1991-2001**
Annual intake	6-22	19-36
Total graduated in 10 years	140	180

Table 7. BEd. and BNS graduates from the University of Botswana 1981-2001.

The Master's of Nursing Science was started in 1996, and has to date graduated at least 50 nurses.

The Kellogg Foundation supported many nurses through the international study grant program. Table 8 shows the number of fellows from Botswana, who benefited at the master's and PhD levels from the program.

Degree	**Numbers Sponsored**	**Numbers Completed**
Master's degree in	15 (MOH)	15
Nursing	9 (UB)	9
PhD's	4 (MOH)	3
	8 (UB)	8
TOTAL	32	35

Table 8. Master's and PhD graduates in Botswana.

The focus in both Lesotho and Swaziland was to upgrade basic academic qualifications for most nurses. Table 9 illustrates very commendable achievements made by nurses from Lesotho, through the Kellogg Foundation study grant program. These achievements were made over a short 15-year period (from 1986 through 2000).

Degree	Numbers Sponsored	Numbers Completed (2000)
BSc.	41	41
MSc.	18	18
PhD	6	5
TOTAL	65	64

Table 9. Nursing education in Lesotho.

Through the University of South Africa (UNISA), a special doctoral program, the D Litt et Phil degree, was implemented with a view to empowering nurse leaders from the five countries. The plan was that these nurse leaders would facilitate the initiation and implementation of nursing education programs aimed at strengthening primary health care in their respective countries.

Some of the factors that added to the success of this program included, among others, the biannual colloquia and opportunities for peer-day discussions. The faculty provided extra student support through consultations with promoters and co-promoters. Table 10 gives a breakdown of the numbers of current PhD graduates from this program.

Leadership development through academic advancement

UNISA Degree (PhD)	Numbers Sponsored	Numbers Completed	Country
	4(-1)	3	Zimbabwe
	4(-2)	2	Swaziland
	4(-2)	2	South Africa
	4(-1)	3	Botswana
	4(-1)	3	Lesotho
TOTAL	20	13	

Table 10. Development of Nursing Leadership in southern Africa (UNISA).

Community Partnerships and Health Personnel Education

The Foundation's entry into southern Africa, with its health programs, coincided with the WHO mandate for member states to undergo health care reforms. In South Africa, the emerging democratic/progressive movement in health care was committed to providing services to all members of civil society. In some subtle way, this move challenged the health service providers and training institutions to revisit and consider the restructuring of their curricula and service delivery modalities that had been in place for many years.

The Foundation's entry into southern Africa, with its health programs, coincided with the WHO mandate for member states to undergo health care reforms.

The seven Community Partnerships in Health Personnel Education projects offered to transform the situation of communities, which were poor, neglected, and marginalized by oppressive systems of education and health. The partnerships would also transform health personnel education as well as health personnel, academics, and their structures and systems in order to be relevant and responsive to the needs of the communities they served.

The role of communities was crucial not only in shaping the training programs but also in articulating their perceptions and their needs, and ultimately in influencing the delivery of health personnel education in a way that enhanced the delivery of primary health care. Academic sites were upgraded to accommodate participation of health sciences students who were then sent out to care for patients in settings located close to the people. In these settings, students worked with other members of the primary health care team, thus promoting interdisciplinary as well as multidisciplinary training of students in settings similar to where they would work once they graduated.

Participating institutions worked to improve the curriculum to make it more responsive to the community, upgrading clinical services and academic facilities to include computer links, increasing staff support at community sites, helping students get to academic sites, promoting interdisciplinary learning, and involving departments and disciplines outside of the health sciences faculty. In four out of the seven CPHPE projects in South Africa, the departments of nursing pioneered curricula changes. Many universities within the partnership recognized the need to respond to societal needs, concerns, and priorities through:

- Diversifying curricula with increased attention to operational and applied research.

- Creating partnerships with government and civil society to meet the challenges of modern society.
- Adopting a holistic approach, which sought to develop students' cultural, recreational, civic, and academic interests. This benefited not only the students but the community as well.

The Kellogg Foundation's investment in health personnel education through the Community Partnership in Health Personnel Education initiative has led to the convergence of three elements:

- The innovative development of academic sites for student learning;
- Major developments in the external politics and social environment; and
- New faculty leadership able to support a serious, faculty-wide investment into building a more appropriate and relevant health sciences curriculum.

This convergence has provided a sturdy platform for the future of community-oriented and community-based health sciences education in the tertiary institutions.

The Community Partnership in Health Personnel Education concept offers increased opportunities for community-based practice and interdisciplinary approaches. Involvement in this initiative translated into community empowerment, increasing knowledge and awareness of health issues, and the need for their active participation in the process of health development. The models have informed public policy and plans for a reformed health care system.

The community partners, through their social and economic development projects, have led to ongoing participatory research and academic sites for the partner institutions, thus providing some degree of sustainability of these grant-funded efforts.

Economic opportunities for community development called for appropriate quality technical training of community partners. Table 11 highlights achievements made in preparing the community for their own sustainable economic development.

Vocation	Core area of development	Number trained
Building construction	Project management	30
Stabilized compressed earth bricks	Infrastructure housing	15
Stabilized compressed concrete paving bricks	Infrastructure	6
Welding of steel window frames and doors	Housing	4
Painters (women)	Housing	15
Etsa Phapang (Agricultural project)	Farming skills Management	20
Bakery	Catering skills Management Marketing Entrepreneurship	25
Youghurt factory	Management Marketing Entrepreneurship	2
Cleaners/Receptionists	Entrepreneurship	8
Low cost housing	Construction	10

Table 11. Community partners in Micro-enterprise development: working toward sustainable development. MUCPP, South Africa.

Table 12 illustrates how responsibilities are shared among the partners, a practice aimed at ensuring long-term sustainability.

	Currently sustained by		
Program	**Academic partner %**	**Service partner %**	**Community partner %**
Job Creation			
• Sewing and knitting	10 (UFS)	-	90
• Brick-making (construction)	30 (Tech)	-	70
• Hydroponics (agriculture)	50 (UFS)	-	50
• Bakery	30 (Tech)	20	50
Youth Development			
• HIV/AIDS and STD	10 (UFS)	20	70

continues

continued

Program	Currently sustained by Academic partner %	Service partner %	Community partner %
• Information Systems	70 (UFS)	-	30
• Training and counseling	-	-	100

Table 12. Community working toward self-reliance and sustainable development. MUCPP, South Africa.

Through the health sciences student bursary programs of the Community Partnership in Health Personnel Education, there was a significant contribution toward transforming human resource development and related programs within the sub-region. The Community Partnership in Health Personnel Education promoted the application of innovative programs for schools of nursing, to educate students in community settings with the full participation of community members. The scope of health education and information was broadened as part of the health care systems' response to problems at individual and community levels.

Field of study	Number of students 1995	1996	1997	1998
Medical	22	23	15	15
Nursing	14	12	18	19
Social Work	7	9	4	2
Occupational Therapy	4	4	3	3

Table 13. Field of study and number of students who received bursaries. MUCPP, South Africa.

Focusing on providing quality primary health care services to underserved communities, innovative approaches included:

- Training nursing students from academic sites and bases within communities.
- Training community health workers.
- Involving and partnering with alternative health care providers to offer primary health care services to communities.

- Strategically identifying and selecting community health workers and helping them acquire skills to mentor nursing students.

Nurse training included a focus on primary health care and provided community-based learning experiences for students.

Year		**1994**	**1995**	**1996**	**1997**	**1998**	**1999**	**2000**
Gender	Male	9	13	13	20	21	19	20
	Female	335	362	370	342	265	274	263
Race	White	318	323	304	268	177	163	127
	African	26	52	79	87	99	120	145
	Colored				7	9	9	9
	Asian					1	1	1
Total number of students		**344**	**375**	**383**	**362**	**286**	**293**	**283**

Table 14. Breakdown of first year nursing students, MUCPP, South Africa.

It was evident to all concerned health professionals, that if the health needs of the majority of the people of South Africa were ever to be addressed, then the systems to train health personnel would require massive restructuring. The Community Partnership in Health Personnel Education initiative thus became a very important and appropriate tool for the Kellogg Foundation's contribution toward health leadership development in South Africa. Though the initiative was introduced into an environment hostile to its goals—disempowered communities, fragmented health services, and traditional institutional-based curricula for health personnel education—after the democratic elections of 1994, the political, economic, and educational scenario changed dramatically, making the environment not only receptive but inclusive of these goals. As it turned out, the seven projects were strategically positioned to benefit from the changing climate, thus opening avenues for access to local funds.

Efforts to support plans, preparation, and assistance of health personnel were intensified. In reorienting health personnel education it became essential to link to relevance to health service requirements, by identifying learning experiences in functioning primary health care systems.

The curricula changes shifted from the dominant hospital-based training model, where diseases are treated, to the community-based model, where diseases can be prevented and healthy lifestyles promoted.

The concept of forging partnerships with communities was embraced, following years of deliberation. It was recognized that if strong working relationships were developed between the training institutions, health service providers, and society at large, this would lead to an empowered and cooperative working environment. Among the institution partners, the nursing departments primarily took the lead in affecting change in student training. The curricula changes that followed shifted from the dominant hospital-based training model, where diseases are treated, to the community-based model, where diseases can be prevented and healthy lifestyles promoted. Immersing health sciences students into community dynamics has effectively eliminated their reservations to serve in disadvantaged and/or rural communities upon graduation. It has encouraged students to interact with communities within home settings, building trust and confidence and thereby rendering learning a process that is mutually beneficial for all stakeholders—communities and health sciences trainees alike.

In the shift toward community-based education, it became necessary for institution partners to promote the notion of home visits, which were not intended as a service to hospital-discharged patients, as was normal practice; but to expose students to healthy people within normal family settings and within their community lives.

The partnerships took very seriously the role of communities and their contributions toward the new community-based education curriculum. At the Border Institute of Primary Health, the process of student entry into communities was managed and led by community health workers and other people of good community standing. The community health workers, along with the nurse educators, facilitated student learning at the family and community levels. The strengths of the community health workers, such as giving health education and identifying problems on home visits, were recognized and used to the benefit of all stakeholders. Though there were no signposts providing guidance in the ensuing revolutionary community-based education curriculum, the community partnerships projects took carefully calculated risks and forged ahead with a host of innovations.

In order to ensure that the new approaches to teaching and working would not pose a barrier to educating students, it became essential to engage faculty members in the reorientation process. An inherent feature of community-based education is the uncertainty it poses particularly to the educator, who is no longer in control of the environment. As much as there is uncertainty, there is also excitement. There are often competing interests between community priorities and service and institutional priorities. As a coordinator, the Border Institute of Primary Health office played a major role in preparing all stakeholders, especially the institution partners, to the challenges of teaching students in households and within communities.

It took time to foster real partnerships—as trust among players was nurtured, as faculty learned to operate and teach across cultural barriers, and as all learned to work as one. At the University of the Free State, for example, it took a series of training workshops on thematic topics such as team building, community-based education/problem-based learning, people-centered development, and orientation to primary health care, to bring aboard many skeptics, especially within the training institutions.

Service Delivery

With the primary goal to improve the quality of health service delivery to the majority of people, support was channeled through the national nurses associations in some instances, while in other instances support went through the secondary-level hospitals. This called for a steady realignment of structures and systems, and the reorientation of health personnel education toward more community-based, community-oriented services.

In the early years of programming within southern Africa, given the poor and/or lack of appropriate infrastructure, the Kellogg Foundation was called upon to assist with the upgrading of existing facilities, and in some instances requests came to help construct new facilities. Thus the Foundation supported the building of a select number of facilities, mainly in cases where it could be demonstrated that such investments would

In the early years of programming within southern Africa, given the poor and/or lack of appropriate infrastructure, the Kellogg Foundation was called upon to assist with the upgrading of existing facilities, and in some instances requests came to help construct new facilities.

lead to significant improvements in health service delivery. It was on the basis of this logic that expansion to the nursing school at Mpilo Hospital in Bulawayo, Zimbabwe, was funded. With extra classroom space, the school was able to train more nurses to serve in rural health centers and clinics.

Within the Community Partnerships in Health Personnel Education initiative, four health centers were constructed. These were to serve as academic teaching sites for the faculty of health sciences of the University of Transkei-Community Partnership Project, in South Africa. These structures helped tremendously in grounding the CBE approach to the training of health sciences students. The presence of these quality health service facilities marked dramatic changes in service delivery for the four local communities.

Another strategic initiative, which catered specifically to communities in the most underserved areas of South Africa, was the Decentralized Education Program in Advanced Midwifery (DEPAM). This training program for midwives was run from McCord's Hospital, a teaching hospital of the University of Natal, in South Africa. The DEPAM was offered as a decentralized course, therefore participants remained in their home institutions during the best part of their instruction, receiving regular home visits from the course instructors, and coming together for eight weeks in the year, split into four teaching blocks. This was a crucial aspect of the program, because it facilitated the building and strengthening of relationships between the course trainees, their employers, and the course instructors. Additionally, the qualified advanced midwife was able to make a smooth transition from being a trainee to practicing as a qualified and recognized health professional with an expanded scope of practice. Since many of the course participants were from rural areas, having an advanced midwife in remote service locations meant the beneficiaries were assured of very high standards of midwifery, thus leading to a reduction of the peri-natal mortality rates.

As expressed by virtually all other grantees, the DEPAM course instructors felt that the relationship they have had with the Kellogg Foundation was enabling and empowering, as it gave room for new innovations within the active grant. Thus when need arose, and amid tremendous political uncertainty and upheaval within the health service, the DEPAM facilitators course was introduced and received full support from the Kellogg Foundation. This meant that the program could be rolled out to more training bases, with McCord's Hospital still serving as the central coordinating site for the program. As of 1999, 48 nurses had graduated from the DEPAM program, and a study conducted by N. M. Khumalo demonstrates that these graduates are safe practitioners, that the DEPAM approach to training was valid, and that significant improvements in peri-natal mortality rates were being recorded.

Community-Based Development Projects

Nurses across the sub-region participated in a number of community-based health development projects, which offered them excellent opportunities to interact with and work closely with members of the community under ordinary settings, away from hospitals or clinics. In Botswana, through the primary health care project, two preschools were established, as well as community food gardens. Today the local city council continues to work with the nurses in running the preschools.

Nurses across the sub-region participated in a number of community-based health development projects, which offered them excellent opportunities to interact with and work closely with members of the community under ordinary settings, away from hospitals or clinics.

The nurses in Swaziland, deciding that nursing wasn't about waiting in clinics and hospitals, made a commitment—which they identified as their social responsibility—to work with the community on the Umcubunguli primary health care project.

Accomplishments

There is no doubt that the relationships fostered between the peoples of southern Africa and the staff of the Kellogg Foundation, and the ultimate impact of the programs supported in the sub-region, have been very positive for nurses and nursing across the five countries wherein the main work was concentrated. There is also no question that the nurses have significantly affected health care internationally, nationally, subregionally, and locally on the community level. Local communities have also enjoyed development projects that employ nursing personnel outside the confines of health centers and clinics, thus nurturing healthy relationships, and developing trust and confidence in one another.

In sum, with the guidance of dozens of dedicated leaders and visionaries throughout the region (see Appendix), the Kellogg Foundation investments in southern Africa have led to the following accomplishments:

1. An increase in the numbers of adequately trained registered nurses and midwives, with capacity to provide quality primary health care services within easy reach of civil society. A key achievement was human resource development for leadership, management, and decision making, and embracing and subtly promoting the notion of change agents.

2. A steady realignment of structures and systems, and reorientation of health personnel education toward more community-based, community-oriented services.

3. Completion of the construction of physical structures (Mpilo Hospital, Zimbabwe), health centers and academic teaching sites (University of Transkei Community Partnership Project, South Africa), which has enabled larger intakes of nursing students into the training institutions.

4. Academic preparation of nurses through graduate programs within the region, such as the regional master's program at the University of Zimbabwe, and the Master's of Science program offered at the University of Botswana.

5. The study grant program, which helped prepare a critical mass of nurses at higher degree levels, thus ensuring that the leadership in the newly established nursing faculties within the national universities in the region would be made up of local nurse professionals.

6. Efforts that have enabled southern Africa to prepare master's and PhD nurses to serve as faculty for nursing education programs and to manage systems in which nurses work.

7. Development of an adequate number of quality programs to prepare nursing leadership at undergraduate and graduate levels.

This effort was best achieved through distance education, bringing instruction to learners at sites remote from the campus of origin (master's programs offered through the Department of Nursing at the University of Zimbabwe, and the PhD program offered through the University of South Africa).

8. The upgrading of the nursing workforce to achieve a critical mass of adequately prepared nurses for community-based primary health care, who were strategically positioned to form the "second generation" of nursing leadership.

9. Financing replacement personnel when employees went on study leave (study grant fellows). In this way leadership and management skills at the home institution were developed, while ensuring uninterrupted coordination of activities within the organization.

10. Nursing staff increasingly gained knowledge and improved their skills and efficiency in service delivery through participating in international conferences and meetings.

11. International exposure through working with international consultants, including study tours and traveling seminars.

12. Training targeted at nurse educators, facilitating their development of community-based problem focused programs, through progressive nursing education (University of Natal). Graduates say that they have made a difference in the lives of their students; that they have helped their students become self-directed learners and critical thinkers. The students have become reflective learners and practitioners who are comfortable with the notion of challenging and questioning issues, rather than accepting things as told by others. Graduates have assumed leadership roles in their places of employment, and have moved into senior positions within the universities' faculties of nursing science, and in government.

13. Skills training of all stakeholders (particularly involving community partners), ensuring improved efficiency and efficacy in health service delivery.

14. Nurses working in close collaboration with community members on community development projects, thus fostering cordial relations with community members, away from hospital settings.

15. Growing recognition and respect for nurses and nursing in the sub-region. Increasingly, nurses are occupying positions of leadership, both in their own countries and internationally.

16. Major curricula changes with a bias toward more community-oriented, problem-based learning approaches at most departments of nursing, and in some cases entire faculties of health sciences.

17. Improved status of women's health professionals; for instance, in Lesotho, the National Director of Public Health Care is for the first time a woman and a nurse.

18. Nurses are increasingly participating in international collaborative research projects, where they are making commendable contributions.

Challenges

While there is good reason to celebrate the gains and successes of all efforts over the years, it is important to note the challenges as well:

Managing change. The creation of a new breed of nursing personnel across the region has raised tension and fear in some circles, both public and private. Society is still not used to having a nursing force that is so highly qualified academically. Nurses now must learn new negotiation techniques to overcome the biases they are often confronted with in the name of their being overqualified for some positions.

Society is still not used to having a nursing force that is so highly qualified academically. Nurses now must learn new negotiation techniques to overcome the biases they are often confronted with in the name of their being overqualified for some positions.

Political and social factors. From a broader view, political unrest and poor governance in the sub-region, an influx of refugees from other parts of Africa, the ongoing food-shortage crisis in southern Africa, compounded by the rising incidences of new and re-emerging diseases, put a huge burden on the already overstretched health services, and make the noted achievements seem miniscule.

Outside influences. While the level of achievement varied, factors such as insightfulness, authority and power commanded by the nursing leadership, the support and commitment of their local institutions, as well as the extent to which their efforts and aspirations were embraced by the national governments, had a direct correlation to the degree of progress made by the individual countries.

Barriers and disincentives. While barriers were broken in some respects, such as in the case of legislatures, which governed the status of national nursing associations, the challenges of unfavorable national policies like forced retirement at age 55 in some countries have negatively affected the early achievements, as many nurses with master's and doctoral degrees are in this age bracket. This fact, compounded by poor rewards for nurses who have higher degrees and poor working conditions, is one of the reasons for the high attrition rate of nurses from southern Africa to the Middle East, the United Kingdom, and North America.

The burden of disease. Today, the southern Africa region records the worst HIV/AIDS rates of any region in the world. This is largely due to the slow political acknowledgement of the seriousness of the problem during the eighties, when national leaders unashamedly made public statements denying that the problem existed. Since the mid-eighties the sub-region has witnessed a sharp rise in both adult and infant mortality resulting from HIV/AIDS-related deaths. Health professionals, as with all other peoples of southern Africa, are also severely affected by the scourge of HIV/AIDS. They too are getting sick and dying in ever increasing numbers. The demands on the services of the health workers are ever increasing, leaving the already overstretched services worse than ever before. The playing field has changed, the dynamics at hand different from what they were even in the recent past. It is for this and other reasons already cited, that achievements of recent years may appear not to make a difference at all to the health status of the broader society.

Since the mid-eighties the sub-region has witnessed a sharp rise in both adult and infant mortality resulting from HIV/AIDS-related deaths. Health professionals, as with all other peoples of southern Africa, are also severely affected by the scourge of HIV/AIDS.

Appendix: Nurses and Nursing Education in Southern Africa: The Prime Movers

Country	Prime movers
Botswana	Dr. N. Ngcongco, Professor S. Mogwe-Kupe, Ms. K.J. Gasennelwe, Ms. M. Kobue, the late Ms. Tlale, Prof. S.D. Tlou, Dr. N. Seboni, Dr. E.S. Seloilwe, Mrs. F. Kelobang, Dr. C.N Pilane, Mrs. K Mmatli, Mrs. O. Mogano, Ms. T. Mothobi, Mrs. Matlhabaphiri, Ms. C Modungwa, M Seretse-Sybia, Mrs. M.K.M. Magowe, Mrs. G. Feringa, Ms. K. Rantona, Mrs. D Lejowa, Mrs. Matlhabaphiri, Dr. C. Pilane, Mr. E.N.Ncube, Mrs. M.K. Magowe
Lesotho	Mrs. Chabane, Mrs. Phakisi, Dr. L.N. Makoae, Dr. M.J. Leteka, Dr. E.T. Makoa, Ms. M.A. Ramokoena, Mrs. C.M. Thakhisi, Mrs. Moji, Mrs. L.Tsekoa, Ms. R.E. Mosala, Ms. Seholoholo, Ms. L. Keketsi, Ms. Tsiane, Ms. Y.M.Thakhisi, Ms. Motseko, Ms. L. Thoahlane, Ms. Moahlodi, Ms. M. Ramphoma, Ms. Thatho
Swaziland	Dr. N.T Shongwe, Ms. L.G. Dlamini, Ms. R.L.L. Manana, Dr. M.D. Mathunjwa, Mr. M.P. Dlamini, Mr. Africa
South Africa	Ms.T. Gwagwa; Prof. B. Robertson; Prof. M. Viljoen, Prof. Uys, Ms. N. Mazaleni, Dr. B. Nzama, Prof. R. Gumbi, Ms. G. Andrews, Ms. M. Ramalepe, Prof. S. Ross, the late Mrs. D. Dhlomo, Dr. D. Hackland, Ms. N.C. Nzolo, Prof. H. Philpott, Dr. M. Adhikari, Mrs. Bolani, Dr. T. Anderson, Mrs. Gumede, Mrs. Zulu, Mrs. Umlaw, Mrs. Hadebe, the late Prof. M. Beukes, Dr. E. Potgieter
Zimbabwe	Dr. R. Ndlovu, Mrs. J. Kadandara, Mrs. C. Chasokela, the late Mrs. C. Rinomhota, Mrs. S. Mupepi, the late Mrs. I Mafethe, Ms. P. McKenzie, Mrs. C. Nondo, Dr. Mapanga, Dr. C. Mudokwenyu-Rawdon, Dr. J.Z.M.Chiware, Dr. E. Makondo, Mrs. A. Makwarabara, Mrs. M. Mandaba, Mrs. T. Makuyana

16 Shaping Direction: The W. K. Kellogg Foundation and the Nursing Profession

Helen K. Grace and Gloria R. Smith

This book has provided three perspectives on the support of the Kellogg Foundation for nursing in three areas of the world: the United States, Latin America and the Caribbean, and southern Africa. In the United States, the Kellogg Foundation's interaction with nursing and the nursing profession has extended over its entire 77-year history. The relationship between nursing in Latin America and the Caribbean has extended over 65 years, and in southern Africa the relationship has been short-lived, less than 20 years. The common ground that has connected the Foundation so closely with the nursing profession has been the key role that nurses play around the world in linking people from various communities to the broader health care system. In the 1930s in rural Michigan, it became clear that nurses were essential if youth were to receive comprehensive health services. Even though physicians and dentists were committed to providing health services to people in the community if they came to their offices, nurses, working with teachers in the schools, and with the families of youth, were essential in bringing the pieces together. Over the 77-year history of the Kellogg Foundation, this central lesson has been learned over again—in the United States, in Latin America and the Caribbean, and in southern Africa.

The common ground that has connected the Foundation so closely with the nursing profession has been the key role that nurses play around the world in linking people from various communities to the broader health care system.

In the early years in the United States, a strong external advisory committee served to shape the direction that funding for nursing was to take. The strong focus on community health in the early years and the important role of nursing in weaving all of the pieces of comprehensive health care for youth in a rural seven-county area of Michigan became a national model and the training ground for nurses in community health. This was in keeping with nursing's focus on community health, and the Kellogg Foundation and nursing worked hand in hand.

Following World War II and through the 1970s, the Kellogg Foundation shifted emphasis from the community to focus on hospitals. In this context, the agenda for nursing within the Kellogg Foundation was caught in a conflict between the needs for adequate numbers of nurses to staff hospitals and the aspirations of the nursing profession to focus upon upgrading the educational entry to nursing practice to the baccalaureate level. This conflict is captured in the controversy over the Foundation's support of associate degree nursing education. Associate degree nursing education addressed the need to produce adequate numbers of nurses to staff hospitals and addressed the concern of nursing educators that nurse training be moved out of hospital settings and into educational institutions. But those who wished to make the baccalaureate degree the point of entry to the profession viewed this support as subversive to their goal, particularly since nurses from two-year associate degree programs and those graduates of baccalaureate programs took the same licensure examinations. From the perspective of those wanting to advance nursing as a profession, nurses educated in two-year programs were viewed as "holding back" this movement, and the Kellogg Foundation became viewed as allied with community colleges, and, because of this support, suspect in terms of support for development of the nursing profession.

In the 1980s, the Kellogg Foundation made a very conscious decision to change its focus from the hospital as the center of health care back to its roots in the community and community-based health care. While in the thirties, the Foundation's focus on the community and the nursing profession were in synch, in the 1980s, as Kellogg returned to the community, the nursing profession moved farther and farther from its roots in community health. Public health nursing had been a distinguishing characteristic of baccalaureate nursing education since its beginning in the United States. But through the sixties and seventies, this component of nursing education had diminished and the focus shifted to specialization, matching the specialization occurring within hospitals. These changes in education were paralleled by changes within the professional support infrastructure, which saw the rapid proliferation of nursing specialty groups and the decreased influence of professional organizations representing nursing as a whole. Within the Foundation the commitment to nursing as an integral participant in comprehensive health care has remained constant for its entire 77-year history. In recent years, nursing in Latin America, the Caribbean, and southern Africa has partnered with the Foundation in very productive ways. But in the United States, this has not been the case. The lack of a voice and vision to unify the nursing profession and to interact with a world external to the profession has become increasingly problematic. Unlike in the early years, when the nursing leaders in the Foundation could link to external nursing leaders who were working toward the same goals, in many ways the Foundation and the nursing profession found themselves moving in disparate directions. Ironically, the success of community-based health programming within the Kellogg Foundation was largely a result of the outstanding leadership of individual nurses out in the highways and byways of communities across the country who were working to improve the quality of life for elderly, to reduce infant mortality and morbidity, to improve the health of families of underserved groups, to work with adolescents in highly creative ways to help them move toward a more promising future, and to become leaders in the revolution that needs to occur to transform the health care system. Indeed, it has become increasingly clear that the nurses who are working in communities around the world know how the system could be transformed, but their voices are not being heard. And despite numerous efforts to make these voices heard, without a unified collaborative effort that mobilizes nursing as a profession to take the lead, no revolution will occur. And as long as nursing remains fixated on building its prestige based on rhetoric and on competition with medicine in the battleground of hospitals, the future looks very bleak.

In recent years, nursing in Latin America, the Caribbean, and southern Africa has partnered with the Foundation in very productive ways. But in the United States, this has not been the case.

Although the Foundation has provided the most extensive support for nursing in the United States of any philanthropy through the years, frequently its relationship with the nursing profession has been strained and sometimes viewed as at cross-purposes.

Although the Foundation has provided the most extensive support for nursing in the United States of any philanthropy through the years, frequently its relationship with the nursing profession has been strained and sometimes viewed as at cross-purposes. As the nursing profession has tried to gain prestige, the Foundation has worked to support improved health care for underserved people in community settings. Instead of being a support for allying with the powerful and prestigious in the health care arena, the Kellogg Foundation has pointed to the importance of joining the cause for the poorest and most powerless segments of society, and in so doing to gain increased recognition and political power.

From this perspective, the case of Latin America and the Caribbean is most interesting. In the early years, the support provided to nursing was minimal, but perhaps the most important aspect was that of building connections and human capital. The support of fellowships that allowed nurses from Latin America to come to the United States to gain advanced education was critically important. Nursing leaders, identified through Kellogg-funded projects, have seized the agenda of community-based health care, and in so doing have addressed significant issues in the nursing profession throughout Latin America: upgrading nursing education, regulating nursing practice and licensure, developing a nursing information system that is fueling research and scholarly practice, and leading with other health professions in providing community-based primary health care throughout the region.

In southern Africa, the history has been much shorter, and the alliance with the nursing profession aborted too prematurely to be able to assess the situation fully. In this instance, the approach taken was much different from that in the United States and Latin America and the Caribbean. The Foundation initially focused its support on nursing in the countries of Botswana, Lesotho, Swaziland, and Zimbabwe. In these countries, the strategy was to work through formal structures: ministries of health, universities, the Nursing Professional Association, and nursing educational programs. This strategy was much more of a top-down approach than in the other contexts. While much was accomplished, the process was more political—fraught with considerable posturing and maintenance of position, rather than a collaborative, as in the Latin

American example. Although the time span was limited, major changes occurred within the educational system, particularly through distance education, which gave nurses who had been prepared through the colleges of nursing—primarily black African nurses—the chance to achieve baccalaureate degrees, putting them on a par with nurses of European background within the region. The large number of fellowships allowing nurses to complete graduate degrees was another significant contribution. And in South Africa, the insistence that nurses were a vital part of multidisciplinary community-based primary health care brought African nurses forward into new leadership roles with their counterparts in other health areas. Time will tell whether the Kellogg Foundation's support in this region over this brief time span will have made lasting contributions toward improved health care through strengthened nursing.

In South Africa, the insistence that nurses were a vital part of multidisciplinary community-based primary health care brought African nurses forward into new leadership roles with their counterparts in other health areas.

From these case studies the complex nature of the role philanthropy plays in shaping social order is quite apparent. The entrepreneurs who established the early philanthropies clearly held strong views about how the social order could be shaped, and wished their private money to be applied to those ends. W. K. Kellogg clearly wanted his money to be invested in people. From his own personal history he valued the experience of moving from being a broom salesman to heading a major cereal manufacturing company. He valued the ability of people to solve problems given resources to do so.

As the philanthropic field has developed over time, public concerns over the use of private monies in shaping public policy have emerged. A particularly nettlesome issue has been the boundaries between charitable purposes of Foundations and their involvement in influencing public policy. At an operational level, those who work in the philanthropic world walk a narrow line between being activists engaged in social engineering, and being passive in responding to the requests emerging from the field. In an attempt to analyze this complex issue, this book has looked carefully at the relationship between a major philanthropic organization, the W. K. Kellogg Foundation, and the profession of nursing and the approaches used to attempt to strengthen nursing.

The Broad Perspective

Foundations generally do not want to be viewed as simply another charity filling the gaps left by public funding sources. They view themselves as agents of social change and a source of "risk capital" for experimentation with innovative approaches to solving societal problems. They see themselves as accountable to the founders of their foundations for wise use of resources to further desired changes. When successful approaches are demonstrated, philanthropies wish to see these changes moved into the mainstream and supported by public and private sources already in place. Many foundations expend considerable resources to disseminate the results of their experimentation and inform public policy debates.

From the perspective of the federal government, charitable giving in this country constitutes a substantial tax write-off. The concern is for how these dollars are spent, and for assuring that this substantial resource is not used for political purposes. Periodically, the laws governing foundations are reviewed, federal investigations are held, and new laws emerge. The relationship between the philanthropic world and the public sector has become increasingly complex over time.

President Hoover visited rural Michigan to see the comprehensive youth program developing under funding from the Kellogg Foundation, and in one of his radio addresses to the nation, highlighted this as a national model.

This is seen in the history of the Kellogg Foundation. While in its early years, the foundation world and the federal government collaborated extensively, but this relationship has diminished over time. In developing his foundation, W. K. Kellogg was strongly influenced by participating in a White House conference on children and youth convened by then-President Herbert Hoover. The Children's Charter, which stemmed from this conference, provided a baseline for development of the early programs of the Kellogg Foundation. President Hoover visited rural Michigan to see the comprehensive youth program developing under funding from the Kellogg Foundation, and in one of his radio addresses to the nation, highlighted this as a national model.

In the early years, the Kellogg Foundation collaborated with federal agencies in several initiatives. With the approach of World War II, the State Department was concerned about building strong alliances with our neighbors to the south, and along with the Rockefeller Foundation, approached the Kellogg Foundation for assistance in addressing this problem. From this starting point, the Kellogg Foundation has provided continuing support for projects in Latin America over the past 70 years. Another example of public-private collaboration at its best was that between the Kellogg Foundation and the Department of Defense to address the problem of production of adequate health personnel for the war. The approach that had been taken, after consultation with medical school deans, was to accelerate educational programs and run them year-round to increase the supply. This had not taken into account the reliance of many medical students upon summer employment to finance their education; without that option they did not have the means to continue. While the federal government recognized the problem, enacting legislation to address it would take at least a year. The Kellogg Foundation rose to the occasion and rapidly implemented a scholarship and loan program in medicine, dentistry, and nursing throughout the United States and Canada. This made an enormous contribution to addressing a national problem.

While these early examples illustrate the efficacy of collaboration between the public and private sectors, this relationship has become more problematic over time with the public sector fearful of the power of the private sector in shaping public policy and the private sector increasingly concerned over the inability of the public sector to readily respond to the need for change. Within this broad context, key players are working to improve interaction between the public and private sectors. This book has looked at one particular field in depth, the profession of nursing, and the relationship of this profession with the Kellogg Foundation over more than a 70-year time span and in three geographic contexts: the United States, Latin America, and southern Africa. From these case studies it is possible to identify lessons to be learned by both the public and private sectors. It is hoped that these lessons will help to inform those who wish to be leaders in particular fields of interest to understand the process involved in acquiring private funding to facilitate their work. It is hoped that these case studies will also be informative for the broader philanthropic world by detailing different approaches in facilitating social change. While this is an ambitious goal, hopefully this will be one small step in that direction.

Doing Philanthropic Work

Those involved in philanthropic work face a major challenge of focusing sufficiently so that philanthropic resources are used most productively to create the "greatest amount of good" possible, and at the same time remain responsive to those seeking support. Foundations are bombarded with requests from a wide array of sources. Without priorities, the work accomplished would be highly fragmented and ineffectual. Too narrowly defining priorities cuts off the majority of those seeking support. Those engaged in the grant-making process continually walk the narrow line of trying to be responsive to the needs of those seeking funds while seeking to demonstrate to a foundation board that the funds granted to projects are being used wisely toward some overarching goal. In many ways the process is one of matchmaking—of finding grantees who are engaged in work that would further some articulated goal of a foundation. Those seeking funding usually see the process as one in which their project is judged on its individual merit, rather than seeing their work within a broader context of the overarching goals of a particular foundation. Having a request turned down is interpreted as a matter of not "being found worthy" of a foundation's support, rather than not "fitting" into its programming priorities.

Those involved in philanthropic work face a majoı challenge of focusing sufficiently so that philanthropic resources are used most productively to create the "greatest amount of good" possible, and at the same time remain responsive to those seeking support.

The work of a philanthropic organization is organized to accomplish the goals of the original benefactor. For example, initially W. K. Kellogg wished his money to be used solely to enhance the welfare of children. Based on a realization of the broad scope of factors affecting the well-being of children, the mission was broadened. In the first stage of operation of the Kellogg Foundation, this work was furthered by directly operating a project in south-central Michigan, the Michigan Community Health Project. Staff for the project were employed by the Foundation and worked in a seven-county area in the region. Responsibility for managing the project was that of the staff of the Foundation. In its earliest phase, the Kellogg Foundation was an *operating* foundation: in this model, its staff was directly responsible for implementing a project that addressed this original intent.

As the resources of the Foundation expanded, the decision was made to expand the scope from the original project both nationally and internationally. This shift required a transition from being an *operating* foundation to a *grant-making* foundation.

A grant-making foundation must focus its efforts so that its resources can be used in the most effective manner possible. But unlike an operating foundation that is directly responsible for the outcome of a particular project, a grant-making foundation operates indirectly through projects it funds. Usually the foundation employs a core staff, who, in most instances, are recognized experts in their fields that match the foundation's priorities. For example, as the Kellogg Foundation began its grant-making phase of operation, nationally recognized experts in the fields of medicine and public health, dentistry, nursing, and hospital administration served as directors of its health divisions. These experts were assisted by advisory committees chosen because of their expertise in these particular fields, drawn from a national cross-section, and including representation from governmental agencies such as the National Institutes of Health, and from the private sector. Divisional directors working with advisory committees on the one hand, and the trustees of the Kellogg Foundation on the other, shaped decisions for making grants. This form of organization resulted in powerful influence from outside of the Foundation in establishing priorities and setting direction for each of the divisions. Each division was given a budget for the year, and operated within those boundaries. Direction-setting during this phase was predominantly from the outside-in, so much so that each division became isolated from the other, which hampered opportunities for doing cross-disciplinary work in the health field.

As the Kellogg Foundation began its grant-making phase of operation, nationally recognized experts in the fields of medicine and public health, dentistry, nursing, and hospital administration served as directors of its health divisions.

In 1965, the appointment of a new general director of the Foundation altered this form of organization. Divisions and advisory committees were disbanded and broad programming priorities were established. Ad hoc advisory committees could be convened to address particular issues, but these committees were clearly advisory. These changes led to a priority-setting process that was much more internally initiated. In this model, program directors were held responsible for being connected to the external world sufficiently to know the major issues out in the field and to judge the relative merit of requests for funding coming into the Foundation. Program directors proposed programming priorities for consideration by the board of trustees and implemented programs that were approved. Under this form of organization, each of the program directors negotiated directly with the general director and ultimately with the board of trustees to gain support for funding of particular projects. No longer were they operating with a budget that provided boundaries, but now were operating within the total context of monies available within the Foundation for funding of all projects. The program directors no longer had strong advisory committees to keep them up to date with developments in the field, and needed to maintain connections externally to stay abreast. At the same time, they needed to be engaged in the internal politics of the organization to assure their fair share of project funding. This entailed a dramatic change in the role of program director, a change that Mildred Tuttle, a program director in nursing, achieved successfully. The transformation was made from a Foundation largely influenced by outside forces via advisory committees to one in which the decision-making process and ongoing management of projects became largely an internal affair.

The transformation was made from a Foundation largely influenced by outside forces via advisory committees to one in which the decision-making process and ongoing management of projects became largely an internal affair.

Other foundations operate differently. A very popular model, particularly among the East Coast foundations, is a middle-ground approach between operating programs and making grants. A typical process is one that focuses attention on a high-priority problem, such as the "nursing shortage." A national expert in the field is identified to lead a series of projects directed toward addressing this problem. The responsibility for management of the total effort is "outsourced" to this expert, working collaboratively with a program director from the Foundation. A series of grants is made, but the overall responsibility for recommending grants and for managing the total effort is largely handled by this outside expert. Usually, these programming initiatives are highly publicized and maintain a high profile on the national funding scene. This model is somewhat similar to that of the original advisory committees within the Kellogg Foundation, but instead of being generalists over an entire field, the structure is tailored to a particular initiative. The board of trustees of the Foundation and the program directors have responsibility in shaping the broad initiatives, but the decision-making and management of projects long-term is "outsourced."

The differing philanthropic processes sometimes can confuse those seeking grants. Frequently, the prospective grantee may have a vague idea of programming priorities of a foundation, but they are driven primarily by their interests in a particular project. On the other hand, the program director in the foundation is acutely aware of the need to "make a case" for the funding of any particular project. The search for a "fit" between a prospective grantee's interests and the programming priorities that are the framework in which a program director operates is an arduous process and largely a covert activity. Program directors spend considerable time in trying to communicate programming priorities. In doing so, they walk a narrow line between being "heavy-handed" and prescriptive, and being open and responsive to prospective grantees. In the model in which the responsibilities for particular program initiatives are "outsourced," the program director is much more of a "middleman," rather than the one directly responsible for deciding which grantees are worthy of funding. The board of trustees' role is also different. In the newer Kellogg model, in which program directors work directly with the board in shaping programming priorities and in proposing individual grants for funding, the board is much more involved in the decision-making process surrounding individual grants. In the model in which programming initiatives are outsourced, the role of the board and the program directors is that of shaping the broad initiatives, but they rely upon an external expert and an advisory structure to deal with the details of each of the individual project grants.

In the newer Kellogg model, in which program directors work directly with the board in shaping programming priorities and in proposing individual grants for funding, the board is much more involved in the decision-making process surrounding individual grants.

No matter what the form of organization, the role of a program director is challenging and complex. The program director needs to be in tune with what is going on in his or her particular field of expertise and understand the priorities within a particular field. A program director also needs to be a part of shaping programming priorities within the context of the Foundation in which he or she works. The director needs to be skilled in building bridges between the high priorities out in the field, and the programming priorities that govern funding within his or her foundation. This role is easier to manage when the field outside the foundation is relatively unified and can agree on priorities. In the case of the Kellogg Foundation in the health fields, medicine, dentistry, and hospital administration were quite clear in their programming priorities, and much of the funding in these fields up until the 1980s focused upon graduate education in these fields, continuing professional education, and training of assistive personnel. The hospital field focused not only on hospital administration education, but also on the development of hospitals. All of these fields were buttressed with strong professional organizations that represented their interests: the American Association of Medical Colleges, the American Dental Association, the American Association of University Programs in Health Administration, and the American Hospital Association.

While there may be deep divisions within each of these fields, the professional organizations deal with this and put aside these differences to present a united front to the larger world.

Nursing, however, has never been able to present itself as united, which has greatly complicated the role for a program director representing this field in foundations. This is particularly an issue with the Kellogg Foundation, given its more than 75-year history of involvement with nursing. How does one avoid being capricious and siding with a particular faction in the field? In the early years, the strong advisory committees representing various perspectives in the nursing field provided some unifying voice. In these early years the nursing profession and the Kellogg Foundation worked hand in hand.

With the disbanding of this structure, program directors from the nursing field had direct responsibility for connecting the external world with internal decision-making processes governing particular programs and projects. Without clear direction from the external world, the program directors' jobs became all the more difficult. Each program director with a nursing background in the Kellogg Foundation strongly believes in the essential role that nursing has to play in improving the health of people. Each has sought to ensure that the role of nurses in health care delivery is strengthened and supported. This position has been supported by the board of trustees and the Foundation leadership for more than 75 years. The Kellogg Foundation remains committed to viewing nurses and nursing as essential to providing leadership within the broader health field. As in the thirties, the challenge of improving health care for communities and families across the world requires nursing to play a leadership role in weaving the pieces together and in making the connections to the real-world needs of people. The problems of today—such as high health care costs, lack of access to health care services, the inappropriate use of technology—can only find solutions when the people are brought in as problem-solvers, not as recipients of whatever the health care system has to offer. And nurses around the world are the connecting link to the people. It is time for nurses to recognize this reality and to join with those in the philanthropic community that would rally in placing resources in their hands and join as partners in this cause.

Each program director with a nursing background in the Kellogg Foundation strongly believes in the essential role that nursing has to play in improving the health of people. Each has sought to ensure that the role of nurses in health care delivery is strengthened and supported.

Index

NUMBERS

A

B

C

D

E

F

G

H

I

J

K

L

M

N

O

P

Q

R

S

T

U

V

W

Z